SUSTAINABLE MEDIA

Sustainable Media explores the many ways that media and environment are intertwined: from the exploitation of natural and human resources during media production to the installation and disposal of media in the landscape; from people's engagement with environmental issues in film, television, and digital media to the mediating properties of ecologies themselves. Edited by Nicole Starosielski and Janet Walker, the assembled chapters expose how the social and representational practices of media culture are necessarily caught up with technologies, infrastructures, and environments. Through in-depth analyses of media theories, practices, and objects including cell phone towers, ecologically-themed video games, Geiger counters for registering radiation, and sound waves traveling through the ocean, contributors question the sustainability of the media we build, exchange, and inhabit and chart emerging alternatives for media ecologies.

Nicole Starosielski is Assistant Professor in the Department of Media, Culture, and Communication at New York University. She is author of *The Undersea Network*, an exploration of the histories, environments, and cultures of transoceanic cable systems, and co-editor, with Lisa Parks, of *Signal Traffic: Critical Studies of Media Infrastructure*.

Janet Walker is Professor of Film and Media Studies at the University of California, Santa Barbara, where she is also affiliated with the Environmental Media Initiative of the Carsey-Wolf Center. A specialist in documentary film, trauma and memory, and media and environment, her books include *Trauma Cinema: Documenting Incest and the Holocaust* and, with Bhaskar Sarkar, *Documentary Testimonies: Global Archives of Suffering*.

SUSTAINABLE MEDIA

Critical Approaches to Media and Environment

Edited by Nicole Starosielski and Janet Walker

NEW YORK AND LONDON

First published 2016
by Routledge
711 Third Avenue, New York, NY 10017

and by Routledge
2 Park Square, Milton Park, Abingdon, Oxon OX14 4RN

Routledge is an imprint of the Taylor & Francis Group, an informa business

Library of Congress Cataloging-in-Publication Data
Sustainable media : critical approaches to media and environment / edited by Nicole Starosielski and Janet Walker.
pages cm
Includes bibliographical references and index.
ISBN 978-1-138-01405-3 (hardback) — ISBN 978-1-138-01406-0 (pbk.) — ISBN 978-1-315-79487-7 (ebook) 1. Mass media and the environment. I. Starosielski, Nicole, 1984– editor. II. Walker, Janet, 1955– editor.
P96.E57S87 2016
070.4'493637—dc23
2015035526

ISBN: 978-1-138-01405-3 (hbk)
ISBN: 978-1-138-01406-0 (pbk)
ISBN: 978-1-315-79487-7 (ebk)

Typeset in Bembo
by Apex CoVantage, LLC

Printed and bound in the United States of America by Publishers Graphics, LLC on sustainably sourced paper.

This book is dedicated to Constance Penley

and for Ariel Nelson

CONTENTS

FIGURES

ACKNOWLEDGMENTS

We dedicate this volume to Constance Penley for inspiring us to observe and research the many ways that "media and the environment influence, structure, and inhabit each other" and to study "the environment in media, media in the environment."

These words are part of the mission statement of the Environmental Media Initiative (EMI) of the Carsey-Wolf Center at the University of California, Santa Barbara, which Professor Penley led as founding director and co-director for more than a dozen years. She pulled the two of us into an amazing array of media research, teaching, and programming projects that she designed to involve marine and environmental scientists, green media industry leaders, middle and high school classes, and all manner of scholars at the nexus of media and environment. We extend our deepest gratitude to Connie, esteemed colleague, teacher, and friend. Our work together in the EMI's research group and GreenScreen course, "Films of the Natural and Human Environment," was the beginning of *Sustainable Media*.

The first in-person conversation among volume participants was enabled by the Society for Cinema and Media Studies in 2014, and we thank the SCMS as well as "Media and Sustainability" panelists Shane Brennan, Alenda Chang, Bishnupriya Ghosh, and Amy Rust—and the insightful SCMS audience—for the thoughts that flowed. We wish to acknowledge, as well, Hunter Vaughan and the members of the Media and the Environment scholarly interest group of the SCMS for their shared commitment to media and sustainability. Our warm thank you to Lisa Parks for the enormous inspiration of her scholarly and creative work, and for joining us for a workshop at UC Santa Barbara.

Janet wishes to acknowledge the influence of a large and vital group of UC Santa Barbara colleagues with whom she has had the good fortune of collaborating on two interdisciplinary environmental initiatives, with thanks to the university and especially Executive Vice Chancellor David Marshall for valuable support. Josh Schimel, Mary Hancock, Stephanie LeMenager, ann-elise lewallen, John Foran, and LeeAnne French were terrific partners, along with Barbara Herr Harthorn, Therese Kelly, Sara Daleiden, and additional CWC colleagues, on "Figuring Sea Level Rise," the 2012–13 theme of the Critical Issues in America series of the College of Letters and Science. Janet thanks Kum-Kum Bhavnani, John Foran, and Crossroads Fellows Corrie Ellis, Zachary King, Sarah Jane Pinkerton, and Christopher Walker for their significant ideas related to "Climate Justice Futures: Movements, Gender, Media," the 2014–15 theme of the Crossroads initiative of Graduate Division and multiple co-sponsors. She also wishes to thank the participants in her "Media and Environment: Climate Justice" seminar for great ideas and probing questions about the limits—if any—of the media in media ecology: Daniel Grinberg, Lisa Han, Jennifer Hessler, Pablo Sepúlveda-Díaz, and Theo LeQuesne, along with the above named Crossroads Fellows. Many thanks to Alexa Weik von Mossner, Peter Alagona, Teresa Shewry, Kathryn Yusoff, Anna Davidson, Susan Derwin, Ken Hiltner, Bruce Caron, and Selmin Kara for exciting invitations and environmentally conscious conversations in various contexts. Nicole Starosielski has been an inspiring interlocutor, both through her scholarship and personally, and I am thrilled to have had the opportunity to realize this project together. I am deeply grateful to my departmental colleagues and, as always, to Steve Nelson and Ariel Nelson for sustaining support.

Nicole would like to thank the faculty and students at UC Santa Barbara who were mentors, colleagues, and sources of inspiration for her work in environmental media. In particular, she would like to thank Janet Walker for suggesting the volume and for being such a generous collaborator. Nicole also thanks her colleagues at New York University's Department of Media, Culture, and Communication for their support, and her students in graduate and undergraduate seminars on "Media and the Environment" for their thought-provoking ideas and conversations. The organizers, presenters, and filmmakers of the "Lines and Nodes: Media, Infrastructure, and Aesthetics" symposium, particularly Chi-hui Yang, provided a genuine intellectual community for the development of the ideas in this book. She is especially grateful to Jamie Skye Bianco for introducing her to sustainability as an ethical way of life.

Warm thanks to Routledge editor Erica Wetter for inviting us to create this book and facilitating the process with enthusiasm, and to Simon Jacobs for crucial support along the way. Many thanks to Maria Corrigan for her superb copyeditorial contribution, to Autumn Spalding for outstanding project management, and to Mia Moran and Reanna Young for seeing the book through to publication. For the brilliant contributors, we are deeply grateful.

INTRODUCTION

Sustainable Media

Janet Walker and Nicole Starosielski

This book is about all of the ways that media and environment are intertwined: from the exploitation of natural and human resources to the installation and disposal of media in the landscape; from people's engagement with ecological issues via film, television, and digital media to the mediating properties of ecologies themselves. As elaborated by Nadia Bozak in *The Cinematic Footprint* and Richard Maxwell and Toby Miller in *Greening the Media*, there is an "inextricable relationship between moving images and the natural resources that sustain them"[1] and "the media are, and have been for a long time, intimate *environmental participants.*"[2] While it is certainly meaningful in this time of ecological crisis to call out "media" and "environment," we too recognize the infinitude of connections between these phenomena: media and environment are mutually constitutive, melding into one another with all the frictions and potential this soldered state entails. Building on recent work in ecomedia studies, this book locates media within a multiscalar resource economy of extraction, production, distribution, consumption, representation, wastage, and repurposing.[3] With this volume, we dedicate ourselves to the critical exploration of the environmental aspects of the media we construct, exchange, inhabit, and seek herewith to influence.

Media screens and transmission towers beam signals out across the world, sculpting disparate geographies as well as spaces of work and entertainment. Motion picture manufacturers and craft workers produce and deploy paint and lumber; Velcro and bungee cords; C-stands and Steadicams; water towers and sandbags; cameras, recorders, and postproduction equipment and software. Images and sounds distributed via monitors and microphones shape individual and collective insights and values, while interactive applications on handheld devices inflect our bodily movements. Media practices include the cultural activities of producers, distributors, users, and activists, be they scientific animators tasked to represent molecular-cellular events or residents seeking to capture visible traces of radiation. Electronics plants deal in vacuum tubes, circuitry, and motherboards, and in electrical and fiber-optic cables; they rely on data centers, industrial parks, assembly workers, and hauling and recycling operations. And all of this involves the generation and delivery of power, be it hydroelectric, nuclear, oil, gas, or solar. These manifold practices and infrastructures are the substance of "medianatures," in Jussi Parikka's evocative term. "Media are *of* nature, and return *to* nature," he states.[4]

As the chapters in this volume demonstrate, the ramifications of media for the life processes of our planet are beyond measure. We write during a time when human activities are impacting the earth and its ecosystems to an unprecedented degree—an epoch variously called the Anthropocene, the Capitolocene, the Chthulucene, and the Anthrobscene, among other names.[5] We write from concerns about complex, interrelated, disturbing happenings: decreasing biodiversity across worldwide patterns of species extinction; planetary-scale warming due to fossil fuel and other greenhouse gas emissions; sea level rise and increased weather volatility bringing drought and flooding; ocean acidification and oxygen depletion; increasing pollution, from the microplastics swirling in the Pacific vortex to the fumes of burning circuitry for salvage. This list too proliferates and is cause for concern.

Environmental problems are affecting individuals, communities, and publics, and unevenly so, according to what Rob Nixon has called "the environmentalism of the poor." Those who are already lacking money and resources suffer more from environmental despoliation.[6] We are moved by Nixon's attention, not only to catastrophic failures that receive spectacular treatment—disastrous oil spills and radiation leaks—but also to "slow violence," "a violence that occurs gradually and out of sight, a violence of delayed destruction that is dispersed across time and space, an attritional violence that is typically not viewed as violence at all."[7]

Although the specific causes of some problems are known, the intersectional character of these myriad violences relegates a great deal to what the philosopher-geographer Kathryn Yusoff characterizes, following Bataille, as "nonknowledge." Embracing nonknowledge "is not simply a matter of admitting we do not know all the difficult answers to living on a dynamic planet," she writes, but rather "part of the pact we make with knowledge" to seek the excluded and experiential in our research and ethics.[8] This is by no means to ignore the obvious—anthropogenic global warming—but rather to embrace the responsibility of developing active responses that are attuned to the complexity of mediating forces. It is in this spirit that Erica Robles-Anderson and Max Liboiron resuscitate the radical legacies of the sciences of ecology and cybernetics, premier scholarly disciplines, they argue, for describing complex systems.

As film and media studies scholars, we write from the conviction that the objects and systems that absorb us are implicated in daunting changes for the worse, and also, we can only hope, for the better. The burgeoning area of ecomedia or environmental media studies is beginning to grapple with the massive amount of energy drawn and expended to power media systems, the extraction of materials to construct media technologies, and the toxicities of use and disposal.[9] Yet at the same time, media provide a means to come to terms with and help ameliorate the ecological harms produced by industrial processes.

This book has three primary objectives. First, to expose the many ways that social and representational practices of media culture are necessarily caught up with media technologies, infrastructures, and ecologies. Instead of assuming a split between the study of how films thematize environmental issues and the study of how media architectures impact the environment, we are interested in cultivating the connections that inhere between media *about* the environment and media *in* the environment. Second, in characterizing, problematizing, historicizing, and theorizing aspects of media and environment, our contributors attend to the media specificity of a broad range of objects and processes typically outside the domain of film and media analysis, from Geiger counters that register radiation to sound waves traveling through the ocean. Third, as they come to grips with the heavy ecological impacts of our current media practices and systems, contributors also tackle the challenge of identifying existing or potential environmentalist media practices. Across the parts and chapters, *Sustainable Media* articulates the enmeshment of media practices (both textual and technological), infrastructures (physical and social), and resources (natural and human), and reaches toward ecological alternatives.

Bridging Practices, Resources, Infrastructures

In addressing this multiplicity of connections between media and environment, we are indebted to ecocinema studies, a robust area of film and media studies that has drawn attention to the sociocultural, geopolitical, and physical aspects of the environment and environmental issues. A cluster of works including Derek Bousé's *Wildlife Films* and Gregg Mitman's *Reel Nature* concentrates on nature and wildlife films, while another group, including books by David Ingram, Pat Brereton, and Robin L. Murray and Joseph K. Heumann, focuses on mainstream movies.[10] Other ecocinema titles, notably Sean Cubitt's *EcoMedia*, with its commitment to systems theory and dialectics, have made substantive historical, theoretical, and philosophical arguments linking green perspectives with a range of conceptual formations including affect, trauma, and queer theories.[11] If "all cinema is unequivocally culturally *and* materially embedded," Stephen Rust and Salma Monani posit in the introduction to *Ecocinema Theory and Practice*, then ecocinema studies offers "engaging and intriguing perspectives on cinema's various relationships to the world around us."[12]

Notwithstanding—or perhaps because of—the prescience and value of ecocinema studies, Adrian Ivakhiv opined in 2007 that more might be done. "For an approach that names itself after the scientific study of the natural world and that overtly looks for connections between culture and material nature," Ivakhiv realized, "it is surprising how little ecocriticism has dealt with the material aspects of cultural production." He argued that while much ecocritical work had thus far focused on philosophy and ideology, there was a need for a materialist account of the "things, processes, and systems that support and enable the making and disseminating of cultural texts."[13] Lisa Parks, also writing in 2007, pioneered the analysis of television "in relation to natural resources and environmental conditions," pointing not only to the massive amount of electronic waste generated in structured obsolescence, but also to the electromagnetic spectrum itself as a natural resource.[14] In the intervening years such studies have begun to proliferate. Concentrated on the ecological impact of media-related things, processes, networks, systems, and infrastructures, Jennifer Gabrys's *Digital Rubbish: A Natural History of Electronics* (2011) and Richard Maxwell, Jon Raundalen, and Nina Lager Vestberg's anthology *Media and the Ecological Crisis* (2014), along with Miller and Maxwell's *Greening the Media* (2012), Parikka's *Medianatures* collection and work on the geology of media (2012, 2015), and Parks and Starosielski's *Signal Traffic* (2015) are significant in this regard.[15]

Sustainable Media continues to explicate the complex practices and dynamic materialities of mediated environments. In so doing, we do not frame our work as moving *from* representational *to* infrastructural analysis (with the concomitant text-infrastructure binary).[16] Distributed throughout *Sustainable Media* are chapters engaged in reconnecting theoretically informed representational analysis of film, television, and new media (including video games, scientific animation, and social media) to matters of science and technology, labor and power, and contaminated environments. From an "ecomaterialist" perspective and with the benefit of archival research into the film's production files, Hunter Vaughan reads the choreography and physical setting of *Singin' in the Rain*'s signature sequence in light of water usage in the homes surrounding the MGM backlot. Amy Rust rereads theories of primary and secondary identification developed by Christian Metz in the 1970s in relation to her own theorization of "Steadicam's ecological aesthetic," by which she denotes "a phenomenal and psychological form of relation" that hovers between "on- and off-screen environments" (148). Working with scientific animations, Bishnupriya Ghosh critiques the notion that "molecular movies" are "representations of primary data" purely and simply, and reveals instead their capacity conceptually to "bring the nonobservable into existence" (233). For these authors, texts are not readable a priori, independent of the environments they mediate, but rather come into being through relational, ecological processes.

With the great benefit of being a collective work, *Sustainable Media* plumbs material substrates for the transmission of messages, the building of technologies, and the inscription of ideologies. Several of our contributors are inspired by what Eva Horn describes as recent "anti-ontological" approaches in media theory, rejecting fixed concepts of what "media" are—whether technologies, communications platforms, or institutions—in favor of seeing media as conditions of possibility for events and processes, "heterogeneous structures" that comprise cultural practices and forms of knowledge.[17] Today, media studies scholars are accounting for the mediating properties of their surrounding ecologies, including fire and ash, water and ice, sun and oil.[18] In this vein, volume authors are concerned with radiation emanating from cellular towers in people's neighborhoods (Mukherjee), minerals that enable the manufacture of electronics and rend physical and political landscapes (Milburn, Gabrys, Parikka), and the pinging signals that permeate the ocean (Shiga). Peter Krapp describes myriad "technologies of measuring and representing, exploring and navigating, remote sensing and telecommunication [that] have produced a planetary imaginary in the high-tech crucible of polar media" (276). *Sustainable Media* chapters take technologies and infrastructures as active, vibrant, and historically specific sets of processes, forces, and intensities, rather than as some sort of discrete or static physicality.[19] Certainly no essay can enfold all the dimensions of our mediated environment, but collectively these chapters probe the historical and political nuances of complex media ecologies, pushing the bounds of what is "media" in environmental media studies and tracing media's environmental footprint in the epoch of the Anthropocene.

(Un)sustainable Media

The title of the book has served as an invitation to the contributors to expose and clarify media's pronounced *un*sustainability, and they have addressed that charge with noteworthy energy. Yet studies of environmental media that focus *solely* on documenting unsustainable and exploitative practices may leave us amidst the detritus and debris of ecological collapse and questioning, "What now?" In choosing not to incorporate the "un" into our title proper, we intend also to analyze media's existing or aspirational sustainability. "The catastrophe of climate change is excessive and will inscribe all earthly space," cautions Kathryn Yusoff. "It is earth writing writ large."[20] The question we pose in response and with which the contributors deeply engage, is how to rethink or even reinvent media as a form of earth *re*-writing.[21]

From the Latin *sustinēre* (*tenēre*, to hold; *sub*, up),[22] "sustain" as a verb can mean to strengthen or support, to uphold or affirm, to continue or be prolonged, to bear without breaking, but then also to undergo or suffer.[23] The adjective "sustainable" from the 1610s was used to mean "bearable," from 1845 "defensible," and from 1965 "capable of being continued at a certain level."[24] In the modern ecological context, sustainability's meanings have proliferated: one researcher counted more than three hundred definitions of the term.[25] In its broadest and most liberal sense it refers to the prolongation and continuation of human and animal life on Earth, a future-oriented concept that considers the ecological impact of present practices on future generations. "Sustainable development," elaborated in a well-known 1987 report by the Brundtland Commission of the United Nations, is defined as "development that meets the needs of the present without compromising the ability of future generations to meet their own needs."[26] Articulating the value of social justice, the Earth Charter advocates "a sustainable global society founded on respect for nature, universal human rights, economic justice, and a culture of peace."[27] The word has considerable traction for environmental activists and in the realms of policy, politics, and economics.

And yet, in the context of global development, sustainability discourse has often served the interests of the privileged. It has been implicated in corporate efforts to privatize and capitalize on the environment, and concomitantly, in greenwashing campaigns that obscure the actual environmental impacts of products and organizations. It has naturalized particular assumptions about the forms of life to be sustained, thereby justifying exploitative practices, exclusionary humanisms, and the expansion of capital along ever-expanding routes. As Daniel Bonevac has observed, most common definitions refer not simply to the sustainability of human or nonhuman life, but to the preservation of particular economic orders and to capital.[28] For example, as Shane Brennan discusses in this collection, internet companies consider data redundancy a sustainable practice. In light of economic primacy, Steve Mentz has argued, "the era of sustainability is over."[29] He argues that we need to consider the postsustainable world, one that does not endorse such fantasies about a recoverable past or a static future state.

Sustainability remains a problematic and contested concept, and we recognize the wisdom of the editors of the journal *Resilience: A Journal of the Environmental Humanities*, who chose this name that "hacks its own brand" in part because "resilience" is distinct from "sustainability," in the latter term's conservative meaning.[30] The *Resilience* editors invoke systems ecologist C.S. Holling, "who brought resilience into the scientific community in the early 1970s" because he favored "destabilization" over "equilibrium as the core of the ecosystem concept. . . . in the process tossing out the Club of Rome's *Limits to Growth* and other arguments for sustainable development." In nominating "resilience," the journal's editors seek to "claim the environmental humanities as a field of abundance and profusion rather than scarcity and crippling precarity."[31]

We proceed with "sustainability," critically conceived, in order to complicate and destabilize existing modes of environmental media practice and analysis. Sean Cubitt holds that "when the weight of ideological individualism is crushing," thus pressuring the rise of "political activism and a turn towards a new politics of nature," then "sustainable media will demand not only a sustainable community of workers but equally a sustainable commons embracing workers and consumers, and beyond that a community of workers, consumers and their environments" (176). In tune with the book's commitment to bridging the perceived gap between texts and infrastructures, Cubitt calls for "an aesthetic approach" that considers "both the sustainability of the media themselves as material practices and their role in mediating between phyla and among humans" (177). Interrogating "sustainable media," then, entails asking what processes, human and nonhuman, are sustained by our current media systems.

The concept of sustainability also helps to decenter the common focus on the documentation of crisis and catastrophe. In her introduction to a special issue of *qui parle* on "Eco/Critical Entanglements," Katrina Dodson observes: "Crisis has long been the defining catalyst of the modern environmental movement, which has gained momentum and legislative traction through its ability to communicate the plight of plant and animal species on a more immediately human scale."[32] But yet, while "impassioned urgency" has its role to play in "finding a way to coexist more ethically with our nonhuman neighbors," Dodson articulates the importance of "taking time to think through the very concepts we begin with."[33] Through its invocation of continuation and prolongation, "sustainable media" considers not merely past and present harms, but also the implications and unfoldings of ways of being carried on into the future. And the concept can help apply the critical brakes to the "fast violence" of the crisis mode by drawing attention to problems that are slow, nuanced, and complex, and by highlighting potential alternatives. A recent and inspiring model of sustainable media studies is Jacob Smith's *Eco-Sonic Media*, a "green" media archeology of audio technologies.[34] Identifying shellac phonograph discs, singing canaries, and radio plays as exemplary of sound media, Smith sees such low-wattage, sustainable infrastructures as offering alternatives to an energy-intensive global media system. They put us into contact with nonhuman nature, he argues, and grant a sense of both localized places and planetary systems.[35] In this volume, Milburn broaches "modding," a fan practice involving the refurbishing or remodeling of gaming consoles, as a way, however small, to minimize "the [video] game industry's calculus of planned obsolescence" (88). Minori Ishida discusses intensive efforts by groups of people communicating online to scrutinize photographs for the effects of radiation and pay attention to topsoil remediation in irradiated zones. Emerging projects of this sort expose adverse ecosystemic effects of unsustainable media and foreground practices and infrastructures through which alternative futures may be imagined, "rewritten," and constructed.

Organizing *Sustainable Media*

By dividing this book into sections, we seek to highlight selected dynamics that manifest across disparate environments, technologies, and social contexts. These are, in Part One, the co-implication of media productions, installations, and circulations with a sprawling set of (so-called) natural resources; in Part Two, the mediations of emergent forms of social ecological engagement; in Part Three, the varied materialities of complex media systems; and in Part Four, the meditational properties and problematics of multiscalar environmental phenomena.

Readers will note that the logic of the sections is not that of traditional media objects (film, television, computer games, social media, etc.), nor have we adopted a geographically referenced *Planet Earth* organization ("mountains," "deserts," "ice worlds," "jungles," etc.). But yet we are awed by the variety of topics, objects, and geographies the contributors have brought to the fore. This volume encompasses post–World War I meteorological measurements, server farms, and *Final Fantasy*

VII; it extends across factories in Taiwan and South Korea, farms in the Catskills of New York state, the streets of Philadelphia, and below the surface of the Pacific Ocean.

In the following portions of this introduction we describe some of the subjects and arguments of the individual chapters. And, in addition, we trace the byways of the book as a whole, mapping interdisciplinary, multimodal approaches to media and environment over time and at this moment of necessary attunement to the power and potential of media with regard to environmental change.

Part One: Resource Media

In this part of the book, we present four chapters strikingly different from one another in their ostensible topics: a Hollywood musical from the 1950s, fiber-optic cables, data backup systems, and a popular environmentally themed video game. What the chapters have in common is their shared dedication to analyzing media's imbrication in endless chains of extraction, energy usage, and disposal, and to revealing the ecological underpinnings and potential resistances to contemporary "resource media."

In *The Cinematic Footprint*, Bozak offers the concept of the "resource image" to evoke "[t]he image and the determinants of its technology" as "the manufactured resources of industrialized culture."[36] Filmic images are made *from* resources and they themselves *are* resources. Moreover, so-called "natural resources" absorbed by media—while physically extant as ore, oil, water, nitrocellulose, and so on—come into being as resources through media fabrication, distribution, and usage. In titling this part "Resource Media," we appreciate the nuances of Bozak's term along with its application to all films (and not just those that are reflexive with regard to the production process), and then extend the proposition to all media modes, practices, processes, and infrastructures. That is to say, adapting Shane Brennan's comment near the end of his chapter, all media are "resource media."

Singin' in the Rain, the memorable Gene Kelly–Debbie Reynolds musical about the coming of sound, might seem an unlikely media object to initiate a book committed to the critical study of infrastructure. But the movie's frivolity—and the pseudo-reflexivity of the "500,000 kilowatts of stardust" with which Hunter Vaughan begins—aptly serve his ecomaterialist reframing of this film that, like all films, exists in and through its inevitable and mostly unfettered use of natural resources. The connection to Culver City residents runs deep, as he shows through an innovative focus on water—systems, towers, rainmaking, waste-making—and with the benefit of archival research in the production files of MGM's Arthur Freed Unit. It seems only natural, then, for readers to move with Nicole Starosielski up the west branch of the Delaware River. From farmers tilling the land and telecommunications workers stringing it with fiber-optic cables to hydraulic fracking's fluid injections into the earth, this chapter realizes and studies the multiple, interdependent conduits—"pipelines," in Starosielski's term—of resources in the western Catskill Mountains. Networks of media distribution, she submits, do not simply spread over environments, insulated from the complex process and flows that permeate them. Rather, as her historical and on-site research demonstrates, they are caught up with tree cutting, groundwater pollution, dairy farming, and with a developing struggle over fracking in the area. Colin Milburn, analyzing the environmental video game *Final Fantasy VII*, recognizes the natural resource and sociopolitical entanglements of energy and media, writing of "the energy consumed by legions of powerful gaming machines" (81) and the "pollution effects" of coltan mining for electronics manufacturing and the dumping of e-waste. "To play a video game, any video game," he writes, "is to contribute however indirectly to the environment hazards represented by electronic media" (81). Shane Brennan, in his discussion of the history of data backup practices, brings to the fore the power-hunger that belies the ephemeral imagery of "cloud" storage.

At the heart of Brennan's chapter are the distinctions he makes between risk perceptions of different eras with regard to data vulnerability (the corporate responses of the 1980s were not those of the Cold War era) and his key insight about the profound incommensurability of *environmental sustainability* with cloud computing's purported *sustainability of data*. Vaughan too critiques industrial practices with purported attentiveness to sustainability, revealing the limits of Hollywood's "complex, if inconsistent, environmentalism" (26). The motion picture industry's "green" ambitions (e.g., environmentally themed movie production and the Environmental Media Association's Green Seal certification, the implementation of "carbon-neutral" production through the practice of offsetting) are qualified, he argues, by the poor rating the industry received on its "Southern California Environmental Report Card."

That said, chapters in this part (and beyond) also point to media practices from which environmental benefit accrues. Milburn shows that the conundrum of an environmental video game that thematizes the exploits and exploitations of a global energy corporation, while contributing through its allure to that very energy economy, has not gone unnoticed by players. He repurposes the term "pwning," an erstwhile typo in the gaming community that came to signify dominating or owning an opponent, to suggest how a "*mistaken owning*" (Milburn's emphasis) might move players to take responsibility for their—or humans'—environmental errors. Alenda Chang (whose chapter we have located in a later part of the volume), suggests that the interactivity of video games (a prime example is the world-making game *Spore*) and their scalar properties may indeed have invigorating properties beyond those of the more passive experience of film viewing.

The development of new environmental media practices is also a topic of discussion. Brennan suggests that the concept of the "resource data file" might reorient media in sustainable ways—if we were to change backup practices by lessening the resources they require. Or, tracking the histories of various pipelines through the Catskill Mountains, Starosielski reveals how technological infrastructures might be leveraged by local cooperative networks and mobilized toward environmentalist ends. The geography of resource media thus includes water systems, data centers, and rural resource sheds—ecologies that are all open to alternate patterns of connection.

Part Two: Social Ecologies, Mediating Environments

To consider sustainable media requires an interrogation of the practices, publics, and socialities that are sustained by particular forms of mediation as well as their unequal distribution of environmental harm. In other words, it requires an understanding of media's varied social ecologies. While this is a shared concern across the volume, in this part we have grouped a few of the chapters where authors tackle directly some of the slow and spectacular,[37] small and enormous, inconvenient and devastating violences that industrial scale energy production, media infrastructures, and sociopolitical frameworks have brought to bear on humans and other forms of life.

Kathryn Yusoff begins her article "Excess, Catastrophe, and Climate Change" with the words of poet René Char:

> Today we are closer to the catastrophe than the alarm itself, which means that it is high time for us to compose a well-being of misfortune, even if it had the appearance of the arrogance of a miracle.[38]

The chapters in this section seem to us to do just that. First, the authors are at pains to describe various "highly technological space[s]" (128), in John Shiga's apposite phrase, of catastrophe or erosion: the coverage areas of radiation-emitting sector antennas in Mumbai (Mukherjee), Japanese television news reports and the Fukushima Daiichi nuclear plant meltdown (Ishida), underwater acoustic media and "eco-crisis" (Shiga), and the Steadicam rig's remediation of "the impoverished and contaminated landscapes" of postindustrial Philadelphia (Rust, 148). Then, the authors' evocative descriptions of environmental degradation or damage—in excess of any would-be alarm—are developed via analyses of historical and institutional contexts and resistance or "technostruggle" (Mukherjee quoting Kath Weston) through to "a well-being of misfortune," as it were. In keeping with the overall philosophy of the volume, media are vividly presented here as more than objects or technological relays among people, places, and things. According to the technically (as well as theoretically) savvy authors of this section, media "continually 'becko[n]' into being" (Shiga, 129, quoting Adey et al.) the "*ecologies*, or *logics of dwelling*" (Rust, 148), and media are key, therefore, to possibilities for reshaping and remediating our world.

Yet media infrastructures, even those of steel and concrete (like cell towers) or visible from space (like the Fukushima Daiichi Nuclear Power Plant), may be weirdly inconspicuous in everyday life. Rahul Mukherjee describes how a woman whose daughter was diagnosed with cancer suddenly saw the radio wave–emitting cell towers that had been there all along. Minori Ishida analyzes a craftwork of invisibility, or perhaps disguise, supported by the "safety myth" that nuclear energy for power generation could be controlled by advancements in industry and technology. While an in-studio news reporter and expert together denied that the plant explosion they were witnessing was an indication of meltdown, members of the public, Ishida reveals—amateur investigators in the age of new media—are using Geiger counters and Internet images to make fallout visible. Disruption or failure, as Mukherjee discusses (citing Stephen Graham), may often be the catastrophic event that lays bare the breadth and banality of media infrastructures, and the environmental problems to which they may contribute if deployed unsustainably.

Mukherjee's chapter is informed by his fieldwork in Mumbai, where he spoke extensively with Professor Girish Kumar, radio-frequency scientist at the Indian Institute of Technology, Powai, and anti-radiation activist Prakash Munshi. As readers will learn, Mukherjee accompanied Munshi to a presentation for the residents' association at the Meherabad apartment building, where, in front of a microwave oven being used for demonstration purposes, the LED lights of a radiation detector turned glowing red. For Mukherjee, the various "cell antenna publics" themselves, with their particular interests often at odds, form a crucial part of Mumbai's "media infrastructure ecology."

Minori Ishida's chapter also involves the author's personal experience, in the face of unseen, unknown, perhaps even unmeasurable danger. Having watched the television news reports of the Great East Japan earthquake and tsunami, Ishida reports the striking contrast she observed in early coverage of the Fukushima Daiichi plant accident. Only later was the meltdown characterized as a "Level 7" accident, on par with the Chernobyl Nuclear Plant disaster of 1986. As Ishida discusses, the Tokyo Electric Power Company (TEPCO) runs the Fukushima Daiichi plant *and* is a major sponsor of commercial television broadcasting. In fact, the company's influence has extended to the publication of certain images in this book, as evidenced in Figure 6.2b.

Among the particles and waves that flow through environments are those that are propagated below the surface of the ocean by an underwater cacophony of detonations, bells, and sirens, and received by all matter of hydrophonic devices. A later development, a form of echo-location known as sonar ping, "acts as a radiant in the ocean environment," Shiga informs us (130). Nor is this a neutral observation. Shiga writes the history of the ocean "as a technical workshop in military, industrial and commercial practices" (128) with ecological implications. The occupation of the ocean by acoustic media, he argues, produces a space of "eco-sonic" disturbance[39] that has been absorbed by and affected a range of human and nonhuman bodies over time, whether U-boat crews during wartime or whales deliberately frightened by sounds for hunting purposes.

Pertinent to their focus on degraded or toxic environments and irradiated or affected bodies, the chapters in this part have in common with one another (and with other contributors as well) an attunement to the disproportionate suffering of those already disadvantaged when environmental circumstances change for the worse. Mukherjee writes of some people who could not afford to move out of the antenna's path while others in a posh area were able to reach governmental dignitaries with their messages. Or, as Amy Rust discusses in her chapter, "individual triumphs" cannot alter the grind of air and water pollution and "unemployment, stagflation, oil shortages, and inner-city decline," (150) neoliberal capitalist machinations that have undermined the gains in social equity and economic and physical infrastructures achieved by New Deal policies. Working-class people *may* be sustained by environmental activism that combats pollution and promotes public health, but, in a worst-case scenario, Rust shows in her historical analysis, this constituency may be shunted aside by conservation programs that reserve nature for more privileged users.

Each of the four chapters in this part makes a point of articulating patterns or potential for resistance to environmental destruction. Reminding us that Rocky's success is one of enduring the beating rather than winning, and that Steadicam's "purported flights from materiality" (150) are grounded in Hollywood's neoliberalization, Rust nevertheless persists in tracing Steadicam's ability to supply an "*ecological ethic*" as well as the more socioecologically ambiguous "ecological aesthetic." Differentiating between hydrophonic and ping-based technologies, Shiga not only exposes the contribution of underwater acoustic techniques to eco-crisis but also revisits older devices and practices, the use of which might enhance the sustainability of ocean ecologies. Mukherjee and Ishida engage with amateur activist efforts to reclaim and repurpose media devices and practices, from microwave ovens and Geiger counters to television images and online social media. All of these chapters reveal the technological mediations that inhere in disparate environments while also thinking through ways that *mediating* as an activity—in a sociocultural and/or critical vein—can contribute to media's ecological potential. All of the chapters contribute brilliantly to *Sustainable Media*'s project to express the profoundly intermedial nature of such entanglements.

Part Three: (Un)sustainable Materialities

As Jennifer Gabrys observes in her chapter in this collection, materialist approaches now pervade media studies. Indeed, as we have discussed, environmental media studies has taken its own materialist turn, with authors documenting the wide range of matter that is harnessed, manipulated, and exchanged in service of human communication. Scholars have followed media objects as they transition (are transformed) from raw materials to commodities to e-waste. Others have carefully documented the effects of media production on particular ecologies, from the water and lumber use by Hollywood studio productions to the energy drain of Internet data centers in rural Oregon.[40] In such research, media are thus always more than content, they are residues of mountains and they will linger in our water systems.

All of the chapters in this volume touch on some aspect of media's materiality—from the pulsations of electrons and photons through Internet infrastructure, to the fluid currents of aquatic media, to the effects of radiation registered on digital cameras. The chapters in this part push us to grapple with materiality in ways that both expand the realm of objects and landscapes that fit under the rubric of environmental media studies and extend the ecological into media studies as a whole. Each oriented toward a different intellectual genealogy, the authors theorize the ecological in various domains of materialist critique and disciplinary practice. In his chapter, Jussi Parikka reflects on the media materialism of Friedrich Kittler, a key inspiration for media archaeology and media studies more broadly. Parikka argues that Kittler's work on the substrate of technical media and its historical ties to science, engineering, and war has set the groundwork for a new kind of media ecology and yet does not go far enough to address media's materiality in the Anthropocene. Introducing the concept of "nonmediatic materialities," which include the minerals that comprise media, the afterlife of media technological waste, and the labor that supports media production, Parikka offers a robust and ethical version of media ecology, one that expands both the temporal scope and ontological assumptions of media analysis.

An ecological media materialism thus always addresses more than just the mere physicality of media infrastructures, technologies, or objects. Yet in her chapter on the Internet of Things, Jennifer Gabrys highlights how this critical approach to materiality remains absent in the perception of computerization and sensorization of objects, even those that mobilize media infrastructures toward environmental ends. Electronic environmentalist projects such as "Trash Track," which places sensors on material waste in order to reveal its widespread trajectories, fail to account for the new distributions of electronic waste generated by the sensors themselves. While Parikka engages the media materialism of Kittler and media ecological research, Gabrys draws insight from new materialist authors, such as Karen Barad and Isabelle Stengers, to argue that things are processual, relational, and plural. Taking a "more-than-empirical" approach, Gabrys suggests that we need to "re-thingify" the Internet of Things. This would mean doing more than composing lists of the "static stuff" that constitute our media technologies, and instead would require an attention to the wider landscape of media's material relations. Gabrys's work, as does Parikka's, argues for a critical and material media studies that is accountable to ecologies beyond the technical.

Included in this wider landscape of material relations are the social ecologies of the labor through which media technologies are produced, distributed, and prodused. These are the central concerns of Sean Cubitt's chapter, which documents yet another aspect of media's environmental impacts in the manufacture of digital technologies. Cubitt details the unsustainable industrial practices implemented in the fabrication of semiconductors: from the exploitation of workers, especially via outsourcing and offshoring, to the ecological impacts of by-products released during production. Grounded by the lineage of historical materialist and political economic theory, Cubitt points out that "it is growth itself, the engine of capital, that opposes sustainability" (165). Media's sustainability is intimately tied to the circulations of capital that underwrite its production and consumption: "to the extent that our dominant technical media bear the stamp of the political-economic regime that gave them birth, they can sustain themselves only as long as capital can sustain itself" (163). This extends not only to laborers employed in global fabrication plants, but also to the produsers that sustain this system. Counter to the oft-ascribed "immateriality" of cognitive capitalism, Cubitt argues that the exchange of signals is likewise a form of material production that contributes to the unsustainability of our media systems, rather than standing as an alternative to it.

Cubitt's chapter is a powerful reminder that media, whether texts, technologies, or infrastructures, are connected to a larger global economic system premised on the exploitation of labor and the indifference to environmental externalities. As Parikka succinctly puts it: "Any talk of sustainability has to critically ask also whether the political economy of contemporary technological culture is what exactly should *not* be sustained" (197). Such approaches highlight the necessity of infusing industrial and political economic research with a consideration of capital's ecological costs, as well as the critical import of unsustainable labor practices in the analysis of environmental media.

Part Four: Scaling, Modeling, Coupling

The chapters in the book's final part, "Scaling, Modeling, Coupling," all deal with a long-standing problematic central to both humanistic and scientific studies of the environment: how to represent and engage with incredibly complex ecological phenomena. Processes such as climate change are both global and local and they enfold a multiplicity of systems and agents as expansive as life itself. How do we begin to develop an approach to the complex dynamics of ecological systems? Alenda Chang offers one fresh possibility: through an analysis of the scientific literature on scale, she argues that we can neither make clear-cut divisions between scales nor simplify interactions among them. Scale, as she notes, "serves equally well as an instrument of revelation and distortion" (216). Taking a critical approach to this literature, she investigates how some media make possible the nuanced and open-ended engagement of ecological processes. While the representational practices of film (and one might add here Google Earth) tend to reduce scalar complexity, Chang situates video games as an ideal site for a new scalar epistemology.

In her chapter, Bishnupriya Ghosh focuses on the intricate interdependencies of humans and viruses, documenting how these processes, far displaced from the realm of direct human encounter, are intensely mediated. Like Chang, Ghosh's chapter foregrounds the need to look critically at scientific discourse, and also takes the scientific visualization of molecular events as her object. In interviews with scientists and animators, she tracks an epistemological shift—from seeing viruses as parasites to entities that we live with symbiotically—that is underway. The "molecular movies," she argues, are a form of environmental media that is helping to reconfigure and repair the host-parasite relationship, a shift necessary for the greater project of "human-microbial planetary cohabitation" (240). While earlier research, and chapters in this volume such as Brennan's study of data backup and Gabrys's work on the Internet of Things, have tracked the immense energy draw and waste produced by digital systems, Ghosh shows how new techniques of digital imaging and manipulation are key to developing multiscalar ecological knowledges and relationships. Throughout the chapters in this part, the authors foreground media's capacities to produce and represent dynamic, historically contingent relationalities, whether via the interactivity of video games or the visualization of human-viral symbiosis.

At the opposite end of the scalar continuum from Ghosh, Peter Krapp offers an historical reading of the production of planetary visions, arguing that polar media, from documentaries of polar expeditions to digital installation art using climate data, have provided "an invisible axis . . . for our world picture" (265). Polar media have redefined how we comprehend the planet, expanding both its spatial and temporal scope, and inflected contemporary climate science in a multitude of ways. Yet these very representations, and the climate science that is supported by it, have been shaped by histories of heroic exploration and military resources. Rather than take the poles as merely a science lab, Krapp suggests, we might think through the polar imaginary and the scales that it normalizes in order to better understand ecosystemic relationships. As do Chang and Ghosh, Krapp looks to the ways that the environment's irreproducibility has generated other forms of data-rich media representations.

While sciences focused on representing such planetary imaginaries have often led to a totalizing, all-encompassing knowledge, Erica Robles-Anderson and Max Liboiron historicize the respective legacies of ecology and cybernetics to reveal their less well-known critique of technocratic ecological knowledge. As Chang found in her discussion of the scientific literature on scale, Robles-Anderson and Liboiron find that these fields, which sought to describe complex systems, rejected linear and causal models in the interest of developing new ways to model complexity. Given the ecological crisis we face today, the authors suggest that ecology's and cybernetics' emphasis on coupling, the process of "hooking systems together to constitute each other's environment" (254), provides a model for rethinking the representation of ecological problems such as the case of ocean plastics. Drawing from a cybernetic imagination, they submit, we should not be trying to assemble a macro-scale view, but rather should be focusing on the ways that systems, whether media or environmental, are co-regulated.

Contributors across the book engage the question of how technologies can represent the complexity of contemporary environmental problems. Tracking the possibilities for ecological visualization, Amy Rust sees in the Steadicam rig a potential to reorganize "encounters between living and nonliving milieus" (148). And Mukherjee and Ishida both home in on intermedial strategies for documenting the imperceptible spread of radiation. The chapters in this part, in particular, bridge media studies with science and technology studies in addressing this question, while charting the contestations against and mobilizations of scientific knowledge in understanding ecological problems. For these authors, understanding environmental media necessitates a critique of the anthropocentrisms and colonial histories embedded in contemporary science, coupled with new forms of technologically mediated relationships between humans and nonhumans.

In describing our logic in constructing the four parts of the book, we also embrace the random accessibility of this established medium. As we have mentioned here and there, whether located in "Resource Media," "Social Ecologies, Mediating Environments," "(Un)sustainable Materialities," or "Scaling, Modeling, Coupling," the authors who have entrusted us with their chapters share a commitment to articulating the challenges to and the possibilities for environmental media. With this book, we hope to connect with readers and advance this necessarily collective project.

Notes

1 Nadia Bozak, *The Cinematic Footprint: Lights, Camera, Natural Resources* (New Brunswick, NJ: Rutgers University Press, 2012), 1.
2 Richard Maxwell and Toby Miller, *Greening the Media* (Oxford: Oxford University Press, 2012), 9.
3 Lisa Parks's and Nicole Starosielski's co-edited volume *Signal Traffic: Critical Studies of Media Infrastructures* (Urbana: University of Illinois Press, 2015) is generative of critical infrastructure studies with Part II, "Resources, Environments, Geopolitics," specifically focused on issues of media and environment. See also, Jennifer Gabrys, *Digital Rubbish: A Natural History of Electronics* (Ann Arbor: University of Michigan Press, 2011); Jacob Smith, *Eco-Sonic Media* (Oakland: University of California Press, 2015); Richard Maxwell, Jon Raundalen, and Nina Lager Vestberg (eds.), *Media and the Ecological Crisis* (New York, NY: Routledge, 2015); Stephen Rust, Salma Monani, and Sean Cubitt (eds.), *Ecomedia: Key Issues* (New York, NY: Routledge, 2016); and Julia Leyda and Diane Negra (eds.), *Extreme Weather and Global Media* (New York, NY: Routledge, 2015).
4 Jussi Parikka (ed.), *Medianatures: The Materiality of Information Technology and Electronic Waste* (Living Books About Life, Open Humanities Press, online version 2012, unpaginated). http://www.livingbooksaboutlife.org/books/Medianatures (accessed November 20, 2015). Frozen version also available (Open Humanities Press, 2011, unpaginated; accessed at the listed site on November 20, 2015).
5 Paul Crutzen and Eugene Stoermer, "The Anthropocene," *International Geosphere-Biosphere Programme Newsletter*, 41 (May 2000): 17–18; Donna Haraway, "Anthropocene, Capitalocene, Plantationocene, Chthulucene: Making Kin," *Environmental Humanities*, 6 (2015): 159–165; Jason Moore, "The Capitalocene, Part I: On the

Nature & Origins of Our Ecological Crisis" (2014), https://www.google.com/search?q=the+capitaloscene+part+I+june+2014&ie=utf-8&oe=utf-8#q=the+capitalocene+part+I+june+2014 (accessed November 20, 2015); Jussi Parikka, *The Anthrobscene* (Minneapolis: University of Minnesota Press, 2014).

6 Rob Nixon, *Slow Violence and the Environmentalism of the Poor* (Cambridge, MA: Harvard University Press, 2011). As Adrian Ivakhiv has observed, "the movement of images across physical and electronic networks also depends on a certain 'social' or 'political ecology' of interactions among producers, distributors, consumers, and others." Ivakhiv, "Green Film Criticism and Its Futures," *Foreign Literature Studies*, 29, no. 1 (2007): 19.

7 Nixon, *Slow Violence and the Environmentalism of the Poor*, 2.

8 Kathryn Yusoff, "Excess, Catastrophe, and Climate Change," *Environment and Planning D: Society and Space*, 27 (2009): 1014. Yusoff references Georges Bataille, *The Unfinished System of Nonknowledge* (Lincoln: University of Nebraska Press, 2001).

9 Maxwell and Miller, *Greening the Media*, 2.

10 For example, Derek Bousé, *Wildlife Films* (Philadelphia: University of Pennsylvania Press, 2000); Cynthia Chris, *Watching Wildlife* (Minneapolis: University of Minnesota Press, 2006); Gregg Mitman, *Reel Nature: America's Romance with Wildlife on Film* (Cambridge, MA: Harvard University Press, 1999); David Ingram, *Green Screen: Environmentalism and Hollywood Cinema* (Exeter: University of Exeter Press, 2000); Pat Brereton, *Hollywood Utopia: Ecology in Contemporary American Cinema* (Portland, OR: Intellect Books, 2005); Robin L. Murray and Joseph K. Heumann, *Ecology and Popular Film* (Albany, NY: SUNY Press, 2009).

11 Sean Cubitt, *EcoMedia* (Amsterdam: Editions Rodopi B.V., 2005); see also Anil Narine (ed.), *Eco-Trauma Cinema* (New York, NY: Routledge, 2015); Alexa Weik von Mossner (ed.), *Moving Environments: Affect, Emotion, Ecology, and Film* (Waterloo, ON: Wilfrid Laurier University Press, 2014); Adrian J. Ivakhiv, *Ecologies of the Moving Image: Cinema, Affect, Nature* (Waterloo, ON: Wilfrid Laurier University Press, 2013); Nicole Seymour, *Strange Natures: Futurity, Empathy, and the Queer Ecological Imagination* (Urbana: University of Illinois Press, 2013); and E. Ann Kaplan, *Climate Trauma: Foreseeing the Future in Dystopian Film and Fiction* (New Brunswick, NJ: Rutgers University Press, 2015).

12 Stephen Rust and Salma Monani, "Introduction: Cuts to Dissolves—Defining and Situating Ecocinema Studies," in *Ecocinema Theory and Practice*, edited by Stephen Rust, Salma Monani, and Sean Cubitt (New York, NY: Routledge, 2013), 3.

13 Ivakhiv, "Green Film Criticism and Its Futures," 18–19.

14 Lisa Parks, "Where the Cable Ends: Television beyond Fringe Areas," in *Cable Visions*, edited by Sarah Banet-Weiser, Cynthia Chris, and Anthony Freitas (New York: New York University Press, 2007), 106. Lisa Parks, "Falling Apart: Electronics Salvaging and the Global Media Economy." In Residual Media, Charles Acland, ed. Minneapolis: University of Minnesota Press, 2007, 32-47.

15 See notes 2–5. See also Jussi Parikka, *A Geology of Media* (Minneapolis: University of Minnesota Press, 2015).

16 Contemporary film theory was born with the insight that "texts" are polysemous, open, and in-process; there is no such thing as a discrete film or program under study that exists apart from reading. According to theories of film and system that received crucial development in the 1970s in the pages of *Cahiers du cinéma*, *Screen*, *Camera Obscura*, and other outlets, a text is a complex relation co-constituted with processes of spectatorship and psychological subjectivity. Moreover, informed by (and cognizant of the frictions between) Marxist and psychoanalytic theories, these theories of texts and filmic systems emphasized material as well as ideological production processes: a text was understood to take shape and "take place" within sociopolitical, ideological, material (including "raw materials"), dynamic, and unbounded relations of production, distribution, and consumption. This is the history we have in mind and seek to build on in this volume. For just a few examples, see Jean-Louis Comolli and Paul Narboni, "Cinema/ldeology/Criticism," *Screen*, 12, no. 1 (1971): 27–38;

John Ellis, "Made in Ealing," *Screen*, 16, no. 1 (1975): 78–127; and Rosalind Coward, "Class, 'Culture' and the Social Formation," *Screen*, 18, no. 1 (1977): 75–106.

17 Eva Horn, "There Are No Media," *Grey Room*, 29 (Winter 2008): 8; Sarah Kember and Joanna Zylinska, *Life after New Media: Mediation as a Vital Process* (Cambridge, MA: MIT Press, 2012); Jussi Parikka, "FCJ-116 Media Ecologies and Imaginary Media: Transversal Expansions, Contractions, and Foldings," *Fibreculture Journal*, 17 (2011): 35, referencing Bruno Latour, "Morality and Technology. The End of the Means," trans. Couze Venn, *Theory, Culture & Society*, 19, nos. 5–6 (2002): 247–260. This essay is also included in Parikka, 2012.

18 Here, too, there are useful precursors. Over twenty years ago, W.J.T. Mitchell observed that

> landscape is a medium in the fullest sense of the word. It is a material "means" (to borrow Aristotle's terminology) like language or paint, embedded in a tradition of cultural signification and communication, a body of symbolic forms capable of being invoked and reshaped to express meanings and values.

W.J.T. Mitchell, "Imperial Landscape," in *Landscape and Power* (Chicago, IL: University of Chicago Press, 1994), 14. For recent scholarship see John Durham Peters, *The Marvelous Clouds: Towards a Philosophy of Elemental Media* (Chicago, IL: University of Chicago Press, 2015); Parikka, *A Geology of Media.*

19 See Diana Coole and Samantha Frost (eds.), *New Materialism: Ontology, Agency, and Politics* (Durham, NC: Duke University Press, 2010).

20 Yusoff, "Excess, Catastrophe, and Climate Change," 1010.

21 Janet Walker has also broached this concept of "earth *re*-writing" in "Projecting Sea Level Rise: Documentary Film and Other Geolocative Technologies," in *A Companion to Contemporary Documentary Film*, edited by Alexandra Juhasz and Alisa Lebow (Chichester, West Sussex: Wiley-Blackwell, 2015).

22 Wikipedia, the free encyclopedia, entry for "sustainability," https://en.wikipedia.org/wiki/Sustainability (accessed August 17, 2015).

23 Oxford English Dictionary Online (OED Third Edition, March 2012), http://www.oed.com.proxy.library.ucsb.edu:2048/view/Entry/195209?rskey=mgod9W&result=2#eid (accessed August 17, 2015).

24 Dictionary.com, http://dictionary.reference.com/browse/sustainable (accessed August 17, 2015).

25 Andrew Dobson, "Drei Konzepte ökologischer Nachhaltigkeit," *Natur und Kultur—Transdisziplinäre Zeitschrift für ökologischer Nachhaltigkeit*, 1, no. 1 (2000): 62–85. Cited in Daniel Bonevac, "Is Sustainability Sustainable?," *Academic Questions*, 23 (2010): 84–101.

26 Gro Harlem Brundtland, et al., *Our Common Future: Report of the World Commission on Environment and Development* (Oxford: Oxford University Press, 1987).

27 Earth Charter, http://www.earthcharterinaction.org/content/ (accessed August 17, 2015). Here we would qualify the discussion by acknowledging the critical scholarship on "rights." See, for example, Wendy Brown's key point in *States of Injury: Power and Freedom in Late Modernity* (Princeton, NJ: Princeton University Press, 1995), that

> The question of the liberatory or egalitarian force of rights is always historically and culturally circumscribed; rights have no inherent political semiotic, no innate capacity either to advance or impede radical democratic ideas. . . . [yet] while the measure of their political efficacy requires a high degree of historical and social specificity, rights operate as a political discourse of the general, the generic, and universal. (97)

28 Bonevac, "Is Sustainability Sustainable?"

29 Steve Mentz, "After Sustainability," *PMLA*, 127, no. 3 (2012): 586.

30 Stephanie LeMenager and Stephanie Foote, "Editors' Column," *Resilience: A Journal of the Environmental Humanities*, 1, no. 1 (January 2014): 4.

31 Ibid., 2. The editors continue:

> We wish to *occupy* resilience for the purpose of thinking and acting together, and we propose this journal as a bridge toward a scholarly language commons, by which we do not intend a common language, where such is conceived in the reactionary sense of English only or the reactionary-progressive sense of consensual compromise.

32 Katrina Dodson, "Eco/Critical Entanglements," *qui parle*, 19, no. 2 (Spring/Summer 2011): 5.
33 Ibid., 7.
34 Smith, *Eco-Sonic Media.*
35 As another example, Leila Nadir and Cary Peppermint's project *Ecologies of Inconvenience* (2015), meditates on the slowness of labor-intensive forms of food production and their associated cultures of sustainability, http://www.ecoarttech.net/about/ (accessed August 17, 2015). Also, in a discussion of "how the printed word *changes*" the environment it may also represent, Maxwell and Miller indicate that "the environmental costs of production for one e-reader (including raw materials, transport, energy, and disposal) far outweigh those of one book printed on recycled paper." Maxwell and Miller, *Greening the Media*, 63. Bozak also probes the assumption that "'Going digital' is more than ever considered a default means of 'going green.'" Bozak, *The Cinematic Footprint*, 12.
36 Bozak, *The Cinematic Footprint*, 67.
37 Nixon, *Slow Violence.*
38 Yusoff, "Excess, Catastrophe, and Climate Change," citing René Char in an epigraph as quoted in Georges Bataille, *The Absence of Myth* (London: Verso, 1994), 132.
39 Referencing Smith, *Eco-Sonic Media.*
40 See Jennifer Holt and Patrick Vonderau, "'Where the Internet Lives': Data Centers as Cloud Infrastructure," in Parks and Starosielski, *Signal Traffic*, 71–93.

PART I

Resource Media

1

500,000 KILOWATTS OF STARDUST

An Ecomaterialist Reframing of *Singin' in the Rain*

Hunter Vaughan

Cut to:

INT. DARKNESS

A large warehouse door opens to reveal a couple in extreme long shot, silhouetted in the light. Cut to a long shot as they—Don Lockwood and Kathy Seldon—enter a Hollywood soundstage.

DON: This is the proper setting.

KATHY: Why, it's just an empty stage.

DON: At first glance, yes, but wait a second.
(Don turns a handle and lights flare on a backdrop: a pink and silver skyline.)

DON: A beautiful sunset.
(He pushes a lever: smoke comes out of a tube on the floor.)

DON: Mist from the distant mountains.
(He throws another lever: red overhead light, the trill of a flute.)

DON: Colored lights in a garden.
(Don takes Kathy's hand and ushers her to a ladder, which she ascends.)

DON: A lady is standing on her balcony, in a rose-trellised bower.
(Don turns a gel light on her: purple.)

DON: Flooded with moonlight. We add five hundred thousand kilowatts of stardust.
(The flute trills as Don throws a number of levers and overhead lights of white, red, and green shower down.)

DON: A soft summer breeze.
(Don turns on an industrial-sized fan and takes a step toward the ladder, pauses.)

DON: And . . . you sure look lovely in the moonlight, Kathy.
(Cut to: close-up of Kathy, the lights in soft focus behind her.)

KATHY: Now that you have the proper setting, can you say it?
(Cut to: medium shot of Don.)
DON: I'll try.

This pivotal scene from Stanley Donen and Gene Kelly's 1952 classic, *Singin' in the Rain*, provides the pinnacle of sincerity in a film that hinges on irony, artifice, and play. An integrated backstage musical set during Hollywood's transition from silent to sound eras, *Singin' in the Rain* is a self-reflexive satire of the deceptions and hypocrisies that fuel the myths of the silver screen. Its multiple semiotic layers and performative innovation have incited a long-standing track record of criticism from many angles: star studies, genre theory, Hollywood historiography, as well as more postmodern analysis of narrative self-referentiality. The film's Hollywood-insider premise gives it a buffer from criticism, a sort of built-in self-analysis that invites us to partake in its irony; however, while many would agree with Sharon Buzzard's generalization that the backstage musical genre "alerts the viewer to the importance of their engagement as informed spectators," we must also consider what is being hidden deeper beneath this flattering guise.[1]

A big-budget genre film from the studio that defined excess, and an unapologetic catalyst for the performance of its star, MGM's *Singin' in the Rain*'s industrial provenance reaffirms the very artifice and manipulation it critiques—even the sacrilegious use of dubbing that is so central to its narrative meaning.[2] In fact, despite its tongue-in-cheek revelation of the apparatus, *Singin' in the Rain* never fully subverts the artificiality of Hollywood, and as such has maintained an ongoing threesome, comfortably in bed with the mainstream *and* its discontents. While the film's narrative transparency has led many to embrace it as subversive, I would agree rather with Carol J. Clover's insistence that we "see the moralizing surface story of *Singin' in the Rain* as a guilty disavowal of the practices that went into its own making."[3] This disavowal is not a condemnation of Hollywood artifice, but a cover-up, a misdirection through which the film invites us to believe in its revelation of artifice, and plays upon our faith in its honesty. But what, then, is it covering—what is the film's "guilty" secret?

This chapter provides an alternative approach to the film, an eco-critical approach that explores the film's rich layering of conflicting discourses, the green ramifications of its material practices, and the larger significance of how it represents our relationship to the natural and the artificial. In light of these concerns, the soundstage scene laid out in my introduction offers great insight into the actual extent of Hollywood's self-consciousness as to the natural resource cost of its fabrications, and becomes emblematic of the complex audience-industry compact designed to justify the exploitation of nature at the service of screen spectacle. In these pages, I will offer an ecomaterialist intervention in conventional film historiography, an alternative narrative of the ecological significance of a film's production, textual meaning, and the network of discourses that extends from film marketing to its critical reception. By "ecomaterialism" I invoke a material turn in eco-criticism, a shift in current eco-critical approaches to focus less on the problems of representation and more on the concrete environmental consequences of film culture, from production methods to marketing discourse. I also hope to contribute to critical paradigms that move "materialism" beyond a unilaterally Marxist understanding, by offering an approach that pierces the concrete and tangible consequences of our cultural practices. Through archival research and production culture fieldwork, we can reveal far more about how, both on- and off-screen, media use nature to produce culture. This should permit us to build a bridge from representationally driven eco-critical analysis to infrastructural approaches as exemplified by Starosielski, Gabrys, and others in this volume while also adding medial specificity and industry studies depth to current attempts to integrate media into environmental studies. Like the other works in this book, my aim is not only to encourage an ecological reframing of media technology, but also to insist that we rethink the environmental ramifications of our daily attitude toward cultural practice.

Water, Water Everywhere

> Of all our natural resources water has become the most precious. By far the greater part of the earth's surface is covered by its enveloping seas, yet in the midst of this plenty we are in want. . . . In an age when man has forgotten his origins and is blind even to his most essential needs for survival, water along with other resources has become the victim of his indifference.
>
> —Rachel Carson[4]

This passage from Rachel Carson's trailblazing 1962 environmentalist sermon on the mount, *Silent Spring*, identified the important ecological role of water—including humanity's dependence and negative impact on it—over half a century ago, and our species' interaction with it is constantly evolving. Current crises of clean and consumable water in some places, and growing concerns about fresh water's general potential scarcity, reflect not so much what nature offers us as what we offer it—what humanity does with and to this natural resource and the atmospheric ecosystem that makes it naturally renewable. Water makes up 70 percent of the human body and, in striking proportionate congruency, nearly three-fourths of the surface area of our planet. We use water for nearly everything we do: from drinking to bathing to developing photographs to cooling Internet-server warehouses, water is involved in—if not central to—most human activities. As such, it is no surprise that water has been deeply integral to the history of film, from its basic production needs to its topics of fascination, its sublime affectual force, and its inquiry into human social practices.

In this chapter I provide a case study to investigate the way that American cinema has engaged with water, from representation to production to a surrounding network of discourse. What are some thematic and resonant ways in which water has been shown on the big screen, and how have humans been positioned in relation to it? How has water been integrated into the production of mainstream films, in terms of its extra-filmic consumption (from the basic hydration of cast and crew to the use of water for cooling postproduction digital memory servers, to California's reliance on hydroelectric power) as well as its textual role as setting and prop, from manufactured oceans to artificial rainfall?

Singin' in the Rain is an ideal starting place for this approach for two reasons: first, its high-profile status at MGM during production, in the trade world upon release, and in Hollywood historical studies and criticism ever since, means that its production process was painstakingly recorded, its marketing strategies and reviews remain accessible, and its handling by decades of scholarship leaves a paper trail reflecting the methods and concerns of the discipline. Second, the film's prototypically high-concept life cycle focuses specifically on issues of natural resource use and natural representation. From the film's title, to the shooting (and critical celebration) of its most iconic scene, to its advertisement slogans and cross-marketing strategies, water is central to every level of the film's magic. I will use this central object in order to discuss Hollywood's use of (and impact on) the environment, and to explore what the underlying connotations of mainstream practices and textual meanings say about our historical, collective views on the relationship between nature and the culture of spectacle. Beginning with the contemporary environmentalist swing in Hollywood, I will expand to a historical perspective of the role of water in film production, and then turn to a close study of *Singin' in the Rain*, moving beyond the representational problems of textual analysis and toward an ecomaterialist study of this film's production methods and marketing discourse.

While Hollywood was historically founded (and remains so) upon the shoulders of excess and waste, it has in recent decades developed a complex, if inconsistent, environmentalism. Discourse on the subject in the trade paper *Variety* reached a peak in 1993, ebbed for a decade, and then experienced a resurgence building up to the climactic release of Davis Guggenheim and Al Gore's hugely successful *An Inconvenient Truth* in 2006. A well-timed 2006 study made by Charles C. Corbett and Richard P. Turco for UCLA's Institute of the Environment and Sustainability, titled "Southern California Environmental Report Card" and derived from a larger study, "Sustainability in the Motion Picture Industry," condemned the environmental footprint of industry practices.[5] This groundbreaking study not only reaches alarming conclusions about the ecological footprint of the moving-image culture industry, but also helps to map out a grid of the various levels on which the industry has an environmental impact, as well as how its contemporary view of environmentally sustainable practice does not promise a progressive change in the near future.

Still, Hollywood was forced to realize that this issue is publicly popular, a cause célèbre for the new millennium and, just as importantly in an almost poststudio age, economical. The Environmental Media Association developed the Green Seal certificate in 1989 to foster a set of best green practices, and the Producers

Guild of America developed its own PGA Green network, followed by environmentalism and sustainability positions at many of the studios. Major productions began to go carbon neutral (though this really only means offsetting otherwise traditional pollutant methods by donating money to independent organizations); and, not to be outdone by anyone, Rupert Murdoch turned NewsCorp carbon neutral in 2010. All of this is to say that Hollywood's green conscience to this point is a marketplace conscience, torn in conflicting directions by the forces of economics, industry, and public image. Recognizing the economic benefits of sustainable practice, Hollywood studios have begun to tighten the efficiency and renewability of their raw material use, foregoing, however, any radical industrial change or empirical critique of the price our planet pays for this culture of excess. To maximize visibility, the industry has cosmetically shifted the aesthetics of the star system, with websites such as Ecorazzi.com taking advantage of photo ops to show icons like Leonardo DiCaprio driving around in a Prius. As such, the discursive machinery of Hollywood manages to deflect popular criticism of the environmental ramifications of film production, and to avoid governmental regulation of its practices.

With similarly benign visibility (and weighted by an intellectual inertia typical of the slow crawl of mainstream criticism), scholarly interest in this topic has maintained its focus on questions of representation: how do films represent nature and our relationship to it; what do films say about environmental issues; how can "eco-films" act as tools of geopolitical change? These are valid and important questions and, while most eco-critical studies of film acknowledge the need to include analysis of production methods, they rarely do; although it is important to address the way we show nature, it is also necessary to move beyond questions of representation to the material concerns and consequences of our cultural practices. In my forthcoming book, *500,000 Kilowatts of Stardust*, I develop an ecomaterialist framework that begins with a film-philosophical approach to representation and moves on to explore questions of the environmental consequences of film production, as well as how these two are tied together in the discursive practices surrounding a film—from its marketing to historical studies of the film done many decades hence. In this chapter I use *Singin' in the Rain* as a case study to introduce this methodology. By focusing specifically on the role that water plays in the film's many layers, I build a larger eco-critical framework for addressing film representation, production culture, and discourse.

Lights, Camera, Water!

Film, like its users, needs water. And, once that water goes through its system, film has to excrete the waste left behind. Water is a necessary element on multiple levels throughout the life cycle of a film, as both an article of representation and an essential part of the chemical process that makes both analog and digital movie magic possible. The recent turn to more sustainable and environmentally conscious practices has elicited some water-based solutions, such as the shift from individual disposable water bottles back to old-school collective water-cooler efficiency, which stands out as a pretty consistent "greening" initiative of mainstream production practices. However, the real problems lie—as they often do—behind the behind-the-scenes, in the processes that are not highlighted in information brochures.

Film's technology and industrial methods have a deep impact on this ubiquitously needed resource, tracing to the production of the raw material itself. Eastman Kodak, which established its monopoly on the patent technology and practices of film stock production in the early 1920s, was not only the second-largest consumer of pure silver bullion (after the US Mint) but also a cavernous abyss for water use and pollution. As Maxwell and Miller document in *Greening the Media*, the Kodak Park plant in Rochester, New York, was propped strategically alongside Lake Ontario, from which it drew *more than twelve million gallons of water daily* for the annual production of 200,000 miles of film stock during the 1920s.[6] This comes to 18.56 cubic feet of water per second, or the equivalent of filling 240,000 bathtubs, or 18 Olympic-sized swimming pools, per day.[7] While these figures should be placed in perspective with those of other industrial production practices that are shockingly high in their required water use (it takes more than ten gallons of water to produce one slice of bread, over 713 gallons go into the production of one cotton T-shirt, and more than 39,000 gallons are required in the manufacturing of a new car), the magnitude of consumption necessary—or, at least, made necessary—by the production of early film stock is noteworthy.

Twelve million gallons per day, beginning nearly a century ago. By the end of the twentieth century, when it was responsible for 80 percent of the world's film supply, Kodak Park was using 35 to 53 million gallons of fresh water per day—at times over quadruple the amount in the 1920s.[8] To put this in perspective: 53 million is 530,000 times the 100 gallons used per day by American residents (and 10 to 20 million times the two to five gallons—seemingly insignificant by comparison—used daily by a resident of sub-Saharan Africa).[9] The water use of half a million viewers all to make a raw cultural material the majority of which will end up as waste on the cutting room floor.

However, it is not only what goes in that must be accounted for, but also what comes out—and where it goes. After being siphoned off of Lake Ontario, the water is run through the plant's elaborate chemical rinsing process and is then dumped into the Genesee River, which extends through Rochester and another 157 miles down through New York and into Pennsylvania. In 1972, the Clean Water Act forced American factories to collect the majority of their wastewater in treatment plants. Regardless, by the end of the century, Kodak's dumping of postproduction chemicals into the groundwater of New York made it the primary source of carcinogenic pathogens in the state, and Rochester was "ranked number one for overall releases of carcinogenic chemicals" from 1987 to 2000.[10]

And this, as they say, is just the tip of the iceberg; or, rather, this is a chunk of what is invisible to the eye, kept out of the spotlight of industry discourse. It is, nonetheless, symptomatic of the media industry's sleight of hand that diverts public and regulatory attention away from the sausage factory and toward the hot dog stand. Beyond the technical details of how the nuts and bolts are themselves made, though, issues of water consumption, usage, and dispersal have been central to the practice and discourse of many of Hollywood's most iconic films. The two most recent Hollywood films to build much of their prerelease momentum and sublime aesthetic upon the grandiose control of water are James Cameron's 1997 epic romance, *Titanic*, and Roland Emmerich's 2004 disaster flick, *The Day After Tomorrow*, each of which illustrates the necessity of an ecomaterialist approach to the complex environmental ramifications of film production and marketing culture.

Garnering fourteen Academy Award nominations and eleven wins, including Best Film and Best Director, *Titanic* indisputably held the critical respect and appreciation of its industry peers and the adoration of the viewing public. A parallel narrative that moves seamlessly between the ill-fated 1912 maiden voyage and a present-day attempt to salvage artifacts from its remains, the film offers a sort of ironic testimony to the cold destructive power of an indifferent water-based nature, explored through a romanticized vision of human exploratory technology and driven by an extra-filmic braggadocio about the excesses of production and the indulgence of spectacle, from the film's ballooning budget to the extravagant crafting of natural resources for its aquatic soundstage filming. In *Greening the Media*, Maxwell and Miller analyze in great depth the contradictions of the film's making, highlighting its outsourced production in Mexico as a testament to the transnational cultural project intended by the North American Free Trade Agreement (NAFTA).[11] However, there was also a very negative "unseen" impact on the local biosphere—unseen in that it was neither part of Fox's marketing campaign for the film nor part of the narrative being told in the trade papers. Popotla, the Mexican fishing town where the film was shot, was cut off from the sea and local fisheries were devastated by a massive movie wall built to keep local citizens away; and, Fox's chlorine treatment of the water on set led to the pollution of surrounding seawater, decimated the local sea urchin industry, and reduced overall fish levels by a third.[12]

Hinging on the narrative premise of a series of water-based natural disasters (tsunami, flood, and ensuing ice age) caused by global warming, *The Day After Tomorrow* unabashedly asserts itself as a warning to human civilization and a champion of the rising tide of environmental science. Aggressively marketed as an environmentalist film, not only for its apocalyptic science fiction warning but also for its production methods, *The Day After Tomorrow* was a carbon-neutral production, which means that the carbon dioxide generated by its production was offset by funding environmental groups and planting trees; it did not actually alter its practices to become more environmentally sound, but merely budgeted an institutional atonement for its pollution. The film is ultimately not an eco-friendly production, and its message is scientifically inaccurate, but it was advertised as carbon neutral and its grandeur of spectacle proved enormously successful on a commercial level, catering to heightened audience fears in an age of increasing uncertainty and unpredictability regarding meteorological events and natural disasters.

These snapshots of *Titanic* and *The Day After Tomorrow*, provided by Maxwell and Miller in the context of their broader infrastructural study, inspire entry into the material complexities of an eco-critical approach to film. In order to unveil the praxis of such superficially progressive texts, it is important to contextualize representational analysis in relation to the practical problems of filmmaking and the discursive channels that spin the textual meaning into a tapestry of ideological appeal and cross-market advertisement. In order to cultivate such an approach, I turn now to a close analysis of the Stanley Donen and Gene Kelly 1952 classic, *Singin' in the Rain.*

Laughing at Clouds: *Singin' in the Rain* and Ecomaterialism

Like any fledgling methodology, the ecomaterialist approach I am etching out here struggles with access and record, attempting to wrest indirect analysis and theoretical conclusion from a black hole of information scarcity. The early 1950s lacked today's perspective on the environmental impact of human industry, and therefore the film studio documentation of natural resource use, treatment, and waste is subtle at best and often gleaned only indirectly. Nonetheless, the role of nature and natural resources in *Singin' in the Rain*—from production to marketing to critical discourse—is profound and can be systematically revealed through an ecomaterialist approach.

Ranked fifth by the American Film Institute (AFI) on its top 100 films of all time, *Singin' in the Rain* has maintained an important place in the upper echelon of film criticism, remaining a signet of studio, stardom, and genre operations in Hollywood. The polysemous nature of its self-reflexive premise makes it enjoyable on numerous levels, and its blanket self-irony invites viewers to participate and even get lost in its layered play of sincerity and spectacle. As demonstrated by the soundstage scene quoted at the start of the chapter, *Singin' in the Rain* explicitly acknowledges the vast amount of natural resources Hollywood uses in order to manufacture an artificial version of the very nature from which its resources are taken, implicating the viewer in a shockingly frivolous representation of film culture's cavalier disregard for environmental sources.

This extreme arrogance burrows to the discursive core of the film's titular meaning: singin' in the rain. Don Lockwood is in love and this makes him impervious to the weather; in fact, more than impervious, it makes him reject the impositions of nature, and contradict them even with a jovial eruption of emotion. The film's eponymous number embodies the classical Hollywood nature of excess and what is not only indifference toward the consequences of this excess, but a pledged capacity to overcome the sovereign status of the natural world. Don is not singing because it is raining; he is singing *in spite of* the rain. His "glorious feeling" and the performative spectacle that articulates it are a direct rejection and subordination of the natural elements: not only is he contradicting the stereotypical emotional effect of rain (sadness), but his interactive play with the rain goes so far as to physically assault it (in my alternate reading), repeatedly kicking the rain and thereby accentuating the power of human ingenuity and the triumph of human force over the meager presence of the elements.

Deemed more of a technical than a choreographic challenge, and absent from the screenplays leading up to the shoot, the iconic sidewalk scene from *Singin' in the Rain* has provided no shortage of Hollywood legend and lore, some of it apocryphal or inaccurate, some of it still unfolding through new scholarship on this film that, over sixty years later, continues to delight audiences and intrigue historians and critics. Despite the repeated accounts and in-depth investigations of very reputable scholars, each one—rife with an intellectual balance of admiration and scrutiny—offering a painstakingly detailed portrait of the production of this scene in particular, not one study of this film raises what seems an obvious and disturbing line of questioning: where did all that water come from, and where did it go? How was its use orchestrated and accounted for?[13]

In a typically romanticized account, the Arthur Freed Unit producing the film has been described as "living almost in a dream world" of limitless budget and material supply, a degree of excess worthy of MGM's reputation and demonstrated by the more than five hundred costumes used in the film.[14] The "Singin' in the Rain" musical number, which originally appeared in *The Hollywood Revue of 1929*, had been pitched without a specific locale in mind, and upon Gene Kelly's revelation—"I thought of the fun children have splashing about in rain puddles and decided to become a kid again during the number"[15]—the right space was sought for what Kelly and Donen envisioned as an innovative form of cine-dance. Peter N. Chumo II claims that the film moves beyond the conventions of its genre, taking this pivotal dance number into the street under the seemingly worst possible weather conditions. The characters are not trapped, he argues, by narrative or generic conventions; nor, I would speculate, are the cast and crew trapped by limitations of natural resource consumption, water use, or power supply, though such concerns (or lack thereof) are undocumented.[16]

Where direct accounts fail to record industry history, we must turn to production culture study to provide us with some insights about the actual way in which artificial rain is manufactured. A very generous conversation with longtime Hollywood special effects coordinator Steve Galich illuminated some of the practicalities of such a shoot, as well as the unique nature of the MGM infrastructure and the studio's relationship to Culver City. For rainwork (or a "rain job"), special-effects teams typically run fire hoses to the "rain pipe" that is held in place by a thirty-ton crane; the water is supplied by large pumps set in place to take water from variant sources, depending on the location of the shoot, including dammed-up creeks, newly installed wells, or water reservoirs. Different cities have different policies in terms of how such practices are regulated. Because of local instances of industrial pollution—such as Lockheed Martin's decades-long chemical runoff into the San Fernando aquifer, uncovered in 1980—regulators now keep a close watch on industrial runoff in the Los Angeles River; as a result, the city of Burbank now requires Disney to catch all runoff and pump it into the sewer. As Galich puts it, in its heyday MGM more or less ran Culver City and laid out its studio lots around a central water tower (prominent and in some ways iconic in older photos of the studio) that could reach its extensive lots and sound stages.

This scene in *Singin' in the Rain* does not, of course, take place on the street in the worst of weather, but was filmed—with highly controlled meteorology—on the East Side Street in Lot 2 of the MGM studio in Culver City, the nighttime sky simulated by draping black tarpaulins across two entire blocks (which also allowed MGM to avoid paying its technical crew overtime). The centrality of heavy rainfall to the scene's metaphorical meaning and affective force necessitated an elaborate blend of engineering and design; and, far from being achieved in one take, the scene's day-and-a-half shoot actually rendered the worst shooting ratio of the entire production, capping off an eight-day run of extreme resource waste. The set design specified seven puddles with two inches of water apiece, which were to be carved out of the pavement on the lot. Over six days of rehearsal, a network of pipes using water directly from the Culver City water system ran six hours per day in order to provide for the complicated orchestration of what would become Kelly's most iconic moment, with the technical crew making adjustments to extend the scene space for Kelly's movement. Six hours a day for six days, just to rehearse; water, running. The star himself was meticulous in planning the scene; for example, twenty minutes was dedicated to practicing the "cascade" effect of standing under the broken downspout, with technicians adjusting the pipes feeding the water to achieve the desired effect.

While archival documents do support the piecing together of this material production history, there is still a glaring hole in Hollywood's record when it comes to documentation of natural resource use: the production accounts lack a listing of the total amount of water used. Although this gap may be telling, it remains difficult to gauge to what extent classical Hollywood productions like *Singin' in the Rain* were aware of their environmental footprint, and popular discourse on green politics and sustainability were some decades on the horizon. Still, the last note on the Assistant Director's Report from the rehearsal of this scene on July 17, 1951, coming at 6:35 p.m., reads: "Company dismissed because crew was about to run into meal penalty—water pressure is low (40 lbs in Culver City) necessitates more rigging."[17] As Hugh Fordin describes it in his book on the Freed Unit's years at MGM, *The World of Entertainment*: "Instead of the desired downpour, all they got was a tired drizzle, and this no matter how high they turned the control valves. All of the studio's tanks were checked."[18] It was discovered, though, that "rigging" was not the problem—the problem was that, at five o'clock in the afternoon, the residents of Culver City would come home and turn on their sprinklers, and thus trigger a drop in water pressure on the lot. That the Assistant Director's Report notes this, and that filming was consequently scheduled for earlier in the day, shows at the very least that the filmmakers and studio were aware that water was in fact a collective natural resource that was in some way finite, limited in its supply and distribution. It is on the basis of tangible hints such as these that I propose this ecomaterialist reframing of the film-about-which-seemingly-everything-has-been-written in the interest of an alternative industrial history, as a counter-narrative analyzing the cost-benefit ramifications of the Hollywood spectacle.

Shower of Profits: The Title Should Have Been *Hollywood*

The role of natural resources and the material environmental implications of Hollywood production practice extend beyond what appears on screen and how it was put there. Those eight days on Lot 2 at MGM set the tone for how water was and is used discursively in the marketing, reception, and subsequent scholarly preservation of *Singin' in the Rain*. In *Movies about the Movies*, Christopher Ames captures the central song's underlying textual significance:

> This particular number belongs to a class of popular songs that preach the power of one's outlook to overcome external obstacles, a significant theme in American popular culture. Thus the rain represents hardship, and Gene Kelly begins the number by dismissing his cab and folding up his umbrella (which he later gives away after using it as a dancing prop). Having fallen in love, Don Lockwood is made oblivious to the rain.[19]

We can see here that, just as this number epitomizes the film's soaring romantic gesture, so does it sum up in a nutshell the film's tacit discourse on nature.

This scholarly focus on the "Singin' in the Rain" number echoes the critical reception upon the film's release: most American reviews praised the film, in particular Kelly's performance—itself encapsulated by this very scene, in which only the technical virtuosity of MGM's sweeping Technicolor cinematography could match the star's lavish performance and the brilliant choreography that melded his tap-dancing with the patter of the raindrops. However, perhaps *the* most influential reviewer of the time saw the title as misrepresentative. On March 28, 1952, *New York Times* film critic Bosley Crowther trained his blistering wit on Hollywood's newest sensation. Though noting, as did most other reviews, that Kelly's performance dominated the film and "by far his most captivating number is done to the title song—a beautifully soggy tap dance performed in the splashing rain," Crowther insists that the title "has no more to do with its story than it has to do with performing dogs." He continues with characteristic aplomb: "But that doesn't make any difference, for the nonsense is generally good and at times it reaches the level of first-class satiric burlesque."[20] As if echoing Crowther's skepticism, Stanley Donen himself remarked on the bad fit of the film's title: "The title of the picture never should have been *Singin' in the Rain*. Look at the picture. It's not about the weather. The theme has nothing to do with rain. The picture is about movies. The title should have been *Hollywood*."[21] The titular number is, indeed, as Ames puts it, "a song about singing in a musical about musicals in a Hollywood movie about Hollywood,"[22] and reviews of the time consistently applaud its self-ironic brilliance: *Variety* estimates that "the fact that Hollywood can laugh so heartily at itself, only adds to the appeal."[23]

But, in response to Donen, one must ask: which Hollywood? The film's level of denotation claims it is about the Hollywood of 1927. Yet, its connotations are heavily drawn to comment on the Hollywood of 1952, in a state of profound conflict between increasing regulation and the need to refine its transparent production of spectacle—which is why the soundstage and the sidewalk scenes are so pertinent. This film, which offers so much insight into the way movies are made, encourages us to admire how much power is used to light a film set, thus asking us to absolve its waste and to justify its excess by paying our price of admission to the resultant spectacle. What Crowther fails to identify—indeed, what Crowther seems admittedly to be blinded to by the film's excessively entertaining spectacle—is that this is not in fact a misplaced title, but a practice in discursive misdirection. The title does not fail to bring to light what the film is really *about*; it is *we* who, being taken in by the film's subterfuge of an engaging story about movies, miss the mark and accept the comfortable meta-pleasure of the film's superficial self-reflexivity. It is we who, despite clear references made in the film to the absurd amount of power and resources wasted to produce the artificial spectacle that is a "movie," choose not to register the larger ramifications of this waste. Jean-Louis Comolli's materialist argument of formal transparency still holds, and is actually buttressed by the audience's participation in the farce of self-reference; but, here, the spectatorial disavowal required for analogical fiction is projected beyond the typical level of denotation to a disavowal of the hidden material consequences of production practices we get to see on the screen.[24] By drawing our attention to the hypocrisies of Hollywood, the film's sleight of hand distracts us from the ultimate hypocrisy of spectacle and waste both practiced and celebrated in *Singin' in the Rain*.

Despite what Donen says, the film is very much about rain, from the manufacturing of artificial rain to the evolving role of Hollywood as a multidirectional convergence marketing entity, as we can discern from the production files, Exhibitor's Campaign Book, print advertisements, and on-location anecdotes. On each of these levels, the film sells itself according to this scene and its poetic metaphor, and sells many other things in conjunction. The playbill for the film goes into extensive production detail about the shooting of the sidewalk scene in order to impress upon industry colleagues and audiences the ingenuity of natural resource manipulation necessary to master the manufacturing of rainfall.[25] The marketing blitz that ensued upon the film's release revolves entirely around the theme of rain, rainwear, and rain-based puns, and MGM even marketed a tie-in "Gay Parasol" for sale.[26] The Exhibitor's Campaign Book includes a "Disk Jockey Stunt for M-G-M Records Album 'Singin' in the Rain'": a press release littered with photographs of female models, clad in designer raincoats and umbrellas, making promotional visits to prominent radio personalities. The text reads: "A pretty young model dressed in shorts complete with raincoat and umbrella appropriately lettered with copy advertising the sound-track album can get you a lot of free radio plugs."[27] Such gratuitously exploitative tie-ins are continued in large full-page sample ads for Milliken's Dacron wool and Rain Shedder raincoats, including pictorial examples of ads in newspapers and magazines, descriptions of their goods and campaigns, and contact information for the respective companies.

The "pretty young model" promised in the "Disk Jockey Stunt" is exemplary of the unabashedly conservative and masculinist gender politics at play in the marketing discourse surrounding the film. While this is not a study of that particular topic, it is certainly noteworthy how this problem is enmeshed in the discursive dependence on rain-based wordplay, imagery, and consumer goods, and the suggested ways for exhibitors to market the film based on its natural imagery, including radio slogans such as:

> It's as gay as a rainbow . . . bright as a cloud's silver lining . . . welcome as the sun after a spring shower! It's M-G-M's glorious-feeling, Technicolor musical—SINGIN' IN THE RAIN with a skyful of stars and a steady down-pour of song hits!

Singin' in the Rain was selling nature as an analogy, and was striking it rich. This is perhaps most abundantly clear in the film's full-page wide-release ad in the meta-trade journal, *The Exhibitor*, which opens in bold letters: "**IT WILL RAIN GOLD AT EASTER! Be ready with open dates to catch the shower of profits.**"[28] Nowhere are love and happiness mentioned, and this is not a discourse about Hollywood, unless it is about turning natural resources into money, one prominent characterization of Hollywood. *Singin' in the Rain*, which cost $2,540,800 to make, grossed $95,000 its opening weekend and brought in $7,665,000 during the entire run of its initial release.[29] Put quite succinctly in the *New York Compass* review of March 28, 1952: "Another MGM Technicolor mu$ical i$ here, folk$."[30]

The dynamic centrality of rain and water use to the film's identity stretches even to the production unit's mode of celebration: the film closed production on November 21, 1951, and as with all Freed pictures was celebrated with a set party. The party was arranged on Stage 28 at MGM, as Freed recounts in Fordin's *The World of Entertainment*:

> We brought the guests into the stage door and in order to get in I rigged up several pipes of rain. So, the only way you could get to the party was

> by taking an umbrella, which we handed out at the door and everybody walked through the rain.[31]

From rehearsal to production, to wrap party to playbill, to multimedia publicity stunts and cross-market tie-ins—in its visual imagery and linguistic punditry, meticulous orchestration of artificial rainfall, and boosted selling of rain jackets, this film is about rain. Though *Singin' in the Rain* stages a thoughtful critique of Hollywood with a number of memorable setups, gags, and numbers, the film's legacy is secured by four and a half minutes of artificial rainfall and beaming white-toothed smile. On every level, the film hinges on the triumph of humanity over nature, the crafted excess of the Hollywood spectacle. You can dig out the puddles until they look perfect; if the water pressure drops, it is human error, and you can choose a better time of day—only a few Inyo County farmers will feel the sting, while millions will enjoy the movie for decades to come. As this film reinforces through its representational meanings, its production practices, and its various layers of discourse, the Hollywood spectacle is not to be limited by the frugality of nature.

Notes

1 Sharon Buzzard, "The Do-It-Yourself Text: The Experience of Narrativity in Singin' in the Rain," *Journal of Film and Video*, 40, no. 3 (Summer 1988): 21.
2 Most notably, there is a montage sequence in which the lovely voice of the heroine, Kathy Seldon (played by Debbie Reynolds), is used to dub over the scenes played by the shrill and tyrannical silent star who is stunting Seldon's success. This was, in fact, done using the voice of Betty Royce, and not that of Debbie Reynolds. See Christopher Ames, *Movies about the Movies* (Lexington: University of Kentucky Press, 1997), 65.
3 Carol J. Clover, "Dancin' in the Rain," *Critical Inquiry*, 21, no. 4 (Summer 1995): 725.
4 Rachel Carson, *Silent Spring* (Boston, MA: Mariner Books, 2002 [orig. 1962]), 39.
5 Charles J. Corbett and Richard P. Turco, "Southern California Environmental Report Card, 2006," UCLA Institute of the Environment and Sustainability, http://www.environment.ucla.edu/reportcard/article1361.html (accessed January 2013).
6 Richard Maxwell and Toby Miller, *Greening the Media* (Oxford: Oxford University Press, 2012), 73.
7 "A Million Gallons of Water—How Much Water Is It?," US Geological Survey (USGS), http://ga.water.usgs.gov/edu/mgd.html (accessed June 2013); "Water Trivia Facts," US Environmental Protection Agency (EPA), http://water.epa.gov/learn/kids/drinkingwater/water_trivia_facts.cfm (accessed June 2013).
8 Maxwell and Miller, *Greening the Media*, 73.
9 "Water Trivia Facts," US Environmental Protection Agency (EPA).
10 Maxwell and Miller, *Greening the Media*, 73.
11 Ibid., 69.
12 Ibid., 70.
13 Peter Wollen's book-length study of the film, *Singin' in the Rain* (London: BFI, 2008), and Earl Hess and Pratibha Dabholkar's monograph, *Singin' in the Rain: The Making of an American Masterpiece* (Lawrence: University of Kansas Press, 2009), both go into great detail on the production of this scene.
14 Hess and Dabholkar, *Singin' in the Rain*, 12, 81.
15 Ibid., 126.

16 Peter N. Chumo II, "Dance, Flexibility, and the Renewal of a Genre in *Singin' in the Rain*," *Cinema Journal*, 36, no. 1 (Autumn 1996): 39–54.
17 Arthur Freed Collection, Box 21, Folder 2, Cinematic Arts Library, Doheny Memorial Library, University of Southern California.
18 Hugh Fordin, *The World of Entertainment* (New York, NY: Doubleday, 1975), 358.
19 Ames, *Movies about the Movies*, 66–67.
20 Bosley Crowther, "Singin' in the Rain (1952)," *New York Times*, March 28, 1952.
21 Stephen M. Silverman, *Dancing on the Ceiling: Stanley Donen and His Movies* (New York, NY: Knopf, 1996), 142.
22 Ames, *Movies about the Movies*, 67.
23 Bron., *Variety*, March 12, 1952.
24 Jean-Louis Comolli, "Machines of the Visible," in *The Cinematic Apparatus*, edited by Teresa de Lauretis and Stephen Heath (London: Palgrave Macmillan, 1985), 121–143.
25 Arthur Freed Collection, Box 21, Folder 1.
26 Pressbook Collection, Folder 73.
27 Arthur Freed Collection, Box 22, Folder 2.
28 Arthur Freed Collection, Box 22, Folder 4.
29 Hess and Dabholkar, *Singin' in the Rain*, 188.
30 F.R., *New York Compass*, March 28, 1952.
31 Fordin, *The World of Entertainment*, 361.

2

PIPELINE ECOLOGIES

Rural Entanglements of Fiber-Optic Cables

Nicole Starosielski

In March 2015, linemen of the Delhi Telephone Company extended a thin fiber-optic cable through a wide valley in the western Catskill Mountains. Next to the winding two-lane highway, the ground was still hard from the cold winter. Farmers had not yet tilled the land for the coming year's crop. Alongside the highway, the west branch of the Delaware River funneled water to an urban hub over a hundred miles south. And beneath Delaware County soil, natural gas flowed freely, yet untapped and unexploited in New York State. With the touch of a button by a network operator, a beam of light shot down the cable's glass fibers and channeled streams of information into the rural environment. The town of Bloomville, New York, had become a node in the high-speed global network.

In many ways, the arrival of fiber-optic cable in rural New York appears as a straightforward case of technological development. It is a story of the expansion of modern infrastructure to the rural fringe, brought by telecommunications companies and government agencies. The New NY Broadband Program map is a typical representation of such projects; in it, the landscape of media access is largely segregated along an urban-rural divide, with disconnected rural expanses signifying an imperative for new development (Figure 2.1). Some of this fiber-optic construction is financed by state programs, including the Connect NY Broadband Program, and tax dollars generated in New York City—facts that only amplify the apparent dependency of rural networks. The Delaware County Broadband Initiative, the latest build-out of this county's communications system, promises to expand Internet access even further to thousands of unserved and underserved users.

To see the construction of Internet infrastructure as merely a top-down process, a gift from external agencies, overlooks the myriad local and ecological forces, both human and nonhuman, that shape fiber-optic networks. It obscures the entanglement of signal distribution with other circulations across the mountains,

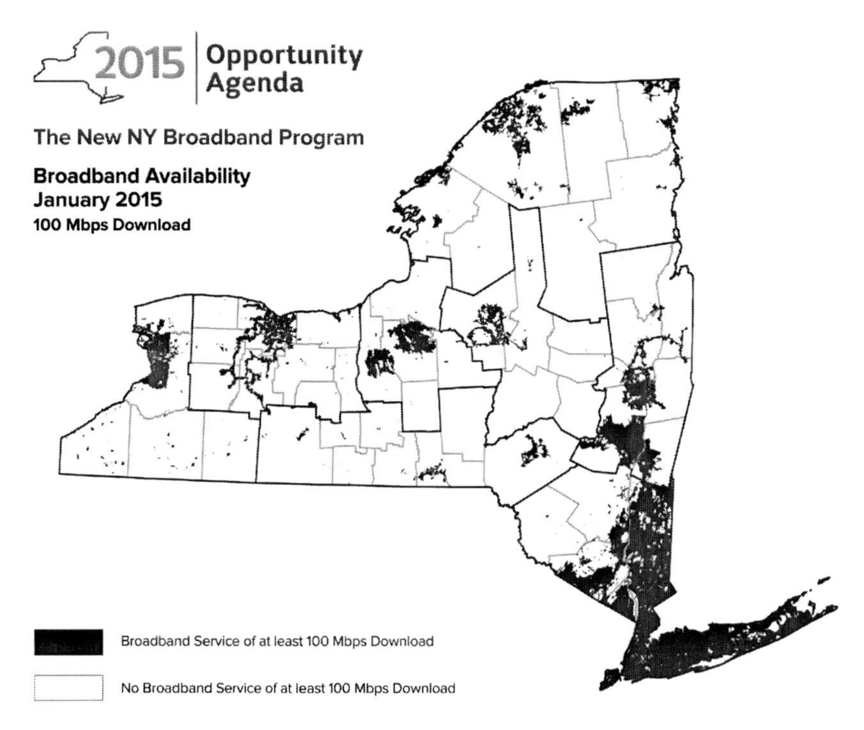

FIGURE 2.1 The New NY Broadband Program represents rural areas as blank spots on the map.

including the supply of electricity, the movement of milk, and the flow of water. "To be entangled," as Karen Barad has eloquently described, "is not simply to be intertwined with another, as in the joining of separate entities, but to lack an independent, self-contained existence. Existence is not an individual affair."[1] Delaware County, with its landscape of ever-present elemental forces, its history of cooperative organization, and its lasting ties to New York City, has long been attuned to such entanglements of existence.

Here, fiber-optic signals are entangled with electricity and tree growth, railways and dairy production, the protection of waterways, and the struggle against natural gas extraction. By documenting these imbrications, this chapter resists the usual narrative of infrastructural expansion. Instead of depicting distribution systems as capillaries that integrate rural areas into global circulations of commerce and culture, infrastructures are ecological forms that materialize within specific social and "natural" landscapes. The sustainability of media, then, is deeply tied to the systems and environments it traverses. In the first section of this chapter, I describe the close connections between the fiber-optic system and the local electrical network. The cable lines of the Delaware County Broadband Initiative follow a set of routes established not by global telecommunications companies but by the Delaware County Electrical Cooperative (DCEC), a legacy of the New

Deal that now provides an alternative to corporate energy provision. As a result, Internet infrastructure is enmeshed with the historical investments and ecological practices of the electrical cooperative.

Yet as they are layered over the DCEC's network, cables bring about a shift from a dispersed and uneven wireless system that requires users to negotiate the topography of mountains and valleys, to the funneling of communications via direct pipeline. Pipelines are a form of distribution that connects individual nodes via linear routes, and significantly for this chapter, insulate signal traffic from the environment. In order to supply a consistent stream of electrons and photons, the electrical system and the fiber-optic network attempt to disentangle these circulations from the surrounding elements, including the growth of trees, the wind through the valleys, and the heavy winter snows. By tracking the entanglements and divisions of such pipelines, this chapter draws attention to the ecological elements of media circulation.

While cables are often linked to electrical networks, in Delaware County both of these pipelines are tied to the transportation of milk to New York City. The county's networks, including the wires of the DCEC and the Margaretville Telephone Company (another partner in the county's broadband initiatives), were first laid to dairy farms. In the second section, I trace how cooperative agricultural networks are both early precedents of and material influences on today's fiber-optic system. These agricultural networks were not intended as a completely independent alternative to external markets, but often leveraged the larger-scale pipelines in the region. Today, digital networks are both used to support small-scale farming in the county, cultivating an ecological alternative to large-scale industrial farming, and to develop new distribution methods for sustainable agriculture—redirecting the circulation of capital from New York City in a way that parallels nineteenth-century practices. Although pipelines are technologies of exploitation and extraction, cooperative organizations can locally manipulate these systems at both their points of access and along their routes.

Underlying contemporary infrastructures, from the power lines that run along the DCEC's poles to digital networks used for sustainable agriculture, are waters that feed one of the most critical pipelines in the state: a series of rivers, reservoirs, and aqueducts that provide drinking water for New York City residents. In order to keep the system in accord with federal regulations, millions of dollars have been channeled into the area via the New York City Department of Environmental Protection. This is the most powerful environmental influence on the region's infrastructures: watershed-related funding has been important not only to the fiber-optic system, but to the ways that digital networks are mobilized. Just north of the watershed boundary, however, is another pipeline, one that conflicts with the watershed's ecological ethos: a planned natural gas pipeline. This system, too, will depend on and shape the possibilities for the area's communications systems. Yet rather than being layered over an existing route, or locally mobilized in cooperative networks, the watershed's zone of influence is being pitted against the proposed gas system by environmentalists who are thereby drawing from one pipeline to combat the expansion of another.

In this chapter, I suggest that we broaden our conception of pipelines beyond the thin metal conduits that transport gas and oil to include fiber-optic cables, electrical networks, railroads, and reservoirs. As they transport resources along a set of linear, insulated routes, these systems are intended to disentangle photons, electrons, milk, and water from the environments they move through. Yet none develop in isolation—they both extend out of past circulations and condition movements of the future. These local and regional entanglements of distribution, what I call "pipeline ecologies," are a critical component of contemporary media's environmental impact. As is documented across this collection and elsewhere, Internet infrastructure is dependent upon the mining of minerals, the unsustainable labor practices in technology production, and the energy-intensive geographies of data centers. However, distribution systems materialize in radically different ways as they extend across the earth's surface, whether through concentrated urban locales, suburban sprawl, mountainous farmland, or the ocean's depths. Regardless of where they are routed, Lisa Parks has argued, media distribution emerges as part of an ongoing spatial and environmental history.[2] Networks are powered by an array of electrical systems, maintained by vastly different laboring bodies, and are entangled with diverse social and ecological practices. While the resource pipelines of Delaware County are certainly underscored by environmentally harmful and exploitative activities, my aim here is to highlight how infrastructural reuse, cooperative modes of governance, and the strategic positioning of systems against one another might redirect these infrastructures toward more sustainable ends.

Wired Mountains

Four miles north of the fiber-optic cable in Bloomville, an index finger slides down a touch screen, signaling a cell phone to refresh its connection to an email server. A body orients to the antenna of a booster system, a domestic technology of amplification that enhances access to cellular transmissions otherwise blocked by mountains and trees. Not unlike the manipulation of television rabbit ears or radio antennae, this is a negotiation with the imperceptible aerial spectrum: a form of what Adrian Mackenzie calls "wirelessness."[3] As a storm moves quickly overhead, bringing intense winds and rain to the valley, it disturbs the atmosphere and generates interference for wireless media. The multicolored circle on Google's mail interface rotates over and over—bodies move in tandem with electronics, searching for the right location to catch the signal.

In this valley, built structures are less concentrated than in the town centers. Farms once staked out wide swathes of land, creating a buffer between architectures lessened only somewhat by the more recent parceling-off of smaller lots. The sparse population, even when clustered in groups of modular homes and trailers, means that there is a high ratio of expense to profit for any infrastructure development, and the fiber-optic line, which costs roughly $35,000 per mile, ends only a few miles outside town. This creates a division in the land. On one side, residents attune easily to global circulations via cable networks. On the other, bodies are oriented differently. Their access to cellular and satellite systems is more overtly mediated by the elements: atmospheric movements, local topography, and altitude shape where and when one can connect. Cell towers are often positioned on hilltops where they can propagate signals easily to users on proximate mountains, but as a result struggle to reach the valleys below. To move along winding county roads is to move in and out of coverage zones. Dense forest interrupts aerial transmissions and snow accumulates on satellite dishes. For wireless users, connection involves confronting an array of ecological forces.

The construction of the fiber-optic network is not about extending the Internet to people who have never encountered it before. It is about insulating network access in order to create predictable and reliable flows *despite* the fluctuation of environmental conditions. This is the structure of a pipeline: it regularizes flow along a series of linear vectors. Unlike cell towers and satellites, which create a wide zone of distribution called a "footprint," pipelines can only be accessed at specific nodes along their route.[4] These are, as Tim Ingold argues, "lines of occupation," throughways that connect "nodal points of power" and segment the earth into distinct territories.[5] The expansion of fiber-optic cables into Delaware County is not simply the introduction of a new technology—it is a renewal of a spatial organization and set of ecological practices established over seven decades before.

The division in the land in which fiber-optic cables are now embedded was made when electricity was first brought to the region. Just as the fiber-optic cable connects Bloomville and excludes farms further up the valley, early investor-owned electrical utilities brought service to the region's town centers. Electrical providers such as the Delhi Electric Company illuminated town streetlights, municipal buildings, businesses and homes in the early twentieth century, but left remote farms off their grid. By 1930, well after many American cities had been wired, less than 10 percent of farms had electricity. Wires were not only expensive to wind through rural terrain, but required an immense amount of labor, both in construction (because a massive number of trees had to be cleared) and in maintenance (because lines could be more easily disrupted by snow and wind). Establishing connections to rural areas was not only a problem of spanning great distances, but a problem of paving reliable pathways through the environment.

Just as the Internet is today accessible outside of rural towns via wireless infrastructure, electrical energy was also available outside of the towns, but depended on kerosene or gasoline-powered generators. Delco-Light Plants manufactured generators that afforded some farmers electrical illumination and power for small machines that would ease the "drudgery" of farming. H.E. Mason & Sons, a company based in Delaware County, advertised electrical plants alongside electrically powered commercial refrigeration appliances. One-cylinder gas engines could also be used to run farm machinery. Generators and engines required a large up-front investment, however, and could not support more than a few appliances. And even though they remained technically independent from the grid, power distribution via generators also tied users to fossil fuels and town centers. To purchase gasoline or contract repairmen, farmers had to traverse the winding mountain roads. As Joseph Tierney reports in his study of electrical adoption during this era, at most farms "it was common for the same engine to be hauled from machine to machine throughout the day."[6] As a result, generator power would be mediated by the weather, topography, and bodily capacities in ways that electricity distributed by wires—a direct pipeline between points—was not.

When the Rural Electrification Administration was established in 1935 to bring electricity to the nation's many farms, it was not intended to support these forms of distributed generation. Rather, as Samer Alatout and Chelsea Schelly observe, it defined electrification as "receiving central station service."[7] Rural electrical networks were conceived as a set of linear routes that would link small communities and remote farms to central plants. Alatout and Schelly argue that this new set of connections between town and country defined the "rural" in ways that were legible to government agencies and enabled new forms of bioterritorial control.[8] In Delaware County, local farmers came together to form the DCEC and were granted funding from the Rural Electrification Administration. In subsequent years, many of the area's farms would be integrated into a single electrical grid.

The resulting network took a particular spatial and ecological form. It did not connect towns already served by investor-owned networks, and instead linked directly between farms—key nodes of both community and government interest. Paving the way for these wires, often along the shortest path over the mountains, was an expansive project that mobilized the muscles of local workers and horses to pull poles, dig holes, and clear the forest. Employing workers was a key objective of the New Deal, and even after the network was established, regular maintenance was required to cut trees that grew under the wires, repair fallen lines after storms, and replace faulty poles. The movements of workers along the cable path stabilized these linear routes of electrical flow and, as a result, fixed users on a grid. This electrical system was a pipeline: an insulated path that funneled bodies and alternating current through the environment.

The fiber-optic lines of the Delaware County Broadband Initiative, as they run along the DCEC's poles, leverage these historical investments in network maintenance and the ecological choices about how it is powered. Since cables route signals along a corridor traversed for decades, a new path need not be cleared through the forest. And the many communications companies that already utilize the DCEC's poles pay yearly fees to a local, member-owned infrastructure company. The DCEC supplies electricity from renewable sources, including both hydropower from western New York and, increasingly, solar power. In recent years, the Cooperative launched an initiative to assist its members in setting up solar energy systems, allowing them to feed surplus energy back into the grid. And in 2015, they proposed the creation of a community solar farm, where members could invest in a solar installation that would be managed by the DCEC. As a result, the fiber-optic signals that pulse down these poles will be powered largely by renewable means and help to sustain the local economy.

There are a number of historical parallels between the expansion of electricity and the expansion of the Internet, including debates about the networking of rural areas, the imperative for government support, and the justification of these systems as essential for economic growth. Yet my purpose in recounting this history is to draw attention to the way that the fiber network here is enmeshed with the electrical grid, and takes on its ecological underpinnings, precisely because of the similarities in the form of distribution. The pipeline's linear routes enable the delivery of reliable capacity at prespecified nodes, situate bodies and technologies within a gridded landscape, and demarcate a split between those on the network and those off it. They offload the need for individual encounters with ecological forces, affording insulation from the environment. The fiber network is not simply an overlay onto an existing infrastructure; it is an entanglement, part of the same spatial and ecological organization of the land.

Cooperative Networks

It would be easy to see pipelines as simple incursions of modernity into rural ecologies as they streamline connections to a vast and centrally controlled grid, create divisions in the landscape, and bring with them technologies that regularize bodily movements and financial transactions. Just as the fiber-optic network today promises to circuit the flow of informational capitalism, the electrical network brought with it promises of modernization and enhanced productivity. With electrification, farmers could use machines to milk their cows and refrigeration appliances to preserve milk, butter, and cheese. The electrification of dairy farms in Delaware County had a dramatic impact on milk production. Shortly after the DCEC laid its first lines, an article reported that the county had obtained a new milk record with the help of the electrical cooperative, producing the most milk in all 71 counties of the New York City milkshed that year.[9]

Just as pipelines are embedded in their physical environments, they also emerge within social ecologies, extending existing relationships between humans and their nonhuman environments. Delaware County's electrical-fiber-optic pipeline, while it does connect users to extensive systems of energy and data circulation, is also locally mediated by cooperative forms of organization that are deeply embedded in the social landscape. While rural areas have often been the site of collective socialities, two distinct features of central New York fundamentally shaped the cooperative relationship to pipeline systems. First, Delaware County's proximity to New York City meant that it was a place where "farmers had to interact earlier and more intensely with capitalist economic and social relations."[10] Second, the initial form of land ownership was a manor-tenant system, in which many farmers worked as share tenants on land they did not own. Far from being a rural area outside of global flows, Delaware County has long been a place where farmers negotiated both the market and its various external mediators, whether landowners, speculators, infrastructure-builders, or tourists, that sought to capitalize on local labor and resources.

In response to these forces a particular breed of agrarian radicalism developed in Delaware County, manifesting overtly in movements such as the Anti-Rent rebellion in the mid-nineteenth century. During this movement, tenant farmers across upstate New York refused to pay their rent and revolted against the landowners. Yet the farmers' position was not one of simple resistance to external influences. As Thomas Summerhill insightfully argues in his history of the area, early tenant farmers at times "appeared to hold two diametrically opposed views: an urge to shield themselves from the market through paternalism's reciprocal obligations and an unbending pursuit of profit to escape dependency."[11] When farming machinery, "often seen as a hallmark of capitalist agriculture," was first introduced in central New York, farmers used it "in ways other than manufacturers intended or hoped, often sharing mowing, reaping, and haying equipment."[12] In the town of Hamden, farmers established a market day to trade goods with each other, rather than to merchants. By developing these local networks, Summerhill points out, the integration of new machinery into existing social formations extended "cooperative, noncash labor exchanges that had worked well in the past."[13]

While agrarian resistance shifted over the centuries, cooperative efforts to locally mediate pipelines of resources and commodities remained. In Delaware County, the railroad was the first major modern infrastructure that routed through the mountains, and like the electrical and fiber-optic cables, it linked local circulations into a larger, industrial network. Like these other systems, railways formed a pipeline: a set of clear-cut routes through the mountains that accelerated movement along a set of narrow paths, required strategies of insulation to decrease environmental interference, and established nodal points of access in the small towns on their routes. Before the railroad was built, the distribution of commodities was much more difficult and was easily influenced by intense environmental conditions. Travelers could move along the rough dirt roads, but snow and rain could make them impassible and the heat would spoil dairy products. Products such as bluestone, wheat, wool, and whiskey could be transported downriver by raft, but this was likewise affected by the height and flow of the water.[14] Railroad operators not only regularized commodity exchange by ensuring the physical path of transportation was clear, but also by circulating cool air in refrigerated cars to preserve perishable products.

The railroad brought about a major shift for the region's many dairy farmers in the late nineteenth century, who up until that point had manufactured cheese and butter (which could be more easily preserved), and distributed it locally to merchants. In 1890, almost one-tenth of the butter produced in New York State originated on Delaware County farms.[15] In the years that followed, farmers began to switch to fluid milk, which could be transported via the rail network to New York City. By 1900, almost half of the milk produced in Delaware County was shipped to market.[16] As this transition oriented farmers toward larger-scale distribution and a cash economy, they were newly subject to the pricing and regulation of milk in the city. New divisions were also carved into the land. Railroads explicitly sought out the milk trade by building creameries on their routes, enabling nearby farmers to enter the milk trade. Farms that were displaced from the railroad remained dependent on the manufacture of butter.

In response, farmers established an array of collective and cooperative organizations that would mediate their new connections to the market economy. They joined county branches of statewide associations, such as the New York State Dairymen's Association. But they also formed their own agricultural societies and clubs to disseminate information, influence the government to protect their interests, and organize the cooperative marketing of their products.[17] Some even developed their own creameries to compete with those constructed by the railroad. Such cooperative initiatives were not intended to shut down the pipeline of milk to New York City, nor to resist their incorporation into the widespread networks of circulation, but rather to redirect these flows toward collective ends.

Although the rail networks were eventually replaced with truck distribution, the modes of cooperative organization developed in this context were enfolded in subsequent negotiations with infrastructural pipelines. While most histories of rural electrical networks begin with the Rural Electrification Administration, the DCEC emerged out of a long history of farming cooperatives that sought to leverage larger-scale circulations through the area. The early network of the Margaretville Telephone Company was also a farm-to-farm system. It is perhaps not surprising then that today the county has a number of independent telecommunications companies and one of the few electrical cooperatives in upstate New York. Even while these companies channel electricity from across the state and signal traffic from around the world, the price of capacity, the decisions about upgrades, and the selection of routes are determined locally.

The Delaware County Broadband Initiative, which will circulate global signal traffic, likewise emerges out of this cooperative history: it is a collaborative project of the DCEC, the Margaretville Telephone Company, and the Delhi Telephone Company, which together appealed to state programs just as the earlier cooperatives had leveraged national investment in rural electrification and the milk market in New York City. The area's history of agrarian resistance is entangled with the fiber network via these companies' infrastructural resources, their historical routing, and their prior investments in the circulation of electricity and milk through the county. Like their predecessors, the communications companies attempt to collectively control what might seem like purely external incursions. Thomas

Summerhill's conclusions about the central New York Grange, another organization that mediated the advancement of industrial society, proves apt here:

> [They] sought answers to social and economic uncertainties, thus, by looking inward rather than out toward the larger society. They ardently believed that they could resolve the complex issues of the day by turning to each other at the local level.[18]

While the fiber-optic pipeline certainly brings social and economic uncertainties, today digital networks are also being used in ways that support local and collective development, especially in sustainable agriculture. The environmental impacts of contemporary agriculture, including the water pollution, methane emissions, and poor treatment of animals, especially in the beef and dairy industries, have been well documented. Counter to much industrial farming, dairy and produce production is Delaware County is not dominated by large-scale agribusiness; the area is populated by small farms, pasture-raised cows, and more sustainable growing practices.[19] This both reflects the area's topography (farms are nestled in between hills and limited by the mountains and forests) and the regulations of the New York City watershed, as will be discussed in the next section. Small-scale organic farmers and businesses have used digital media to market themselves, using the higher capacity of fiber-optic cables to circulate images and video that illustrate the labor of organic farming. Others have used digital networks to develop new modes of local distribution as an alternative to corporate grocery chains. In Hamden, where the nineteenth-century market day once took place, the Lucky Dog Food Hub allows small-scale growers to list their crops online and buyers in New York City to place orders. Delaware Bounty is another organization that enables farmers to sell their products directly to consumers in the area via digital networks. Regional websites FarmLink and FarmNet connect farmers so that they can share information, including real estate and prospects for employment. As they integrate agricultural activities into an informational grid, such systems evidence the digitization of rural life. Yet many of these activities are not undertaken via large-scale sites such as Craigslist or eBay, but are instead facilitated by cooperative networks, which attempt to manage the movement of signals, goods, and resources locally, even as they rely on urban markets.

Regulated Flows

While the circulations of light across Delaware County's fiber-optic cables are entangled with the flow of current through electrical lines and the distribution of food along roads and railways, they are also entangled with streams and rivers that flow down the mountains. Delaware County's waters feed one of the most important of the state's pipelines: a series of reservoirs, tunnels, and aqueducts that supply drinking water to nine million people in New York City and its surrounding areas. The New York City water supply is the largest unfiltered drinking water system in the United States, and its two largest reservoirs—the collection points of the system—are located in Delaware County. The system's construction was a massive undertaking, and numerous books have documented the project's monumental engineering, its architectural innovations, and its broad ecological and social ramifications.[20] As Matthew Gandy argues, it sculpted the Catskill Mountains into "a life-sustaining circulatory system" for the city of New York.[21]

As is true for many pipelines, this channel south was significant not only in its life-sustaining capacities—its effect on users of the system—but also for its substantial ecological transformations. Local lands were seized by eminent domain, farms and homesteads were destroyed, and valleys were flooded. Twenty-three communities were displaced, nine of which were in Delaware County. Similar to the electrical system's creation of a stable zone of access and the railway lines' transformation of Delaware County into a milkshed, the water pipeline turned the mountains into a watershed and a zone of resource extraction, solidifying an "imperial relationship between the city and the country."[22] The environmental implications of the water supply system extended far beyond the initial clearing of the landscape. Varied forms of regulation, monitoring, and control were established to ensure the flow of clean water regardless of local pollution and variable weather conditions. At points, this even entailed direct interventions such as a "rainmaking" project in the 1950s that infused the skies with dry ice and silver iodide.[23] These regulations intensified in 1990s after the Environmental Protection Agency passed a new set of quality standards for unfiltered water systems. In order to avoid the construction of a multibillion-dollar filtration system, New York City was tasked with increasing and enforcing the regulation of watershed activities. The city proposed changes, such as the establishment of a 500–1,000 foot barrier between septic systems and all watercourses, which would have "essentially obliterated farming," as well as existing inhabitations of the county's river valleys, without offering any funding to help achieve the Department of Environmental Protection's goals.[24]

The story of the New York City water supply system most explicitly demonstrates what is true of all of the pipelines in this chapter: they are not simply technical, but social and cultural systems; they are not merely routes for "natural" resources, but are ecological forms in themselves; and as these channels emerge in particular places, they can be leveraged toward local ends. As had been characteristic in the cooperative networks of the past, Delaware County residents banded together across communities to contest New York City's regulations. Following a contentious struggle, the city was forced to direct funds toward economic development and environmental protection within the watershed counties. This was a novel mode of environmental regulation. "In place of a relatively centralized, ossified, and nonparticipatory regulatory system," Matthew Gandy writes, the watershed was overseen "by a complex and dynamic jigsaw puzzle of different interest groups."[25] Further, he observes, this shift decreased the city's power and granted new agency to a broad coalition of local forces. Organizations including the Catskill Watershed Corporation and the Watershed Agricultural Council were formed, each of which routed New York City money to local environmentally-friendly development, including the free replacement of septic and wastewater systems, grants for education and sustainable agricultural projects, and assistance with local farm management, among many others. These organizations and their funds helped to create media and publicity for small-scale growers who aligned with watershed interests, especially via the Internet. The Pure Catskills initiative, funded by the Watershed Agricultural Council, released digital videos that documented farm work and social media campaigns to promote local, sustainably grown produce. The close ties between the city, the watershed, and sustainable agriculture is neatly summed up in their infographic (Figure 2.2). The image depicts New York City funds cycling down through these organizations to farmers and foresters who will steward the land, and finally, to the drinking water of New York City residents.

In this context, the fiber-optic network is understood as a substitute for environmentally damaging forms of industrial development, since it will presumably foster businesses and other activities that can survive given the watershed regulations. In their letter to the Federal Communications Commission, Delaware County telecommunications companies reported that "[e]conomic growth is often hampered by the strict environmental constraints as a result of a significant amount of land located within the New York City Watershed."[26] The network is articulated, here and elsewhere, as critical to the expansion of tourism, as a necessary support for small businesses, and as an opportunity to further develop sustainable farming practices. While recognizing both the imperial nature of the water system and the reworking of environmentalism as green consumption, I want to direct attention to the ways that the water pipeline has been leveraged by residents, gained traction in the existing cooperative networks, and now shapes the conditions for Internet cables. In turn, fiber-optic systems are layered into a grid of environmental management, viewed as an infrastructure that will sustain New York City's drinking water.

The water system has not only been leveraged to support local interests, but has been mobilized against other pipelines, and is now a key component in the region's struggle against natural gas exploitation. Beneath Delaware County's mountains and dairy pastures are two of the largest natural gas reserves in the United States, the Marcellus and Utica Shales. The shales extend east, beyond the county's border into Pennsylvania, where hydraulic fracturing has wreaked havoc on the state's water supply. Fracking, as it is called, injects fluids deep into the earth's surface, using extraordinarily high pressures to crack the rock that bounds natural gas reserves and create an opening through which the gas can escape. As they take the place of underground reserves, fracking fluids seep into aquifers and contaminate local water supplies.

Although New York State recently declared a moratorium on fracking, this does not preclude the construction of gas pipelines or facilities. Several pipelines

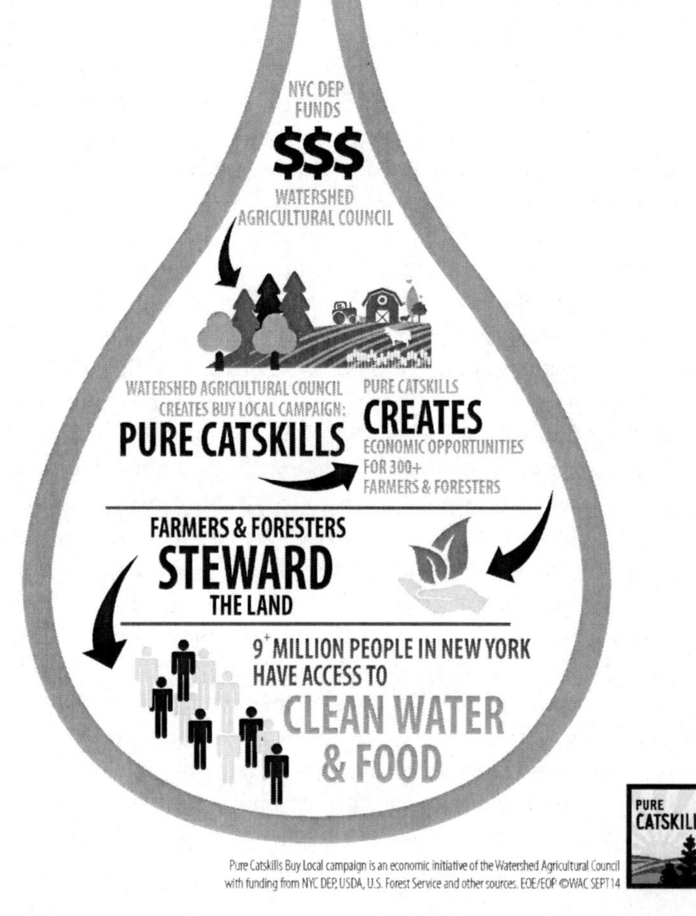

FIGURE 2.2 Pure Catskills links investment in the watershed, agriculture, and conservation.

have been proposed to transport Pennsylvania's gas though Delaware County to New York City. In 2012 the Constitution Pipeline was proposed by Williams, one of the leading natural gas processing companies in the United States. While the current route, as of summer 2015, appears at first glance to run alongside an interstate highway, the planned system actually extends across the nearby hills, where fracking would be preferable to the company. Subsequent extraction systems could easily follow the pipeline, just as the fiber-optic system runs along the electrical network. Plans have been developed for valve and metering stations along the pipeline, where natural gas could be exported to local residents. Given the miniscule market for this gas in Delaware County, paired with the exorbitant costs of the interconnection stations, these nodes are clearly meant to direct gas both ways. The energy companies that have planned the Constitution Pipeline, betting that New York's regulations will shift, are developing infrastructure for the expansion of fracking in the region, should politics, environmental sentiment, or the economy shift gears.

While the current moratorium may or may not be repealed, the water supply system, one imperial infrastructure in Delaware County, has produced a political and geographic boundary that is now being mobilized to keep another pipeline out. The Constitution Pipeline's route runs slightly north of the watershed, at some points only miles outside its bounds. The danger of the pipeline exists not only in the opening of a "gas-shed" akin to a milkshed or a watershed, but in its very installation. As it is currently designed, the pipeline will pave a direct route through the forest and necessitate the removal of almost a million trees. While the electrical cooperative cuts down young trees that continue to grow under its lines, the decision of the gas companies not to use the interstate highway corridor will result in the destruction of a much older forest with a significant role in filtering and stabilizing the water supply, which those opposed to the project argue will affect the watershed. Local investments in the New York City "foodshed" are similarly pitted against the gas pipeline: Food Not Fracking, an alliance of producers, distributors, and consumers of sustainable food, protested with a sign reading "Protect New York City's Food Shed." In these cases, the large-scale pipelines of food and water distribution, and their associated "sheds," are harnessed in service of environmental resistance.

The long-standing cooperative history of the area also infuses this resistance. Far fewer landowners and farmers have agreed to sign over their land to the pipeline than is typical, although the company may still use eminent domain to claim their lands.[27] At local hearings staged by the Federal Energy Regulatory Commission (FERC) for the Northeast Energy Direct Pipeline, a second proposed pipeline along the same route, residents voiced widespread opposition to both projects, often reflecting on the power dynamics of the process and the corruption of the FERC: "The profit is not ours," one participant argued, "the damage that is left behind is ours."[28] Given that the gas companies plan to reverse the flow north to Canada for export overseas, the feeling of exploitation from afar is particularly acute. Anne Marie Garti, local resident and lawyer for StopThePipeline.org, points out that the resistance here has deep roots: "this area is where the anti-rent wars took place," she says, "part of the everyday culture is that this is an area where we shot the sheriff."[29] The protest against the gas pipeline thus gains momentum both from the county's long history of cooperative resistance, and from their position in New York City's varied resource-sheds.

Such pipeline alignments and oppositions are not without contradictions. Although the fiber-optic system is currently envisioned as a way to support environmentally-friendly development, Williams will also rely on communication infrastructures to maintain its system. Along many rural transit routes, networks are not robust enough to provide reliable communications between the crews in the field and the control center, and Williams will establish microwave towers to help its employees monitor the environment for "suspicious" activity and unauthorized local excavation work, as well as to conduct regular equipment inspections. Not only does the gas company monitor the environment around the pipeline, but it also needs communications networks to monitor the gas that flows through it. A SCADA (supervisory control and data acquisition) system will detect the amount, pressure, and temperature of the gas in the pipeline, and transmit this information back to an operations center. Data is collected about system operations at stations along the route, and the company can also use these networks to balance input and output at any given point. Such control mechanisms are absolutely essential to running the network effectively, and as a result reliable data communications, including fiber-optics, are critical.

Williams has a relatively long history with fiber-optic systems. For their 10,200-mile Transco Pipeline, the company used established rights-of-way along their pipeline to lay a fiber-optic cable that supports only their network. And just over twenty years ago, the company was the first to string optical cables through decommissioned gas conduits. In the early 1990s, they owned one of the largest fiber-optic networks in the United States. The *New York Times* reported, "gas and glitz are a perfect fit."[30] Williams may eventually lay a dedicated line along the Constitution Pipeline, if it is built, but this will not be accessible to local Internet users. Initially, however, Williams plans to rely in part on regional telecommunications networks. It is likely that as they send their data monitoring gas flow back to Houston, this information will travel along the fiber-optic infrastructure in Delaware County, perhaps even the systems funded in part by the Catskills Watershed Corporation. In this case, the ecology of the county's information pipelines will extend deep underground.

Entanglements

Several miles down the valley from Bloomville, New York, a large, red dairy barn has been repurposed as a telecommunications facility (Figure 2.3). In place of cows, feed, and farm equipment, the land now houses piles of wooden telephone poles, satellite dishes used to receive television signals, and brightly colored fiber-optic cables. Here, the entanglement of Delaware County's media infrastructure is visible from the roadside; agriculture and communications are coextensive, lacking any self-contained, independent existence. Less overt are the many ways that media

FIGURE 2.3 Cable and satellite farm.

Source: author.

infrastructure extends from preexisting resource networks, follows well-established routes through the forest, and is intimately linked to efforts to generate sustainable systems. By drawing together what might seem to be disparate histories—of electrical networks, agrarian resistance, and the water supply, all of which tend to be written about in isolation—I recast these infrastructural connections as places where routes, capacities, and practices emerge in and through the ecologies they appear to bypass.

Media distribution, whether via cables or satellite networks, data centers or post offices, always takes a particular ecological form. In Delaware County, the fiber-optic network, the electrical system, and the circulation of milk and water have all been developed as pipelines, a kind of media ecology that creates channels along which materials can flow unhindered by environmental circumstances. Pipelines offload the necessity of managing these circumstances from users onto operators; they striate the environment into areas of access and exclusion; and they establish zones of extraction, whether milksheds, watersheds, food-sheds, or gas-sheds. As critical technologies of capitalist expansion, enabling products, information, and finances to transit easily around the world, pipeline systems have been environmentally unsustainable forms. Understanding how cable systems function as pipelines directs attention beyond their carrying capacity to their wide-ranging social and environmental impacts.

Instead of rehearsing such critiques, I have focused here on the ways that pipelines might be leveraged toward more sustainable ends. As in the case of the electrical network, pipelines can be used as a foundation for media distribution that decreases the environmental costs of construction and operation, whether the amount of energy used or the number of trees cleared. As in the case of Delaware County's numerous cooperatives, they can be leveraged to sustain local practices and laborers, including small-scale, ecologically sound agriculture. And, as in the case of the watershed and potential gas-shed, they can be leveraged against one another to defer ecological harm. This is not to idealize the local or the rural, since the cases themselves complicate such distinctions, nor is it to deny that Delaware County is a place without its own inequities and forms of exploitation. Rather, it is an attempt to move beyond the documentation of environmental catastrophe and toward an understanding of where and how ecological alternatives might be found.

Acknowledgments

The author would like to thank Jamie Skye Bianco, Glenn Faulkner, Diane Galusha, Anne Marie Garti, Mark Kennaugh, Ray LaFever, Kristan Morely, Marc Schneider, Bucky Soule, Chris Stockton, and Wayne Marshfield for their assistance in researching this project.

Notes

1 Karen Barad, *Meeting the Universe Halfway: Quantum Physics and the Entanglement of Matter and Meaning* (Durham, NC: Duke University Press, 2007), ix.
2 Lisa Parks, "Where the Cable Ends: Television beyond Fringe Areas," in *Cable Visions*, edited by Sarah Banet-Weiser, Cynthia Chris, and Anthony Freitas (New York: New York University Press, 2007).
3 Adrian Mackenzie, *Wirelessness: Radical Empiricism in Network Cultures* (Cambridge, MA: MIT Press, 2010).
4 Lisa Parks, *Cultures in Orbit* (Durham, NC: Duke University Press, 2005).
5 Tim Ingold, *Lines: A Brief History* (London: Routledge, 2007), 81.
6 Joseph Tierney, "From Fantasy to Reality: The Impact of Rural Electrification on the Dairy Farms of West-Central Wisconsin" (bachelor's thesis, University of Wisconsin–Eau Claire, 2011), 15–16.
7 Samer Alatout and Chelsea Schelly, "Rural Electrification as a 'Bioterritorial' Technology Redefining Space, Citizenship, and Power during the New Deal," *Radical History Review*, 107 (Spring 2010): 127–138.
8 Ibid.
9 "Delaware County Co-Op, Delhi, N.Y., Helps in Obtaining Milk Record," *Delaware County Electric Co-operative, Inc.: From 1941 to Present.* Presentation by Wayne Marshfield, July 2015.
10 Thomas Summerhill, *Harvest of Dissent: Agrarianism in Nineteenth-Century New York* (Chicago: University of Illinois Press, 2015), 3.
11 Ibid., 21.
12 Ibid., 103.
13 Ibid., 104.
14 Tim Duerden, *A History of Delaware County New York: A Catskill Land and Its People 1797–2007* (Fleischmanns, NY: Purple Mountain Press, 2007), 34.

15 Summerhill, *Harvest of Dissent*, 149.
16 Ibid., 150.
17 Ibid., 170.
18 Ibid., 215.
19 Greta Gaard, "Toward a Feminist Postcolonial Milk Studies," *American Quarterly*, 65, no. 3 (September 2013): 595–618; Elizabeth Grossman, "As Dairy Farms Grow Bigger, New Concerns about Pollution," *Yale Environment 360* (May 27, 2014); Bruce A. Scholten, *US Organic Dairy Politics: Animals, Pasture, People, and Agribusiness* (New York, NY: Palgrave Macmillan, 2014); Denis Hayes and Gail Boyer Hayes, *Cowed: The Hidden Impact of 93 Million Cows on America's Health, Economy, Politics, Culture, and Environment* (New York, NY: W. W. Norton, 2015); United States Environmental Protection Agency, "U.S. Greenhouse Gas Inventory Report: 1990–2013" (2015), http://epa.gov/climatechange/Downloads/ghgemissions/US-GHG-Inventory-2015-Main-Text.pdf (accessed May 5, 2015).
20 Diane Galusha, *Liquid Assets: A History of New York City's Water System* (Fleischmanns, NY: Purple Mountain Press, 1999); Gerard T. Koeppel, *Water for Gotham: A History* (Princeton, NJ: Princeton University Press, 2000); Matthew Gandy, *Concrete and Clay: Reworking Nature in New York City* (Cambridge, MA: MIT Press, 2002); Kevin Bone (ed.), *Water-Works: The Architecture and Engineering of the New York City Water Supply* (New York, NY: Monacelli Press, 2006); David Stradling, *Making Mountains: New York City and the Catskills* (Seattle: University of Washington Press, 2010); David Soll, *Empire of Water: An Environmental and Political History of the New York City Water Supply* (Ithaca, NY: Cornell University Press, 2013).
21 Gandy, *Concrete and Clay*, 23.
22 Stradling, *Making Mountains*, 14.
23 Ibid., 174.
24 Jeffrey Baker, "Transcripts of *Behind the Scenes: The Inside Story of the Watershed Negotiations*," 23, http://www.cwconline.org/behind_the_scenes.html (accessed November 27, 2015).
25 Gandy, *Concrete and Clay*, 68.
26 Dean Uher, "Letter to Marlene H. Dortch, Office of the Secretary, Federal Communications Commission, re: WC Docket No. 10–90, Expression of Interest," *Federal Communications Commission*, March 7, 2014.
27 Interview with Anne Marie Garti, July 7, 2015.
28 Public comments to the Federal Energy Regulator Commission regarding the Northeast Energy Direct Pipeline. Oneonta, NY, July 15, 2015.
29 Interview with Anne Marie Garti, July 7, 2015.
30 Allen R. Myerson, "For a Pipeline Company, Moving Data Pays Better," *New York Times*, May 23, 1994.

3

MAKING DATA SUSTAINABLE

Backup Culture and Risk Perception

Shane Brennan

> A young man seeking eternal security for his data came to the master of backup and asked: "What must I do to keep my data safe forever?"
>
> The master replied: "Much data is lost due to a lack of knowledge. You must know the true way of backup. For he who knows it, and follows it, will never lose his data!"

This dialogue begins Terje Ronneberg's *The True Way of Backup*, an online how-to manual written as a collection of pseudo-religious parables. Each chapter focuses on a different aspect of an effective personal data preservation strategy: totality, regularity, continuity, locality, longevity, security, and integrity (Figure 3.1). The chapter on totality, for example, teaches the importance of making comprehensive backups of one's "entire computer" so that no "seemingly insignificant" files are overlooked.[1] These data files are personified as defenseless animals requiring constant protection: "Just as a good shepherd cares for all the sheep in his flock," explains the "master of backup" to his disciple, "it is wise for us to back up every file in our care." Accompanying the text is a slightly pixelated, black-and-white etching of Jesus cradling a lamb. By infusing data management advice with religious imagery, backing up is elevated beyond a simple computational task. It becomes a set of shared rituals, even a way of being in the digital world.

Along similar, though more secular lines, an organization called World Backup Day aims to generate a sense of individual responsibility and broad solidarity in the fight against data loss. "I solemnly swear to back up my important documents and precious memories on March 31," reads a pledge on the group's website. "I will also tell my friends and family about World Backup Day—friends don't let friends go without a backup."[2] Here, backing up is aligned with other global causes and days of international observance such as Earth Day, celebrated just a few weeks later. Together, World Backup Day and *The True Way of Backup* evidence

FIGURE 3.1 Terje Ronneberg, *The True Way of Backup*, 2008.

Source: screen capture by the author, accessed February 28, 2014.

a culture of backup in which computer users everywhere are expected to copy all of their important files on a regular basis to multiple platforms, including online or "cloud" remote server–based storage. Only then, supposedly, will our precious data outlast a multitude of disasters—from fires and floods to errors both human and hardware—and we may achieve the common goal of digital continuity, regardless of the energy costs involved.

This chapter investigates the logics that underpin this culture of backup and their environmental ramifications. As personal backup practices have shifted from more localized forms of data storage, such as the external hard drive, to cloud services, a different idea of backup as a risk-mitigation strategy has emerged. The geographic redundancy of the cloud—in which several copies of a file are spread to at least one, far-off data center—is seen as a way of making information more "sustainable," in the sense of maintaining it over the long term against various kinds of disruption. But this digital "sustainability" is achieved by multiplying the amount of data stored in energy-consuming server farms, leading to more carbon emissions and making the overall system less environmentally sustainable.[3] By contributing in significant ways to climate change, ironically, the proliferation of cloud backup is helping to create more severe storms, floods, and power outages—the very things from which it strives to protect data. As these risks become more immediately visible, digital backup practices respond by creating even more redundancy, constituting a feedback loop. And yet, the dematerializing rhetoric of the cloud, as well as the use of "sustainable data" to describe information that sticks around, have worked to culturally obfuscate this pressing environmental media circuit.

Backup in general—not just digital backup—is a way of mitigating the risk of disruption by designating a secondary object that can stand in for the original if lost or damaged. Thus, the conceptualization of backup reflects historically specific ideas of risk. The chapter's first section outlines how Cold War–era mentalities shaped backup's cultural formation.[4] In this period from the 1950s through the 1980s, two distinct backup strategies were mobilized across US space programs, missile defense systems, and underground architectures built to protect sensitive documents. Backup was configured as both a mode of secure containment against external threats and a method of creating redundancies to protect against more proximate hazards. Appearing in late-1980s corporate culture, this second strategy responded to growing anxieties about information continuity by instituting regular duplication programs. Since no containment system could ever be perfectly secure, given the number and variety of disasters at play, the routine creation of copies came to be seen as an essential part of backing up.

This turn toward habitual redundancy set the stage for the subsequent development of cloud backup services, aimed first at businesses and then at consumers. These services automate and facilitate the process of making data copies and distributing them to multiple, off-site locations across the network. In the second section of this chapter, I argue that the cultural appeal of personal cloud backup over external hard drives is symptomatic of a broadening of risk perception in what Ulrich Beck has called the "risk society," a moment in which we become aware of and must learn to manage a range of mostly intangible, human-caused dangers. The prevailing attitude in the risk society, Beck notes, is "*negative* and *defensive* . . . one is no longer concerned with attaining something 'good', but rather with preventing the worst."[5] This stance carries over to the realm of personal data management, leading to an intensified, hyper-redundant form of backup to the cloud—essentially the regime prescribed in *The True Way of Backup*. Responding to these digital anxieties, promotional materials for cloud backup services, along with the visual-spatial metaphor of the cloud itself, construct an image of resilient data in which copies of one's files are circulated to sites out of harm's way, outmaneuvering many local and regional threats.

However, this "safe" data comes at a cost when viewed with an eye toward energy consumption and CO_2 emissions. In order to reveal the deep tension between sustainable data and environmental sustainability, and connect large-scale energy systems to everyday individual media practices, the chapter's third section introduces the concept of the "resource data file." Inspired by Nadia Bozak's "resource image," I define the resource data file as a way of seeing a digital artifact, which may or may not be an image, in terms of its carbon footprint or impact on the environment. The resource data file is what expands with the increased redundancy of contemporary backup culture. Furthermore, in line with

Mél Hogan's work on Facebook data centers, this section extends the analysis of backup in order to bring into focus a larger area of environmental media research sensitive to the cultural logics and practices that activate media infrastructures in sustainable or unsustainable ways. Returning to the link between contemporary backup culture and global climate change, I conclude by discussing the imaginary of a backup Earth, or the notion that it may one day be possible (and desirable) to duplicate our planet's "data" and store the copy off-world. In short, this chapter examines the cultural formation of cloud backup as both a concept and a widespread digital practice, the stakes this has for the environment, and why those stakes have remained largely invisible to us, even as we have become more aware of other, nondigital aspects of our carbon footprints.

Answering these questions requires bringing together two different ideas of sustainability—digital and environmental—and asking what their surprising incompatibility can tell us about how to reconcile our digital preservation habits with the vital need to preserve our collective habitat. For all of us computer users, this means we might be able to invent new ways to store, retrieve, and back up information more sustainably. And for the field of environmental media studies, it means that, in addition to work on the resource consumption and waste streams of physical media infrastructures, there must be intersecting research on sustainable cultures of media use, including—but by no means limited to—the culture of backup.

Implementing "Shadow Systems"

As a concept and practice, backup is embedded in media history while also reaching beyond it. From floppy disks to flash drives, nearly every mode of data storage has been conceived as a way of backing up at one time or another. Indeed, media are often assessed in terms of their qualities as backups (their cost, storage capacity, durability, or volatility, continued accessibility or potential obsolescence, etc.), and older recording media sometimes persist as backups to newer media based on these qualities (magnetic tape, for example, has remained a viable large-scale backup medium due to its relative low cost, high capacity, and archival durability). While backup develops relationships among media, and its study suggests a corresponding need to reconfigure media histories, the definition of what constitutes a backup medium might well include things not commonly thought of as "media." Consider, for instance, the gas-powered electric generator that becomes a backup to the grid when installed by a concerned homeowner in anticipation of a future blackout. Even humans can be thought of as backups within certain social structures; in sports, backup players substitute for injured teammates, and in some military and governmental command structures, a chain of succession designates emergency replacements.

If almost anything can be a backup, then what defines it as such? Though diverse, backups share a similar strategic placement and intended purpose within an organizational system. This is to say that backups are always relationally and imaginatively produced: they are created when a relationship is established between something that might fail and something else—the designated backup—that can take its place in an emergency, ensuring the relative continuity of the overall system. Thus, backups are always constituted through, and contingent upon, the imagination of potential disaster, from the isolated hard drive crash to the broad-sweeping hurricane that disables power grids, data centers, and other infrastructures. Concentrating in places where breakdowns are seen to be especially likely, undesirable, or both, backups mediate cultural understandings of risk and vulnerability, offering a way to study how these perceptions get materialized in various infrastructural systems.

Before proceeding to consider how historically specific ideas of danger have yielded different backup strategies, it is worth noting that this definition of backup as a perceived standby or reserve only came about in the mid-twentieth century. Up until that point, one might have "backed up" someone in an argument ("to stand behind with intent to support") or complained about the "back-up" of water in a clogged storm drain (or any other accumulation behind an obstruction).[6] A Google Ngram search for the nonhyphenated noun "backup" shows a steady increase in the frequency with which it appears in the English-language Google Books corpus after 1950. Delving into archival US newspapers from this period, one finds a growing number of references to backups as reserve objects, individuals, or entire systems designed to anticipate and mitigate disruption caused by the loss of their primary counterparts.

As this meaning of backup took shape, it was influenced by a Cold War culture of risk perception. In both state-sponsored military projects and corporate information storage facilities, backup was oriented toward threats from outside the boundaries of the nation-state: namely, a possible Soviet nuclear attack and the high-stakes environment of outer space. Backups were used to isolate and contain that which was endangered, or seemingly so—whether astronauts, military data, business records, or the nation-state itself—within layered, protective systems. Some of the most visible backups from the 1950s to the early 1970s were those employed by NASA to ensure the success of early manned space missions. The Mercury capsule (1959–1963), for example, was "literally jammed with repetitive systems built in to assure operating reliability."[7] In 1970, *Newsday* ran an article titled "Apollo Safety Key: Its Backup System," which chronicled NASA's "methodological concern to leave nothing unduplicated."[8] It explains how "the agency has chosen to build back-up mechanical systems into the spaceships themselves, in an attempt to provide on board a sort of already plugged-in shadow system in the event that primary units should fail." On the ground, "backup astronauts" trained alongside first-pick crewmembers, ready to stand in for them at the last minute.[9]

The idea of backup computing emerged within the same mid-century risk climate. As Paul Edwards describes, the Semi-Automatic Ground Environment (SAGE) air defense infrastructures built by the US military in the 1950s combined redundant electrical generators and concrete enclosures—structures that echoed the Cold War's "political architecture of the closed world"—with backup computers for analyzing radar data.[10] Each SAGE center, he explains, had "two identical computers operating in tandem, providing instantaneous backup should one machine fail."[11] In these systems, Cold War anxieties about national security and threats to human life became entangled with fears about the unreliability of computing and the vulnerability of digital information.

Around the same time, data backup techniques responsive to Cold War anxieties were developing in the corporate sector. Iron Mountain Atomic Storage (founded in the early 1950s) harbored copies of sensitive business documents from nuclear threats inside a disused iron mine.[12] Companies offering similar facilities—what amounted to fallout shelters for data—proliferated in the following decades. Although storage media varied, from paper records to microfilm to magnetic tape, the underlying backup strategy of secure containment, informed by a "closed world" culture of risk perception, remained relatively consistent. Another company called DataPort stored backup data records in the bedrock of Lower Manhattan beginning in the mid-1980s. "DataPort would be a safe place for diamonds, rubies, or gold bullion," suggests the *Washington Post*, but "it will contain nothing but thousands of reels of magnetic tape—in effect, duplicates of the memories of modern corporations."[13]

Tung-Hui Hu articulates the links between this practice of protecting media in secure bunkers, sealed off from the outside world, and a cultural imagination of disaster as stemming from the threat of a foreign enemy.[14] While this idea of security has roots in the Cold War era and elsewhere, it continues—in some notably direct ways—in cloud infrastructure. As Hu explains, "many highly publicized data centers" repurpose old military bunkers:

> Inside formerly shuttered blast doors meant for nuclear war, inside Cold War structures . . . and inside gigantic caverns that once served as vaults for storing bars of gold, we see the familiar nineteen-inch racks appear, bearing servers and hard drives by the thousands.[15]

Such defensive architectures, he argues, serve to reinforce the idea of risk as externalized, resulting in the construction of even more data bunkers.[16]

In the 1980s, however, the imagination of data vulnerability was also starting to broaden beyond large dangers, such as a nuclear strike, that must be kept out at all costs. "DataPort and other storehouses across the country," the *Post* continues, "cater to their clients' growing nervousness about their dependence on computers that are vulnerable to fire, flood, sabotage, theft, or simple human error." Corporate backup providers began to see themselves as responding to a slew of everyday, even mundane, hazards to information continuity. In contrast to the foreign attack that could, at least in theory, be mitigated by structures of containment, these dangers threatened to permeate physical boundaries, or even emerge *from within* sociotechnical systems themselves. The random accident, breakdown, or natural disaster was starting to seem a lot more likely—and equally destructive to data—than the prospect of nuclear war. Backup practices adjusted in turn, emphasizing increased redundancy in addition to isolated storage in a single location.

This is not to suggest that the "bunker mentality," as Hu calls it, disappeared; in fact, he shows that it persists in the design of some contemporary data-storage installations.[17] But by the late 1980s, redundancy had begun to invade the corporate imagination of backup once dominated by fortification. "Backup, backup, backup," chants a computer security consultant in a 1989 issue of the *New York Times*.[18] She goes on to advise businesses concerned about data longevity to "declare a 'backup holiday'" in which they direct their employees to "make backups the rest of the afternoon," foreshadowing the first World Backup Day in 2011. If data were not truly safe anywhere, given the number of micro-disasters that could occur near or inside even the most secure facility, then one should create many copies on a routine basis and store them in a few, off-site locations. This can be read as an early instance of a contemporary cloud backup logic—a logic that relies on habitual redundancy and geographic dispersion to hedge against everyday threats—gaining traction before an online storage infrastructure had been fully realized.

Geographies of Risk and Redundancy

Until the advent of high-speed Internet connections, remote backup involved physically transporting data copies to an off-site storage location. Remote Backup Systems (formerly Quantum Tech) was one of the first companies to offer businesses software-driven backup over the network in 1987, followed by many similar services throughout the 1990s. In the next decade, home broadband connections allowed companies to market remote online—rebranded as "cloud"—server-based data storage and backup as a desirable commodity for individual users, not just corporations.[19] "It used to be that we kept our data on our (actual) desks," observes Andrew Blum. Now, "the 'hard drive'—that most tangible of descriptors—has transformed into a 'cloud,' the catchall term for any data or service kept out there, somewhere on the Internet."[20] This distributed form of backup has been positioned as a more reliable alternative to the still quite prevalent, localized mode of backup on a personal drive.[21]

Local backup recalls the secure containment seen in Cold War systems. The user's important files are kept in a specific place, perhaps on a hard drive in a locked desk drawer. Cloud backup, in contrast, claims to make data safe through geographic redundancy, or the practice of placing copies in multiple locations. To be clear, the Cold War–era backup mentality favors the storage of information in stationary, often isolated locations and bunker-like facilities; data files "in the cloud" are, of course, still stored in fixed and often heavily fortified sites (cloud data centers), but the image of the cloud encourages us to think of these sites as unfixed and immaterial, distant while being nowhere in particular, informing a different vision of data at risk. As one *Popular Mechanics* article explains, backup to a secondary drive protects data from a primary drive crash. But what if files are lost,

> not because of a misaligned read/write head but because of a flood or a robbery or a house fire that wiped out both drives at the same time? Preparing for such a disaster calls for a different level of redundancy and a solution that includes backup to the cloud.[22]

In addition to being a matter of cost and convenience, the growing popularity of cloud backup reflects a collective sentiment that the cloud—and more precisely the dematerialized, geographic redundancy it represents—is a safer way to store information in the contemporary world.[23] It reflects, in other words, a change in the spatiality of risk perception, as the threats to digital continuity we imagine inform where and how we choose to back up our data.

This shift from local to distributed backup is not primarily technological, since most information in the cloud still exists on hard drives.[24] The critical difference is where those drives are located from the perspective of the user. Instead of being visible and close at hand, they are in distant server farms that most users can neither imagine nor locate with any precision. In this way, an image of data resiliency is maintained, for if one cannot clearly imagine how data centers look, sound, smell, or feel, it is also difficult to imagine their material breakdown. The cloud's geographic uncertainty, moreover, supports the illusion that our data is somehow placeless and, therefore, immune to regional disruptions.

Cloud backup, in short, depends on a logic in which potential dangers are everywhere, and thus all data should be replicated and stored in several places at once. In *The True Way of Backup*, the "master of backup" imparts this very lesson with a natural metaphor: flowers, he explains,

> store their most important information in seeds, and those seeds [are] spread far and wide, by wind, water, and wings of birds. So too, you must spread copies of your backups far and wide, then your data will also survive disasters like fires, floods, droughts, and snow.[25]

Ever since Darwin observed that seeds can be "transported by sea to distant islands,"[26] they have been understood as a mode of dispersal for bio-information, one that resonates with a model of backup that spreads data copies across space, in a sense "germinating" the network. The wind, water, and birds represent online backup and cloud storage services like Dropbox, iCloud, and Google Drive, which facilitate this dispersal. While keeping a local copy on the user's hard drive, these services replicate a file—including multiple versions saved incrementally—and place copies in one or more data centers, separated by perhaps hundreds, if not thousands, of miles and often strategically located in regions at low risk of natural disasters.[27] Even if one site goes off-line, a copy will be retrievable from somewhere else.

This geographic redundancy is a logical response to contemporary risk perception. As Beck argues, risk was previously experienced as a variety of clear and present "personal risks."[28] But contemporary risk culture is defined by a growing awareness of globally diffuse and far-reaching, yet largely invisible and unpredictable, threats to human well-being. Many of these dangers are the products of earlier periods of industrial modernization, he contends, but have only recently come into focus. It is important to note that this doesn't mean that life today is "riskier" than in previous centuries. Rather, Beck points to a change in risk perception as a culturally constructed form of knowledge. With the shift to the cloud, backup logics have also changed in response to this new risk consciousness.

In contrast to the well-defined perils of the Cold War, pollution, radiation, and other toxins "travel on the wind and in the water," writes Beck. "They can be anything and everything . . . along with the absolute necessities of life."[29] If contemporary threats travel by wind and water, then our data must behave like the seeds of flowers and move by the same means; the dispersal of risk is matched by an equal and opposite dispersal of data copies. Describing the disaster recovery processes of cloud infrastructure, Hu likens such risks as "cascading computer failures" to "*contagions* that haunt the cloud . . . one cloud server fails to sync with another, one network's router misfires, causing a chain of errors that ripple through all other interconnected networks."[30] Backup strategies premised entirely on secure containment may not be enough to ward off dangers that, according to Beck, "pass through all the otherwise strictly controlled protective areas of modernity."[31] By dispersing data copies to more than one "bunker," cloud backup increases the odds that at least one copy will escape these network "contagions."

Risks expand socially as well as geographically in the "risk society," with broad populations (including those responsible for creating them in the first place) forced to deal with their consequences. But as Beck acknowledges, many threats disproportionately affect the poorest and most vulnerable, and "abilities to deal with risks, avoid them or compensate for them, are probably unequally divided."[32] The rhetoric of World Backup Day similarly frames data loss mitigation as a near-universal responsibility, departing from the top-down logic of older state- and corporate-led backup systems. Everyone is charged with protecting his or her own data, even though access to the cloud and digital backup, like many other kinds of security and technology, is far from universally distributed.

Cloud backup not only responds to contemporary risk perception, it also actively shapes it through the cloud imaginary. "Risk theorists," argues Ursula Heise, "have paid relatively little attention to the role that particular metaphors, narrative patterns, or visual representations might play in the formation of risk judgments."[33] While she considers how depictions of the planet and narratives of toxic exposure mediate understandings of environmental risk, I propose that the cloud—at once a visual trope and story about what happens to our data on the

Internet—performs a similar function for conceptions of digital safety or vulnerability. As *The True Way of Backup* illustrates, the cloud is a way of imagining how data copies are spread to sites out of harm's way, as if by a weather system. This adds a sense of dynamism and an element of risk avoidance to the "light and airy image of the digital cloud," which, as Allison Carruth observes, "directs one's attention away from the materiality of information."[34] Part of the attraction of the cloud metaphor, in addition to making digital infrastructure seem magically "green" and virtual, is the image of airborne mobility that assuages fears about data impermanence.[35]

In their marketing, cloud backup services emphasize the presence of certain risks from which the cloud's geographic redundancy provides good protection, such as relatively isolated disasters that impact a specific area, including the user's immediate surroundings. For example, a page titled "Always Safe" on the Dropbox website reads: "Even if your computer has a meltdown, your stuff is always safe in Dropbox and can be restored in a snap."[36] In the accompanying cartoon, a stick figure is unfazed by the spontaneous combustion of his PC, which may soon engulf all external backup drives stored nearby. Instead of insulating data from threats outside, cloud backup anticipates disasters that spring randomly from every machine on which our information depends. Against this image of localized danger, the company presents the iconic blue box hovering in mid air, an abstraction of its cloud storage service. In a second cartoon, the box has become a kite with a tail comprised of a text document, film reel, and photograph (Figure 3.2). Devoid of moving parts, it runs no risk of breaking down or bursting into flames. Tethered only by a thin kite string, the box floats on the air currents of the cloud, dodging any terrestrial danger.

While the cloud imaginary accentuates unpredictable, environmental threats to data, it simultaneously downplays certain political risk factors that the physical placement of data centers may actually intensify. The cloud connotes a broad and untargeted dispersion of data across space, yet servers are concentrated in facilities that are deliberately located in relation to state borders and regulatory contexts.[37] For instance, Apple opened its first iCloud data center in China in 2014 with state-owned China Telecom, allowing the company to provide faster services—including cloud backup—to its local customers. But the move also shifted this infrastructure into Chinese jurisdiction, raising concerns about potential government surveillance and censorship.[38] Cloud backup has even been scaled to the boundaries of the nation-state: "Just as computer users back up their laptops in case they break or are lost," reported the *Economist* in 2015, "Estonia is working out how to back up the country, in case it is attacked by Russia," by duplicating government data and software to cloud servers abroad.[39] At the same time, cloud storage can be used to subvert legal and territorial limits. In 2012, The Pirate Bay (TPB), a peer-to-peer file-sharing site, uploaded its entire operation to several commercial cloud-hosting providers, making it more resistant to, or backed-up against, the disruption of police raids. "Moving to the cloud lets TPB move from

FIGURE 3.2 "The Dropbox tour."

Source: screen capture by the author; accessed February 25, 2014.

country to country, crossing borders seamlessly without downtime," the group explained.[40] Both in terms of where information is stored and how it moves across transnational networks, geopolitical conditions are still important to the conceptualization and implementation of backup in the cloud era, just as they were during the Cold War.

Backup is also no less material today than in previous decades. As Stephen Graham attests, digital media infrastructure remains absolutely reliant "on other less glamorous infrastructure systems—most notably, huge systems for the generation and distribution of electric power."[41] Moreover, the geographic spread of data centers may make it more difficult to pin down their energy use.[42] While the cloud imaginary provides a framework for thinking about environmental risks to data and their evasion, its conceptual dematerialization makes the infrastructure of digital backup and preservation seem environmentally benign.[43] But in reality, networked data centers, built to handle ever-larger amounts of information, including backup copies, entail very real resource expenditures and CO_2 emissions.

The Unsustainability of "Sustainable" Data

"One might sustain a belief, or sustain an argument" in the nineteenth and early twentieth centuries, notes digital preservation researcher Kevin Bradley. "By the early 1980s, 'sustainable' had begun to be associated with concerns regarding the environment. Since then the word has continued to expand in its usage."[44] Bradley defines the more recent formulation of "digital sustainability" as providing "the context for digital preservation by considering the overall life cycle, technical, and socio-technical issues associated with the creation and management of the digital item."[45] This approach recognizes that information can never be truly permanent but can only be *sustained* through an active process of building and continually maintaining costly, energy-intensive infrastructures that preserve "valuable data without significant loss or degradation."[46] A crucial paradox arises, however, since "digital sustainability" is essentially at odds with—and its implementation may very well undermine—environmental sustainability, which "creates and maintains the conditions under which humans and nature can exist in productive harmony."[47] The adoption of "sustainability" by digital preservation discourse works to obscure this conflict, making it seem as if ensuring the "life cycles" of data is akin to, and in line with, protecting biological life and planetary ecosystems.

Although some data centers are powered in significant ways by renewable energy like wind and solar,[48] this does not account for the "embodied energy" that went into building these infrastructures and the electricity used to access remote servers on the user end, much of which still comes from the burning of fossil fuels.[49] In addition, data centers rely on industrial backup systems such as massive arrays of polluting emergency generators, which must be tested regularly.[50] These generators are designed to mitigate the risk of disruption from a power failure and meet clients' demands for 24–7 access. Entire data centers are also "backed up" in the form of duplicate physical and digital architectures—known as disaster recovery sites—increasing energy costs even further. If the main center goes off-line, the primary data center can "fail over" or shift operations to the recovery facility, which is usually located in a different geographic region and connected to a different local power grid so that it is unlikely to be affected by the same disruption.[51]

Understanding this profound conflict between cloud backup's logic of increased redundancy (copying data across server facilities supported by material backup systems) and environmental sustainability requires looking at what I call the "resource data file." As indicated at the start, this is an extension of Bozak's "resource image," a concept that frames cinematic images in terms of the natural resources used, and carbon emissions generated, in their production. The resource data file similarly exposes the biophysical materials involved in the creation, dissemination, storage, and backup of digital media. By duplicating each data file, personal cloud backup services—as well as those used by corporate and governmental organizations—effectively multiply the carbon footprint of the resource data file, making it increasingly unsustainable. A large percentage of cloud infrastructure, note Sean Cubitt, Robert Hassan, and Ingrid Volkmer, is used "for [the] back-up of locally produced and maintained files, implying an increase, not a decrease in the quantity of storage and therefore the amounts of energy required both to store and transport files."[52] Indeed, Dropbox assures its users that "files are backed up several times. The primary copy on your computer's hard drive is synced to your Dropbox account online, and that copy is backed up again for safety . . . across several data centers."[53]

Bozak's resource image also points to the cultural perception of moving images as a kind of utility, endlessly piped in to our homes through cable networks.[54] With the growing availability and falling price of cloud storage, online backup can similarly be mistaken for an abundant resource in a way that negates its material costs to the environment. This, in turn, may lead to an overconsumption of cloud backup, or what might be called the "abuse" of the resource data file.[55] In this mode, users back up everything constantly and on multiple platforms in accordance with the teachings of *The True Way of Backup*.

As such, thinking in terms of the resource data file highlights the role of cultural perceptions, imaginaries, and everyday practices in activating physical media infrastructures in potentially unsustainable ways. Along similar lines, Mél Hogan examines "the potential environmental costs of our everyday obsession with self-archiving" on Facebook as well as our expectation of instantaneous access.[56] "The cost of such instantaneity," she argues, "is that almost all the energy that goes toward preserving that ideal is, literally, wasted," since most servers idle in standby mode—itself a form of backup—ready to meet sudden demand.[57] Her analysis reveals Facebook to be a wasteful archive, a paradoxical system that consumes exorbitant amounts of energy in order to "sustain" information and our ability to access it.

Like online "self-archiving," backing up to the cloud has become a digital obsession for some, one that activates these same material infrastructures but multiplies the amount of data requiring storage. Taken together, these habits constitute what Carruth calls a "*micropolitics of energy*—defined as the planetary ramifications of minute individual practices that are fueled by cultural values of connectivity and speed and that rely, above all, on the infrastructure of server farms."[58] Backing up, uploading, and other cultural processes of media-infrastructure activation, as I term them, warrant a branch of environmental media research that examines not only the sustainability of material technologies and infrastructures, but also the sustainability of the cultural forces that help shape them. Yes, a data center may be power-hungry and inefficient, but this is in part necessitated and justified by our unsustainable "digital demands."[59] The field of environmental media research, in other words, must address sustainable cultures of media use alongside and in dialogue with concerns including data center design, electronic waste, and the question of how media become tools for "environmental communication."[60] At the same time, the term sustainability should be subject to interrogation given its general overuse and specific adoption into digital preservation discourse.[61] Taking another cue from Bozak, a resource media framing, inclusive of both the resource image and the resource data file, would keep the focus more squarely on energy and carbon.

What might a low-carbon or more sustainable backup culture look like in practice? This new way of backup might aim to back up selectively or, alternatively, with restraint, resisting backup overconsumption. But it is difficult to know what constitutes acceptable versus excessive redundancy, in part because there is a lack of transparency about precisely how many times our files are replicated online and where they end up. If future empirical studies could determine the percentage of the cloud that is used for storing backup copies, this could be read alongside existing work that addresses the significant energy drain of data centers to estimate backup's carbon footprint.[62] With this number in mind, our desire for data semi-permanence and near-perpetual access achieved by hyper-redundancy could be more accurately weighed against its price in CO_2 emissions and its role in climate change. Only then, we might start to realize, as Cubitt, Hassan, and Volkmer find, that "storing increasingly detailed multiple copies and drafts on multiple hard drives as back-up may no longer be possible."[63]

Even without this measurement, other steps could be taken to shift the conversation around online backup. A user-driven "anti-redundancy" movement, perhaps a World Deletion Day, could reposition comprehensive backup as the digital equivalent of driving an SUV, while fostering pride in having a small data carbon footprint. This movement may find affinities with nondigital efforts to reduce redundancy and thereby cut emissions, such as ride-share programs, mixed-use urban development, or green spaces that double as community and environmental resources. And it might generate demand for "eco-friendly" backup technologies, perhaps a lossy (or incomplete) form of backup (inspired by the MP3, as theorized by Jonathan Sterne) in which certain data or versions of data, unlikely to be missed by the user, are automatically withheld from replication.[64] Amazon already offers a Reduced Redundancy Storage option that still "stores [digital] objects on multiple devices across multiple facilities . . . but does not replicate objects as many times as standard Amazon S3 storage."[65] While the company frames this as a cost-saving measure, it is easy to imagine something similar pitched to users who wish to minimize their digital carbon footprint.

Conclusion

As a way of defining organizational relationships in response to perceived risks, backup is a historically specific concept and media practice. This means that studying changes in how and where we choose to implement backup systems can reveal a considerable amount about how we collectively imagine our world under siege at different historical moments. The culture of remote and distributed, hyper-redundant backup is quickly increasing the resource footprint of our data, powering a vicious cycle among backup practices, risk perception, and environmental unsustainability. The key paradox is this: to protect our data from the threats associated with climate change—such as storms, floods, and drought-induced wildfires—we multiply and distribute our data across a vast network of server farms; but this multiplication, in turn, necessitates the expansion of physical infrastructures, consuming more space and energy and generating emissions that further destabilize the climate. The logic of cloud backup can thus be seen as contributing to some of the very environmental threats it aims to defend against. In other words, the sustainability of data is at odds with the sustainability of the greater environment, a primary life-support system for which no backup exists. In closing, I will mention two visions for the future of backup and the environment, one in which backup is framed as the ultimate "solution" to global catastrophes, and another that rejects this logic as undercutting active and collective responses to the climate crisis.

Climate change and globalized risk perception have already radically expanded what is seen as requiring backup. The Svalbard Global Seed Vault, for example, stores duplicate copies of crop seeds in a Norwegian Arctic mountainside in order to be able to "restart" agriculture after a large-scale disaster, recalling both Cold War bunkers and high-latitude data centers.[66] The *Atlantic* called it "the world's agricultural hard drive," and, according to its director, "the seed vault is a kind of safety backup for existing seed banks and their collections."[67] But a group known as the Alliance to Rescue Civilization (ARC) takes this idea even further by proposing to back up the entirety of human civilization in a dedicated lunar facility. This backup would include a copy of "our cumulative scientific and cultural treasure chest" and "enough human beings (and supporting species) to repopulate Earth," thus hedging against global disasters including climate change, nuclear war, and large asteroid impact, explain cofounders William E. Burrows and Robert Shapiro.[68] As they write:

> Even if Earth were turned into a vast field of devastation, humanity and its achievements would survive. Think of it as backing up the planet's hard drive and keeping the "disk," constantly updated, in a secure location. Many of the possible disasters would affect our entire planet, so the logical location for such a haven would be off of it, in a base on another world.

Burrows and Shapiro employ the same seed-dispersion metaphor as *The True Way of Backup*, arguing that we must "insure against catastrophe by spreading seeds to other worlds." Echoing digital preservation discourse, they note that the backup facility must be "sustained over the long haul" and "maintained indefinitely since the hazards will always be with us." But what are the environmental costs of building and continually maintaining this vast new infrastructure? Given present energy technologies, the backup of Earth's "data" would threaten the sustainability of *this* Earth, exacerbating the very climate crisis it hopes to outlast.

ARC extends the logic of cloud backup to the planetary scale, until off-site backup becomes off-world backup and geographic redundancy becomes a kind of astronomical redundancy. This extreme "backup solution" imposes a deeply problematic artificial separation between the "data" of human civilization and the "hardware" of the planet: its oceans, soils, atmospheres, and ecosystems. Since the former can, in theory, be copied and saved elsewhere, the latter becomes expendable. Why safeguard Earth's hardware if the contents of the planet's hard drive could simply be "reinstalled" after an apocalypse? The world may end, but the world-as-data will survive. In this way, ARC redirects focus from taking immediate action in order to preserve the environmental resources that exist toward creating an energy-intensive, whole-Earth backup. Representing this same idea in a subtler way, the logo for World Backup Day consists of a blue Earth encircled by a counterclockwise arrow, as if we could back up the world.

An opposing message about the relationship between backup and the biosphere has emerged from the climate justice movement. Instead of framing backup as a

FIGURE 3.3 Climate Reality Project, from the group's Facebook page.

Source: http://facebook.com/climatereality/photos, accessed May 17, 2014.

radical solution to ecological collapse, it urges us to reject the fantasy of a fully backup-able world, which can lull us into a false sense of security, promoting inaction. This imperative is represented in a graphic produced by the Climate Reality Project and shared on social media: superimposed on the "Blue Marble" Apollo 17 photograph—an icon of environmentalism made possible by the many backup systems of the Space Race—are the words "THERE IS NO EMERGENCY BACKUP PLANET" (Figure 3.3). A similar statement appeared in the September 2014 People's Climate March in New York City.[69] One participant's sign read "THERE IS NO PLANET B," accompanied by a red-orange heat map of a future world transformed by climate change. To fight for a habitable climate, these texts suggest, we must rein in our imagination of backup while paying greater attention to the planet's capacity.

Just as the image of the cloud obfuscates the large-scale energy and carbon processes that, by fueling climate change, make the world gradually less safe for our information and for us, even the remote possibility of an emergency backup planet distracts from the irreversible mass extinction and ecological devastation already well underway. It deceives us into thinking there is a separation between Earth's natural "hardware" and the cultural and biological "data" it supports, that one can be saved without the other. But running counter to this fantasy of a global backup system is the stark realization that we have only one planet—and if it "crashes," it won't be "restored in a snap."

Acknowledgments

I wish to thank Nicole Starosielski for her invaluable feedback on multiple versions of this text, and Janet Walker for making many helpful suggestions throughout the editorial process.

Notes

1 Terje Ronneberg, *The True Way of Backup*, http://www.backup.info/files/index.htm (accessed November 18, 2015).
2 See http://worldbackupday.com.
3 As Sean Cubitt, Robert Hassan, and Ingrid Volkmer note, a large portion of cloud infrastructure is dedicated to supporting redundant files. "Does Cloud Computing Have a Silver Lining?," *Media, Culture, and Society*, 33 (2011): 154. Facebook, for example, "ensures user content is always available and retrieved quickly by storing lots and lots of copies of every media file in its data centers," explains *Data Center Knowledge*. This includes primary or "hot" data centers, as well as secondary, lower-power, "cold-storage" facilities, which hold additional copies of media files that are less often accessed by users. Yevgeniy Sverdlik, "How Facebook Cut 75 Percent of [the] Power It Needs to Store Your #tbt Photos," *Data Center Knowledge*, May 8, 2015, http://www.datacenterknowledge.com/archives/2015/05/08/cold-storage-the-facebook-data-centers-that-back-up-the-backup (accessed November 18, 2015). According to an April 2014 Greenpeace report, the Internet accounts for approximately two percent of worldwide CO_2 emissions, "on par with emissions from the global aviation sector." And "data centers will be the fastest growing part of the global IT sector energy footprint as our

online world rapidly expands; their energy demand will increase 81% by 2020." Gary Cook, Tom Dowdall, David Pomerantz, and Yifei Wang, "Clicking Clean: How Companies are Creating the Green Internet," April 2014, 9–10, http://www.greenpeace.org/usa/Global/usa/planet3/PDFs/clickingclean.pdf (accessed November 18, 2015).

4 This analysis would fit into a much longer history of backup, stretching, perhaps, from the copying of ancient manuscripts by hand to the data collection and storage of NSA surveillance, which has yet to be written in full.

5 Ulrich Beck, *Risk Society: Towards a New Modernity* (Los Angeles, CA: SAGE, 1992), 49.

6 "To back up," in "back, v." (definitions 8 and 22). OED Online. September 2015. Oxford University Press. http://www.oed.com/view/Entry/14335?rskey=mLQ305&result=3&isAdvanced=false (accessed November 18, 2015).

7 Marvin Miles, "Grissom Flight Spurs Changes," *Los Angeles Times*, July 30, 1961.

8 Thomas Gordon Plate, "Apollo Safety Key: Its Backup System," *Newsday*, April 15, 1970.

9 Andrew F. Blake, "Backup Astronaut Prepares Himself," *Boston Globe*, April 10, 1970. John L. Swigert Jr. took over for primary pilot Thomas K. Mattingly on Apollo 13 after he was exposed to German measles. Rudy Abramson, "NASA to Study 2,000 Safety-Critical Parts," *Los Angeles Times*, March 10, 1986. While backups were extolled during the Space Race, the *Challenger* shuttle disaster was linked to an undersupply of backup systems. As Charles Perrow observes in *Normal Accidents: Living with High-Risk Technologies* (New York, NY: Basic Books, 1984), complex "high-risk systems," including space missions and nuclear power plants, are prone to cascading failures and thus require more backups: "buffers and redundancies and substitutions must be designed in," he writes; "they must be thought of in advance" (94).

10 Paul Edwards, *The Closed World: Computers and the Politics of Discourse in Cold War America* (Cambridge, MA: MIT Press, 1996), 106.

11 Ibid., 104.

12 One of the company's first clients was East River Savings Bank, which stored backup copies of deposit records on microfilm. See Iron Mountain, "History," http://www.ironmountain.com/Company/About-Us/History.aspx (accessed November 18, 2015). For more on this topic, see Brian Michael Murphy, "Bomb-Proofing the Digital Image: An Archeology of Media Preservation Infrastructure," *Media-N*, 10, no. 1 (2014), http://median.newmediacaucus.org/art-infrastructures-hardware/bomb-proofing-the-digital-image-an-archaeology-of-media-preservation-infrastructure (accessed March 2, 2015), and Tung-Hui Hu, *A Prehistory of the Cloud* (Cambridge, MA: MIT Press, 2015), especially 96–97.

13 Peter Coy, "Nervous Firms Burying Data in Vaults: Thousands of Reels of Magnetic Tape Stored as Precaution Against Computer Failure," *Washington Post*, January 26, 1986.

14 See Chapter 3, "Data Centers and Data Bunkers: 'The Internet Must Be Defended'" of Hu's *A Prehistory of the Cloud*, in which he explores the possibility that "some version or mutation of sovereign power is present in the cloud," specifically in its architecture of physical "data bunkers" (94).

15 Ibid., 91.

16 Ibid., 98.

17 Ibid., 82, 80–81. See, for example, Westland Bunker, http://westlandbunker.com (accessed November 18, 2015). "An isolated underground fortress designed to withstand direct nuclear attack has become the perfect home for mission critical IT and business operations," reads the website for the data center outside Houston.

18 Peter H. Lewis, "Friday the 13th: A Virus Is Lurking," *New York Times*, October 8, 1989, F12.

19 Amazon launched its S3 cloud storage service in 2006, and Dropbox in 2007.

20 Andrew Blum, *Tubes: A Journey to the Center of the Internet* (New York, NY: HarperCollins, 2012), 230.

21 "Personal Plans," http://www.carbonite.com/en/cloud-backup/personal-solutions/personal-plans (accessed November 20, 2015). "Your files are automatically, continuously backed up," explains online storage provider Carbonite. "If disaster (or spilled coffee) strikes, you can recover files from the cloud, in just a few clicks."
22 Seth Porges, "Back Up to the Cloud and Prevent a Data Loss Disaster," *Popular Mechanics*, September 18, 2012, http://www.popularmechanics.com/technology/how-to/a8015/how-to-prevent-a-data-loss-disaster-11992705 (accessed March 15, 2014).
23 "Gartner Says That Consumers Will Store More Than a Third of Their Digital Content in the Cloud by 2016," *Gartner*, June 25, 2012, http://www.gartner.com/newsroom/id/2060215 (accessed November 18 2015). In this report, research firm Gartner predicts that although "on-premises storage will remain the main repository of consumer digital content . . . its share will progressively drop from 93 percent in 2011 to 64 percent in 2016 as the direct-to-cloud model becomes more mainstream."
24 "Failure Trends in a Large Disk Drive Population," *5th USENIX Conference on File and Storage Technologies* (February 2007): 17, http://www.research.google.com/pubs/pub32774.html (accessed November 18, 2015). Google estimates that "90 percent of all new information produced in the world is being stored on magnetic media, most of it on hard-disk drives."
25 Ronneberg, "Locality," *The True Way of Backup*.
26 Charles Darwin, "Does Sea-Water Kill Seeds?," *Gardeners' Chronicle*, May 26, 1855, 356.
27 Microsoft Taiwan, "Delivering Reliability in the Microsoft Cloud," YouTube, November 30, 2012, https://youtu.be/owlYan_LkVQ (accessed November 18, 2015). "No one can prevent natural disasters from occurring," explains a Microsoft spokesperson, but "we can architect a resilient cloud infrastructure with geo-redundant backup for recoverability and data integrity." See also http://www.greenhousedata.com/data-centers/low-disaster-risk (accessed November 18, 2015). In a map of the US, titled "Is Your Data Center Safe?" and created by Green House Data, the low-risk zone appears to be near the Colorado-Wyoming border. Or it might be in Utah. Hu notes that, in addition to the "practical considerations that determine a data center's placement," such as inexpensive land and electricity, "security also plays a role." He cites a data center industry report by Utah's Economic Development Corporation, which claims that the state "logged the fewest federal disaster declarations—tornados, earthquakes, and so on—out of any state," as well as having a low "vulnerability to attacks." Tung-Hui Hu, *A Prehistory of the Cloud*, 109.
28 Beck, *Risk Society*, 21.
29 Ibid., 41.
30 Hu, *A Prehistory of the Cloud*, 132 (emphasis added).
31 Beck, *Risk Society*, 41.
32 Ibid., 35.
33 Ursula Heise, *Sense of Place and Sense of Planet: The Environmental Imagination of the Global* (Oxford: Oxford University Press, 2008), 137.
34 Allison Carruth, "The Digital Cloud and the Micropolitics of Energy," *Public Culture*, 26, no. 2 (2014): 339.
35 For a much closer reading of the cloud metaphor, see Carruth, ibid., especially her investigation of "virtual infrastructure."
36 "The Dropbox Tour," http://www.dropbox.com/tour (accessed November 18, 2015).
37 "These data centers produce a cloud that transcends national borders, but they are also rooted in specific geographies," Hu argues in *A Prehistory of the Cloud* (122).
38 Gerry Shih and Paul Carsten, "Apple Begins Storing Users' Personal Data on Servers in China," *Reuters*, August 15, 2014. http://www.reuters.com/article/2014/08/15/us-apple-data-china-idUSKBN0GF0N720140815#qztUfaSV1IvHFIz0.97 (accessed November 20, 2015). Google, in contrast, has moved servers out of mainland China

"after refusing to comply with Chinese government censorship." Also see Jennifer Holt and Patrick Vonderau, "'Where the Internet Lives': Data Centers as Cloud Infrastructure," in *Signal Traffic*, edited by Lisa Parks and Nicole Starosielski (Urbana: University of Illinois Press, 2015), 71–93.

39 "How to Back Up a Country: To Protect Itself from Attack, Estonia is Finding Ways to Back Up Its Data," *Economist*, March 7, 2015, http://www.economist.com/news/technology-quarterly/21645505-protect-itself-attack-estonia-finding-ways-back-up-its-data-how (accessed March 7, 2015).

40 Ernesto, "Pirate Bay Moves to the Cloud, Becomes Raid-Proof," *TorrentFreak*, October 17, 2012, https://torrentfreak.com/pirate-bay-moves-to-the-cloud-becomes-raid-proof-121017 (accessed March 6, 2015). This also enables a backup system for the site: "If one cloud-provider cuts us off, goes offline or goes bankrupt, we can just buy new virtual servers from the next provider."

41 Stephen Graham, "When Infrastructures Fail," in *Disrupted Cities: When Infrastructure Fails*, edited by Stephen Graham (New York, NY: Routledge, 2009), 5.

42 Cubitt et al., "Does Cloud Computing Have a Silver Lining?," 152.

43 Carruth, "The Digital Cloud and the Micropolitics of Energy," 339, 342. As anthropogenic climate change causes more extreme weather, however, the "light and airy" imaginary of the cloud may give way to something more ominous and menacing.

44 Kevin Bradley, "Defining Digital Sustainability," *Library Trends*, 56, no. 1 (2007): 156.

45 Ibid., 151.

46 Ibid., 157.

47 "What is Sustainability?" United States Environmental Protection Agency, http://www2.epa.gov/sustainability/learn-about-sustainability#what (accessed November 18, 2015).

48 Apple boasts data centers "powered by 100 percent renewable energy." See http://apple.com/environment (accessed September 2014).

49 See Barath Raghavan and Justin Ma, "The Energy and Emergy of the Internet," *Hotnets* (2011): 1–6. According to the Energy Information Administration, roughly 67 percent of the electricity generated in the US in 2013 came from fossil fuels. See "Frequently Asked Questions: What is U.S. electricity generation by energy source?" http://www.eia.gov/tools/faqs/faq.cfm?id=427&t=3 (accessed November 18, 2015).

50 James Glanz, "Power, Pollution and the Internet," *New York Times*, September 22, 2012. http://nyti.ms/18nZqnG (accessed November 18, 2015). "The pollution from data centers has increasingly been cited by the authorities for violating clean air regulations," notes Glanz.

51 Several companies, including Recovery Point and Ongoing Operations, specialize in providing such continuity and disaster recovery services.

52 Cubitt et al., "Does Cloud Computing Have a Silver Lining?," 153.

53 "Does Dropbox Keep Backups of My Files?," http://www.dropbox.com/help/122 (accessed September 2014).

54 Nadia Bozak, *The Cinematic Footprint: Lights, Camera, Natural Resources* (New Brunswick, NJ: Rutgers University Press, 2012), 2.

55 Ibid., 158.

56 Mél Hogan, "Facebook Data Storage Centers as the Archive's Underbelly," *Television & New Media*, 16, no. 1 (2015): 5.

57 Ibid., 7. Cubitt et al. also cite a study "according to which only 6 percent of server capacity is in constant use, and nearly 30 percent is entirely unused," but the "sealed-unit construction of server farms" (e.g., in shipping containers) makes this difficult to track ("Does Cloud Computing Have a Silver Lining?," 152). "Online companies typically run their facilities at maximum capacity around the clock, whatever the demand," writes Glanz. "As a result, data centers can waste 90 percent or more of the electricity they pull off the grid" ("Power, Pollution and the Internet").

58 Carruth, "The Digital Cloud and the Micropolitics of Energy," 343–344.

59 Hogan, "Facebook Data Storage Centers as the Archive's Underbelly," 6. As Carruth argues, "the neglected question of how *personal and individual* uses drive the cloud's expansion, and hence energy requirements, is of particular concern" ("The Digital Cloud and the Micropolitics of Energy," 358).
60 Bozak, *The Cinematic Footprint*, 4.
61 Peter Marcuse, "Sustainability is Not Enough," *Environment and Urbanization*, 10, no. 2 (1998): 104. In his critique of "sustainable urban development," Marcuse argues:

> The acceptance of sustainability, at least in principle, in the environmental arena by virtually all actors has led to the desire to use such a universally acceptable goal as a slogan also in campaigns that have nothing to do with the environment but where the lure of universal acceptance is a powerful attraction.

62 Estimates for data center power consumption vary, but all are staggering. Hogan cites a 2011 Greenpeace report: "Internet servers consume upward of one and a half percent of our global electricity. . . . this means that if the Internet (i.e., cloud computing) were its own country, it would rank fifth in global electricity use" ("Facebook Data Storage Centers as the Archive's Underbelly," 5). Glanz puts the worldwide data center energy tab at "30 billion watts of electricity, roughly equivalent to the output of 30 nuclear power plants" ("Power, Pollution and the Internet"). Also see Mark P. Mills, "The Cloud Begins with Coal," *Digital Power Group*, August 2013, http://www.tech-pundit.com/wp-content/uploads/2013/07/Cloud_Begins_With_Coal.pdf (accessed September 25, 2014).
63 Cubitt et al., "Does Cloud Computing Have a Silver Lining?," 155.
64 Jonathan Sterne, *MP3: The Meaning of a Format* (Durham, NC: Duke University Press, 2012), 2. "The technique of removing redundant data in a file is called *compression*," Sterne explains. "The technique of using a model of a listener to remove additional data is a special kind of 'lossy' compression called perceptual coding."
65 "Amazon S3 Product Details," http://www.aws.amazon.com/s3/details (accessed September 25, 2014).
66 In 2013, Facebook opened a data center near the Arctic Circle in Sweden, taking advantage of hydropower resources and natural air-cooling.
67 Ross Andersen, "After 4 Years, Checking Up on the Svalbard Global Seed Vault," *Atlantic*, February 28, 2012, http://www.theatlantic.com/technology/archive/2012/02/after-4-years-checking-up-on-the-svalbard-global-seed-vault/253458 (accessed September 25, 2014).
68 All quotations from William Burrows and Robert Shapiro, "An Alliance to Rescue Civilization," *Ad Astra*, 11, no. 5 (September/October 1999): 18- 22.
69 See The People's Climate March, http://www.peoplesclimate.org. (accessed November 18, 2015).

4

"THERE AIN'T NO GETTIN' OFFA THIS TRAIN"

Final Fantasy VII and the Pwning of Environmental Crisis

Colin Milburn

There is a common refrain among fans of the 1997 Japanese video game *Final Fantasy VII*: "This game pwns."[1] Developed by Square (now Square Enix) for the Sony PlayStation and Windows, the game has attained legendary status: "This game PWNS ALL."[2] For many gamers, it represents not only the pinnacle of the long-running *Final Fantasy* series, but the climax of video game culture at large: "Final Fantasy VII frigin pwns anyone and anything!!!"[3] By now, the widespread adoration of this game has even become a cliché, a stereotype of geek zealotry: "I dont care what you say, ff7 pwns all."[4]

The gamer vocabulary of "pwning" signifies the domination of an opponent—owning, conquering. But it also represents a quality of excellence, brilliance, and delight. According to gamer lore, the verb "to pwn" originated in the gaming community itself, born from a typographic error.[5] One popular theory points to a multiplayer map in the 1994 game *Warcraft: Orcs & Humans*, allegedly the result of a developer's hasty misspelling. Another common theory holds that the term arose during a deathmatch session of the 1996 game *Quake*, beginning with a player's slip of the keyboard when proudly announcing that an adversary had been "owned." Instead of being ignored or overlooked in the reckless pace of the game, the mistake was called out and then wildly embraced.

Despite such origin stories, the notion of pwning as coeval with gaming is dubious. Various creative abuses of the term "own" had been prevalent in hacker leetspeak since at least the 1980s, without suggesting a specific connection to video games. But in their desire to establish the mythic roots of pwning among games and gamers, these folk etymologies suggest the strong communal value of a term that can mean both a smackdown and a reckoning, a mark of virtuosity as well as virtue. For as much as it might indicate a glorious owning, it also suggests a condition of accountability: an owning that owns up to its mistakes, its failures.

In a ridiculously ironic way, after all, the recycling of the mistaken "p" signals a reclaiming of error, even if nothing more than a typographical error, yet expanded into a hallmark of gamer identity. It takes seriously the error as belonging to *us*—our group, our community—owning it and then finding ways to transform it, rendering it surprisingly productive of other ways of thinking, with pleasure, in common.

It is a term that recognizes, in its playfulness, that one cannot simply undo past wrongs, but instead discovers renovated potential precisely by playing through, learning from the blunder, and responding to such unexpected risks of goofing around with technology: a *mistaken owning* now troped into a way of taking charge of the game, taking responsibility.

In this regard, then, pwning might be understood as an ironic ethical concept, a ludic if not ludicrous ethics. Especially in the context of a game such as *Final Fantasy VII*, widely recognized as a richly layered allegory of environmental crisis and the planetary impacts of technological development, the ironies and paradoxes of pwning are front and center.

As video games increasingly dominate our global media ecology, the environmental implications of game technologies themselves become ever more tangible. The engineering of game hardware relies on scarce resources, including rare earth elements caught up in far-reaching geopolitical conflicts. At the same time, the growing ubiquity of powerful console systems and high-performance PCs puts more and more demands on the worldwide energy infrastructure. These environmental costs of gaming have often registered in the ludic operations of video games themselves.[6] This chapter shows how *Final Fantasy VII* recursively implicates game technologies in environmental risk and ecological despoliation. It is a relentlessly self-reflexive game that provides players with conceptual and affective resources to address the consequences of their own recreational pleasures. While encouraging them to love their hardware, it simultaneously galvanizes some players to love responsibly and to take a stand for sustainable media.

A Bird in the Hand

There is a scene in *Final Fantasy VII* where our intrepid band of heroes, led by the ex-corporate soldier Cloud Strife, encounters a nest of monstrous baby birds. The game prompts the player: "What should we do?" If the player chooses that Cloud and his companions should leave the nest in peace, Cloud's childhood friend Tifa responds: "Right! That was admirable of you." If, however, the player chooses to pilfer the nest, the mama bird immediately descends to defend her young. The player is then locked in mortal combat with the mama bird, with no other choice remaining but to destroy her or lose the game. As soon as the mama bird dies, the heroes receive the treasure in the nest: a bundle of magical Phoenix Downs for restoring life to fallen characters. The scene concludes with Cloud silently staring at the nest of newly orphaned birds, scratching his spiky head, as if recognizing that he and his friends have now doomed these young creatures as surely as they have destroyed the mother, all for the sake of gaming advantage.

The scene has provoked considerable debate. Some players see it purely as a tactical moment, another opportunity to maximize utility: "I always took [the Phoenix Downs] . . . I say if it only helps to take them then take them. Don't give in to cute little 10001001, come on its just a computer animated bird." For these players, the game is simply a game, an algorithmic system. To think otherwise is to be duped by fiction: "always take them [the Phoenix Downs]. There's no reason not to except for role-playing purposes." Others take the representation more seriously while still advocating a hard-nosed financial calculus: "I always take the phoenix downs because they are useful and the items are expensive. So it is money saved. Yeah it is sad that momma bird dies, but hey in real life it happens also." Yet even for these pragmatic players, the scene often induces a role-playing effect, an affective response, precisely because it shows the collateral impacts of the most profitable course of action: "i always take them. it saves money and all that, exp [experience points] and such. do i feel guilty? sort of."[7]

While many are perfectly content to live with this choice and its instrumental rewards, others find it intolerable: "I only ever took them the first time I ever played it, then promoptly felt guilty for ages after when I killed the parents :gasp: So Ive never taken them since xD, I just can't bring myself to do it, even though I know they would come in handy. . . . I look at the little chicks and Im like awwwwww. I think about it for afew minutes then come to the conclusion that I just can't do it." It is a common experience: "I took it once, then guilt ate away at me and I ended up restarting the game. Guess I just felt bad for the little birdies." Some even translate the moment into a moral imperative: "I have never taken the phoenix downs from the nest. If I did i could not live with my self literaly the birds are so cute . . . Please dont take them."[8]

It is a strangely poignant moment in a game that, up to this point, has consistently rewarded the player for ransacking every hidden treasure and slaying all manner of other creatures. Following the model of earlier games in the *Final Fantasy* series—and like most role-playing games (RPGs) in general—*Final Fantasy VII* structures its gameplay around "level grinding," the process of gaining power, money, and equipment by fighting monsters and wildlife, becoming stronger and more experienced with each victory. The baby bird scene merely highlights the fact that the entire game, characteristic of RPGs as a genre, encourages us to plow a path of destruction across the planet in order to level up. More than one player has noted the irony: "RPG characters are always trying to fix the world (while killing as much flora and fauna as it takes to reach level 99)."[9]

It is a particular irony in the context of *Final Fantasy VII*, which focuses on the efforts of Cloud and his scrappy companions to save the planet from the exploitations of Shinra, Inc. Shinra is a ruthless technology corporation that controls the global energy infrastructure. Running its own private army called SOLDIER, Shinra dominates the political economy of the world. The Shinra business plan is based on constructing powerful reactors to extract Mako energy from the geological depths of the planet. Mako is a natural resource, converted into fuel by the reactors. In concentrated form, it is also the source of numerous strange phenomena that are commonly considered "magic" (though some characters point out that any mysterious power is just unexplained science: "It shouldn't even be called 'magic'").

Eager for profit, Shinra supplies more and more Mako to feed the energy demands of the human population, especially in the high-tech city of Midgar. At the outset of the game, Shinra's extraction of Mako has reached a crisis point, threatening the integrity of the planet, the vitality of its ecosystems past and present (the "Lifestream"). The Mako economy has also created enormous social disparities. In Midgar, the poor live in the bottom tier of the city, below the surface, no other choice but to inhabit an urban stratum filled with pollution from the Mako reactors: "The upper world . . . a city on a plate . . . people underneath are sufferin'! And the city below is full of polluted air. On topa that, the Reactor keeps drainin' up all the energy."

Riding the train that connects the different sectors of the city, Cloud suddenly observes the path dependencies of existing technological infrastructures and socioeconomic orders, the extent to which choices made in the past lock in certain futures that often seem impossible to change: "I know . . . no one lives in the slums because they want to. It's like this train. It can't run anywhere except where its rails take it." A train on a track—a metaphor for what Martin Heidegger called the technical ordering of destining, that is to say, technological enframing.[10] It indicates the mode of existence in the modern world, the manner in which all things are challenged forth as standing-reserve, component resources in the relentless drive of technologization. This theme is reinforced throughout *Final Fantasy VII*; indeed, the head of the Shinra Public Safety division—that is, the Shinra military—is even named Heidegger.

The game puts the player in charge of a group of characters actively resisting Shinra and the environmental crisis it has created. Cloud and the other protagonists are members of the militant ecological group AVALANCHE, whose tactical ops focus on blowing up the Mako reactors. AVALANCHE's activities—lauded by some, vilified by others—foment an eschatological discourse on environmental justice, as suggested by scattered graffiti in Midgar:

> Don't be taken in by the Shinra.
> Mako energy will not last forever.
> Mako is the life of the Planet and that life is finite.
> The end is coming.
> Saviors of the Planet: AVALANCHE

To the extent that the game presents its core conflict as an epic struggle for the fate of the planet, with Cloud and AVALANCHE fighting a guerrilla war against the militarized corporate power of Shinra, the fact that the heroes must procedurally depopulate the wildlife of every region they visit would perhaps seem an instance of ludonarrative dissonance—a radical disjunction between gameplay mechanics and narrative content.[11] Yet this is exactly the point. The game draws attention to the ways in which its own conventional gameplay design, its random encounters, and accumulative leveling structure allegorize a general predicament: to play a video game, any video game, is to contribute however indirectly to the environmental hazards represented by electronic media.

Final Fantasy VII reminds us, precisely in the conflict between its narrative of environmental heroism and its ludic insistence on random acts of animal slaughter, about the degree to which video games and other technologies of entertainment exacerbate global ecological problems.[12] In recent decades, the production of gaming hardware has escalated the environmental and geopolitical pressures associated, for example, with the mining of coltan and other mineral resources for electronics manufacturing. These pressures include the pollution effects of mining practices and the depletion of wildlife habitats, as well as humanitarian crises in coltan-rich regions such as the Democratic Republic of the Congo. Likewise, as more and more obsolete gaming consoles and computers are discarded as e-waste each year, dumped into landfills or shipped in container barges to countries with less-than-stringent disposal regulations, they add to a steady release of toxic chemicals into the atmosphere, the oceans, and the soil.[13] At the same time, the energy consumed by legions of powerful gaming machines around the world is enormous—and getting worse.[14] By all measures, video games and other computational media leave a sizeable carbon footprint.

Final Fantasy VII pointedly emphasizes these connections. For example, a citizen of Midgar advises Cloud about their media dependency on Mako: "If you knock out Midgar's power, then all of its computers and signals are going to be knocked out too." When playing the game, whether on a personal computer or a PlayStation console, we are reminded that, as gamers, we are contributing to the environmental crisis—even as the narrative of the game charges us, at least fictively, to do something about it.

Runaway Train

Barret, the leader of AVALANCHE, addresses the technological condition of the world with a recurring metaphor—his personal motto: "There ain't no gettin' offa this train we on! The train we on don't make no stops!" The figure of the runaway train suggests the deterministic force of industrialization and the path dependency of the energy economy, the acceleration of petroculture. The game depicts energy consumption speeding up as the result of new technologies, especially in the transition from coal-powered systems to the Shinra-controlled Mako reactors. However gradually at first, the fate of the planet was set once humans began to extract machine fuel from the remains of living things (symbolized as a fluid "Spirit energy" that flows into the underground Lifestream). As the scientist-mystic Bugenhagen says,

> Every day Mako reactors suck up Spirit energy, diminishing it. Spirit energy gets compressed in the reactors and processed into Mako energy. All living things are being used up and thrown away. In other words, Mako energy will only destroy the planet.

Recognizing that technological decisions of the past have become self-reinforcing, Barret repeatedly admonishes his companions that they must fight to change the system in its entirety: "But you gotta understand that there ain't no gettin' offa this train we on, till we get to the end of the line." His metaphor points to the material infrastructures underpinning the social order, while also emphasizing that we are all in it together, for better or worse. Barret's insistence that we are all passengers on the same train distributes responsibility to everyone, in salient contrast to the polarized structure of the game's playable narrative, which identifies the corporate greed and unethical scientific experiments of Shinra as the primary threats to environmental sustainability.

Certainly, Shinra represents the worst excesses of high-tech global capitalism. Even as the existing Mako sources threaten to run dry, destabilizing the integrity of the entire world, Shinra strives to locate the so-called Promised Land, a hidden region of vast Mako reserves, and drain it completely. Irritated by the AVALANCHE insurgents and their strikes against the Mako reactors in Midgar, Shira responds by detonating the supporting structures of the Sector 7 plate, which divides the gentrified upper-level of the city from the slums below. The slum area is crushed by the collapse of the massive plate—a shocking display of corporate force that callously destroys the lives of thousands of people simply to stifle a handful of rebels.

Moreover, it is Shinra's work on the military applications of Mako and the company's unethical biotechnology research that shifts the "slow violence" of the energy economy into a climax crisis: the existential threat represented by Sephiroth and his summoning of the Meteor, a potential extinction event.[15] Sephiroth, the most powerful member of SOLDIER, is the product of Shinra's experimental research, engineered from human and alien materials, as well as heavy doses of Mako. He is both the pinnacle of Shinra's military science program and its foreclosure, for upon discovering his synthetic origins, Sephiroth turns against Shinra and the entire biosphere. Another figure of the runaway train, Sephiroth is an uncontainable force produced by technoscientific choices of the past—and now driving the entire planet headlong toward catastrophe.

Despite that Shinra is a war-mongering, criminal corporation, most people of the world are willingly duped by its propaganda ("Shinra's Future Is The World's Future!! Mako Energy For A Brighter World!!"). In the complacent town of Kalm, for example, an old man says, "Thanks to Shinra, Inc. developing Mako energy for us, everything's more convenient now." A woman says, "I'd hate to think of what life'd be like without Mako energy. . . . Yeah, Mako energy's made our lives much easier. And it's all thanks to Shinra, Inc." Even awareness of the looming crisis cannot shake this attitude. "I hear that the natural resources near the reactors are being sucked dry," says one villager. But he nevertheless concludes, "We're better off with them bringing in the Mako energy." This widespread feeling, according to the president of Shinra, cannot even be threatened by the rising prices of Mako: "It'll be all right. The ignorant citizens won't lose confidence, they'll trust Shinra, Inc. even more." The citizens of the world are happy to accept exploitation as long as they can avoid changing their way of life.

But no one is innocent. Even Cloud bears the evidence of complicity in his own flesh. His glowing blue eyes indicate that he was literally showered in Mako to enhance his capabilities as a corporate warfighter: "That's the sign of those who have been infused with Mako . . . A mark of SOLDIER." Barret, likewise, sports a prosthetic gun-arm, a constant remembrance of the catastrophic destruction of his hometown of Corel. Barret had urged the people of Corel to give up coal and allow Shinra to build a Mako reactor near the town ("No one uses coal nowadays. It's the sign of the times."). But disaster falls: "There was an explosion at [the] reactor. Shinra blamed the accident on the people. Said it was done by a rebel faction." The Shinra army burned the town, killing Barret's wife and shooting off his arm while he tried to save a friend. Replacing his ruined arm with a gun, Barret became leader of AVALANCHE to alleviate his own culpability: "But more than Shinra, I couldn't forgive myself. Never should've gone along with the building of the reactor." But as Tifa says, "We were all fooled by the promises Shinra made back then." The mistakes are ours, we must own them, we are all passengers. We are all playing the game.

Whenever his companions start to lose track of the technopolitical objective—that is, the goal of this game—Barret rallies them with the familiar refrain: "C'mon, let's think about this! No way we can get offa this train we're on." The insistence on the technically determined pathway also self-reflexively points to the narrative of *Final Fantasy VII* itself, the software of the game and its predetermined range of possible choices. For the player of *Final Fantasy VII*, just as for the characters in the story, the only option is to play through to the end or to quit entirely. It is a reminder that our own participation in the game is coterminous with the journey of Cloud and friends to defeat Shinra, a journey that shows how they have always been as responsible as everyone else in maintaining the status quo, the Mako economy and the corporate systems accelerating its expansion.

The central drama of *Final Fantasy VII*, after all, is about discovering the extent to which even those who resist the prevailing systems of control are likewise products of those same systems, mystified by the conditions of high-tech living to overlook everyday failures of responsibility, including their own failures. It is a drama about accepting the fact that we are all puppets of the technopolitical regimes we inhabit, for only by owning up to this can we begin to find ways of turning our puppet condition to advantage, to game the game.

No Strings Attached

The big twist in *Final Fantasy VII* is that Cloud is not who he thinks he is—and in more ways than one. He has played out an elaborate fantasy in his own mind, creating a fictive backstory to avoid grappling with his inability to achieve his ambitions. It blinds him to the fact that he is also being controlled by an external agency: "There's something inside of me. A person who is not really me." He discovers that he is a pawn, a doll, a marionette: "I'm . . . a puppet?"

This puppet condition is the result of a series of failures:

> I never was in SOLDIER. I made up the stories about what happened to me five years ago, about being in SOLDIER. I left my village looking for glory, but never made it in to SOLDIER . . . I was so ashamed of being so weak; then I heard this story from my friend Zack . . . And I created an illusion of myself made up of what I had seen in my life . . . And I continued to play the charade as if it were true.

Haunted by anxieties of weakness, Cloud's personality is further destabilized when Professor Hojo of the Shinra science division tries to turn him into a copy of Sephiroth: "You are just a puppet. . . . An incomplete Sephiroth-clone. Not even given a number. That is your reality." Cloud is rescued by his friend Zack, an accomplished member of SOLDIER, but the nefarious experiment nearly destroys him: "I was a failed experiment." Cloud's mental stability crumbles completely when Zack dies in combat. Cloud then constructs a fake history for himself. His persona as an ex-SOLDIER resistance fighter is merely an avatar, a puppet-identity without substance. He is a shell, a fictive version of himself secretly manipulated by Sephiroth ("I wasn't pursuing Sephiroth. I was being summoned by Sephiroth.").

By owning up to being a failed hero and a failed experiment, conceding his puppet condition and working to move beyond it, Cloud gains the upper hand: "I never lived up to being 'Cloud,'" he tells Tifa. "Maybe one day you'll meet the real 'Cloud.'" He does not accomplish this alone, of course, but with the help of friends. Tifa guides him to reconstruct an identity from the stream of his memories tinged with fantasies of Zack's exploits: a composite fiction that ultimately becomes real life. "I'm . . . Cloud . . . the master of my own illusionary world. But I can't remain trapped in an illusion any more . . . I'm going to live my life without pretending."

This process represents the pwning of error, taking responsibility for the failed experiment. Cloud was a victim of Shinra's military research program, certainly—but only because he had already volunteered for the Shinra ranks in a misguided effort to join SOLDIER. He was a puppet of the military-petroleum complex from the beginning. Only by recognizing his personal contribution to the onrushing environmental calamity (under Sephiroth's influence, he even hands over the Black Materia for summoning Meteor) can he change the course: "I'm the reason why Meteor is falling towards us. That's why I have to do everything in my power to fight this thing . . . There ain't no gettin' offa this train we on!" Taking charge of his own mystification, he actually becomes the virtuous warrior he was not supposed to be. Cloud's pwning of the environmental crisis, even in going forward "without pretending," thus depends on an elaborate role-playing game, adopting the role of the so-called "real Cloud" who emerges from the other side of his own puppet condition, the real Cloud who turns out to have been a hero in sufferance all along.

These narrative twists allegorize the gameplay situation itself. The player, having puppeted Cloud throughout the game, occupies two roles simultaneously: both the puppet, identifying with Cloud, controlled by Sephiroth's will—which is to say, controlled by the game software and its narrative that inexorably drives us through the action—and the puppeteer, the controlling agency behind the console in whose hands the action lies. It is a recursive allegory that valorizes the capacity of RPGs to create an ecologically responsive subject, an eco-warrior fighting for the planet by taking responsibility for past mistakes. Playing the role of Cloud is literally the means by which Cloud works though failure to become better than himself. We, as players of Cloud, are invited to take the same initiative.[16]

In this way, *Final Fantasy VII* fashions itself as an instrument of ecologically responsible technopolitics, an instrument of change. Over and again, its narrative recursively emphasizes the transformative power of games, the subversive potential of role-playing and other forms of ludic recreation. The secret AVALANCHE hideout in Midgar, for example, is actually hidden beneath a pinball machine. Or consider the Gold Saucer: a vast pleasure dome, an amusement park offering a wealth of playable mini-games, including video arcades, chocobo races, basketball, VR battles, and more. The Gold Saucer bodies forth the culture of fun and games as such. On the one hand, it represents gaming as distraction from real environmental crisis, a temporary escape from the problems of modern life. When the AVALANCHE crew enters the Gold Saucer, hot on the trail of Sephiroth, Aeris says, "Wow! Let's have fun! I know this isn't the right time to do this. I wish we could just forget everything and have fun!" The Gold Saucer is an alluring diversion, beguiling the AVALANCHE team to waste time while the fate of the world remains at stake. But on the other hand, the Gold Saucer section shows how games also provide clues, tools, and skills for addressing the material conditions of the present.

After all, when exploring the entry hub to the Gold Saucer, Cloud finds the following poster: "Many attractions await you here at Gold Saucer. You will be moved and excited, thrilled and terrified! Led from one zone to another . . . unlike anything you've ever experienced!" And at the bottom of the poster: "Shinra." In other words, the Gold Saucer and its ludic pleasures have been developed by the very same company controlling the energy economy and driving the planet to disaster. If Cloud therefore learns that games contribute to petroculture and environmental despoliation, we as players must also reckon with the idea that the gaming console in our hands is a node in a much larger system of technopolitical forces and material flows. Certainly, some *Final Fantasy VII* players become acutely attentive to such connections: "You are aware how god-awful video games are for the environment right?"[17] Or as another player explains, "*Why are we still so dependent on oil?* Because we need the electricity to play Final Fantasy VII, of course. Speaking of which, am I the only one who thinks that it's a bit hypocritical . . . it's kind of annoying to get an environmental message from a video game that essentially wastes electricity for the sake of entertainment."[18]

But this shocking revelation of the impurity of gaming is, of course, afforded by gaming itself—a type of double agency. It is for this reason that our first exploration of the Gold Saucer also introduces the playable character of Cait Sith. Like Cloud, Cait Sith is a self-reflexive figure of the puppet. Cait Sith has a robot moogle body, apparently piloted by the intelligent cat riding on the robot's head. Cait Sith joins the AVALANCHE team to fight Shinra and Sephiroth. But it turns out that Cait Sith is a double agent. The intelligent cat is actually a cybernetic avatar teleoperated by Reeve, a Shinra employee who runs the Urban Development Department back in Midgar: "This [Cait Sith] body's just a toy anyway. My real body's at Shinra Headquarters in Midgar. I'm controlin' this toy cat from there." A remote-controlled toy, Cait Sith was set up in the Gold Saucer to infiltrate Cloud's group and undermine the resistance from inside.

Yet as an effect of role-playing as an eco-warrior, Reeve starts to identify with the insurgency. Speaking through the Cait Sith puppet body, he confesses to the AVALANCHE group:

> Alright, yes, I am a Shinra employee. But we're not entirely enemies. . . . Something bothers me. I think it's your way of life. You don't get paid. You don't get praised. Yet, you still risk your lives and continue on your journey. Seeing that makes me . . . It just makes me think about my life. I don't think I'd feel too good if things ended the way they are now.

Reeve switches sides, turning against his employer and embracing the Cait Sith role. He emerges as the virtuous defender of the planet he had not planned to be. In playing along, even from the compromised position of Shinra middle management, Reeve discovers hope—and becomes otherwise.

If at First You Don't Succeed . . .

Hope—even faced with the scope of the problem, recognizing the difficulty of solving a crisis while contributing to it, the double agency of pwning says to play through, try again. Bugenhagen reflects on this conundrum:

> Cloud says they are trying to save the planet. Honestly, I don't think it can be done. For even if they stop every reactor on the planet, it's only going to postpone the inevitable. Even if they stop Sephiroth, everything will perish. But . . . I've been thinking lately. I've been thinking if there was anything

> WE could do, as a part of the planet, something to help a planet already in misery ... No matter what happens, isn't it important to try?

For Bugenhagen, this entails responsible innovation: not a relinquishment of modern science and technology, but an obligation to find other ways, different technopolitics. It turns out that Bugenhagen had also formerly been a Shinra employee, and his deep understanding of the planet was not shaped in isolation from Shinra technologies but in relation to them. Living in the ecotopian village of Cosmo Canyon, Bugenhagen has put Shinra's machines to more sustainable purposes: "Wrapped up in the planet's strange notions surrounded by Shinra-made machines ... Science and the planet lived side by side in that old man's heart."

It is a vision of sustainable media, conjured forth through the mechanisms of unsustainable media. The provocation that a source of danger could itself become a saving power, that the conditions of modern technologization might be the means of their own transformation—a Heideggerian paradox, to be sure—is baked into the structure of *Final Fantasy VII*. For winning this game means using a high-tech system (the console, for example) to work through crisis and achieve a different outcome. Several centuries after the collision of Meteor and the Lifestream, Nanaki—once known as the experimental test subject Red XIII—and his cubs overlook the ruins of Midgar that have grown verdant with new life. The coda suggests that human civilization successfully let go its reliance on Mako, allowing the planet to restore itself.[19] It is the result of the playable events in the game—a consequence of playing through, committing to the challenges. In this manner, *Final Fantasy VII* discloses its own final fantasy of recuperation and rehabilitation. The failed experiment of the global video game industry, with all its contributions to environmental problems, here represents itself as affording something better in the long run ... by putting the future in the hands of players.

Responding to these recursive thematizations of responsibility, fans of the game endlessly discuss its potential to elicit critical thought about the conditions of our technological society. Of course, some prefer to overlook the environmentalist motifs and simply enjoy the fun of slaying monsters. Others are overtly skeptical about the green politics of the game, much to the dismay of those more committed to its ethical twists. But for many players around the world, *Final Fantasy VII* has encouraged them to consider, for example, "how deeply the fights for economic democracy and environmental sustainability are intertwined."[20] It has helped them to conceptualize the stakes: "We're all aware of the idea that Mako energy and all that is a metaphor for the way we mistreat our planet but parts of Final Fantasy VII seem to hint in an even stronger way towards the effects of climate change/global warming."[21] For some, engaging with the game on this level can lead to exasperation:

> The ideological strife here [in the game] mirrors an obvious—and also often ignored—issue in the real world. Fossil fuel consumption, global

> warming, and ongoing pollution are problems that have gone on relatively unchecked since the Industrial Revolution and have continued to increase in severity as technology and a growing population demand more use of coal and gasoline. . . . Gaia is dying, the people are loosely aware of it, and pretty much no one is doing anything about it. In many ways, the same thing is happening on Earth.[22]

Yet for others, the figural aspects of the game entice them to imagine differently, turning erstwhile fantasy into proactivity.

In online discussions about the lessons of *Final Fantasy VII*, ranging from the tongue-in-cheek to the sincere, numerous players have attested to the impact of gaming on their own ecological sensitivities. As one player has put it, "I'd also like to thank Final Fantasy VII, EcoQuest I and II, and Chrono Cross for turning me into a raging environmentalist." According to another player, "FFVII changed the way I think about the planet, I can tell you that much lol. Hurry! Let's all rebel against giant companies and industries because they're polluting the planet!" Another has said, "It made me newly aware about matters like ecology, the need to fight for your rights without hurting anyone else, and most of all it made me think a lot about bioethics." And another: "it made me question a lot of things, including the way we live." And another: "It made me realize that the whole world needs improving . . . and I've gotta help."[23]

It is certainly the case that these insights, these ethical urges and emergent dispositions, are shaped by experiences with a game, or rather, an entire media ecology that is not remotely sustainable in its current form. Ironic, yes. But such irony often cultivates perspicacity. Consider the Spanish console modder MakoMod, who made headlines in 2015 by transforming an old PlayStation console into a physical replica of Midgar. This Midgar PS1 mod is a fully functional gaming machine that also features small details of narrative significance: the Mako reactors, the poisoned terrain beneath the city, the train system, and Aeris's reclaimed church where she grows flowers in the midst of industrial ruin. It reaffirms the game's allegory of environmental crisis and responsibility, identifying the unsustainable city of Midgar with the PlayStation itself: the city is the console, the Mako reactors draw power from a real electrical outlet. The mod does not obscure the material and geopolitical histories that converge at the site of the console, but instead brings them to light and recycles them. In this regard, it also represents a small gesture of resistance, pushing back against the game industry's calculus of planned obsolescence by refurbishing the vintage hardware, giving it new life.[24]

This modding experiment and others like it, modest though they are, can be considered affirmative exercises of remediation, homebrew efforts to reclaim the toxic infrastructures of the present. Some players have likewise suggested that the massive fan-recreations of Midgar in *Minecraft* are comparable to Aeris's flower garden in *Final Fantasy VII*, insofar as they represent DIY attempts to revamp the contaminated city of Midgar and all it signifies from the bottom up. As one player puts it, "if everyone grew flowers in Midgar, it'd be a better place . . . she [Aeris] tried to change that place for the better."[25] Expressing a shared desire for change, such practices of media appropriation confront the givenness and intractability of our high-tech modalities, indicating in miniscule ways that other futures are yet possible: alternate tracks for this train we are on.

Problem and solution, poison and remedy, video games propagate pharmacologically in the ecosystems of our world. Even while contributing to environmental crisis, they also animate a playful and experimental attitude toward technology, a sense that innovation pathways can always be modified, transmogrified. As mechanisms for working through error, overcoming failure, they afford visions of a future reloaded, played with renewed virtuosity thanks to skills learned from past experience.

In a word, pwned.

Acknowledgments

Thanks to Jordan Carroll and Marty Weis for their help with this chapter.

Notes

1 Sour Grape, in response to lynm pahcuh, "FFVII vs. FFVIII," *Square Insider*, March 23, 2005, http://squareinsider.com/forums/topic/19295-ffvii-vs-ffviii/.
2 Baka Neko, in response to cosmic999, "Does Anyone Still Play This?," May 25, 2008, http://www.neoseeker.com/forums/1169/t1161488-does-anyone-still-play-this/.
3 KylesKingdomHearts, "Final Fantasy VII Frigin Pwns Anyone and Anything!!!," *YouTube*, October 10, 2009, https://www.youtube.com/playlist?list=PLE20DDFFD08C15371.
4 Legendary Nick, "Final Fantasy 7," *Urban Dictionary*, May 28, 2005, http://www.urbandictionary.com/define.php?term=Final+Fantasy+7&defid=1285455.
5 jack, "Pwned," *Urban Dictionary*, November 17, 2003, http://www.urbandictionary.com/define.php?term=pwned&defid=354944.
6 Colin Milburn, "Green Gaming: Video Games and Environmental Risk," in *The Anticipation of Catastrophe: Environmental Risk in North American Literature and Culture*, edited by Sylvia Mayer and Alexa Weik von Mossner (Heidelberg: Universitätsverlag Winter, 2014), 201–219. On games as mediating environmental and technological perception, see Alenda Chang, "Games as Environmental Texts," *qui parle*, 19 (2011): 57–84.
7 Sora_lion_heart, in response to Bambi, "Birds Nest on Mt Corel," *Final Fantasy Forums*, October 25, 2008, http://www.finalfantasyforums.net/threads/24277-Birds-Nest-on-Mt-Corel; Siara_Sendai, in response to chaosinwriting, "The Treasure in the Baby Bird's Nest on the Mt. Corel Train Tracks," *GameFAQs*, 2012, http://www.gamefaqs.com/boards/130791-final-fantasy-vii/64181986; MadMonkey and poker king46 in response to Bambi, "Birds Nest on Mt Corel."
8 Bambi, "Birds Nest on Mt Corel"; Sohryuden666 and joeyfinalfantasygod, in response to Bambi.
9 Caostotale, in response to Cadtalfryn, "Games with Environmental Themes," *Destructoid*, August 14, 2010, http://forum.destructoid.com/showthread.php?19383-Games-with-environmental-themes.
10 Martin Heidegger, "The Question Concerning Technology," in *The Question Concerning Technology, and Other Essays* (New York, NY: Garland, 1977), 3–35.

11 The concept of ludonarrative dissonance was coined by the game designer Clint Hocking. For discussion, see Tom Bissell, *Extra Lives: Why Video Games Matter* (New York, NY: Pantheon Books, 2010).
12 See Nick Dyer-Witheford and Greig De Peuter, *Games of Empire: Global Capitalism and Video Games* (Minneapolis: University of Minnesota Press, 2009), 222–224; Sean Cubitt, "Current Screens," in *Imagery in the 21st Century*, edited by Oliver Grau and Thomas Veigl (Cambridge, MA: MIT Press, 2011), 21–362; Nadia Bozak, *The Cinematic Footprint: Lights, Camera, Natural Resources* (New Brunswick, NJ: Rutgers University Press, 2012); Richard Maxwell and Toby Miller, "'Warm and Stuffy': The Ecological Impact of Electronic Games," in *The Video Game Industry: Formation, Present State, and Future*, edited by Peter Zackariasson and Timothy L. Wilson (New York, NY: Routledge, 2012), 179–197; Sy Taffel, "Escaping Attention: Digital Media Hardware, Materiality and Ecological Cost," *Culture Machine*, 13 (2012), http://www.culturemachine.net/index.php/cm/article/view/468; and Jussi Parikka, *A Geology of Media* (Minneapolis: University of Minnesota Press, 2015).
13 Jennifer Gabrys, *Digital Rubbish: A Natural History of Electronics* (Ann Arbor: University of Michigan Press, 2011).
14 Pierre Delforge and Noah Horowitz, *The Latest-Generation Video Game Consoles: How Much Energy Do They Waste When You're Not Playing?* (New York, NY: Natural Resources Defense Council, 2014); Amanda Webb, Kieren Mayers, Chris France, and Jonathan Koomey, "Estimating the Energy Use of High Definition Games Consoles," *Energy Policy*, 61 (2013): 1412–1421; Louis-Benoit Desroches, Jeffery Greenblatt, Stacy Pratt, Henry Willem, Erin Claybaugh, Bereket Beraki, Mythri Nagaraju et al., "Video Game Console Usage and US National Energy Consumption: Results from a Field-Metering Study," *Energy Efficiency*, 8 (2015): 509–526.
15 See Rob Nixon, *Slow Violence and the Environmentalism of the Poor* (Cambridge, MA: Harvard University Press, 2011).
16 On ways that games compel learning through error, see Jesper Juul, *The Art of Failure: An Essay on the Pain of Playing Video Games* (Cambridge, MA: MIT Press, 2013). On ways that characteristics of game avatars affect perceptions and behaviors of their players, see Nick Yee, *The Proteus Paradox: How Online Games and Virtual Worlds Change Us—And How They Don't* (New Haven, CT: Yale University Press, 2014).
17 Pseudonym2, in response to Fightgarr, "Environmentalism and Games," *Escapist*, January 15, 2009, http://www.escapistmagazine.com/forums/read/9.83914-Environmentalism-and-Games.
18 LegendofLegaia, in response to Ultima_Terror, *GameFAQs*, 2011, http://www.gamefaqs.com/boards/605802-dissidia-012-duodecim-final-fantasy/61004655.
19 While the ending is ambiguous, the sequel film *Final Fantasy VII: Advent Children* (2005) retcons in favor of this interpretation.
20 Jon Hochschartner, "'Final Fantasy,' Capitalism, and the Environment," *People's World*, October 24, 2013, http://peoplesworld.org/final-fantasy-capitalism-and-the-environment.
21 Jiro, "FFVII Preempting Climate Change Movement," *Eyes on Final Fantasy*, December 24, 2013, http://home.eyesonff.com/archive/index.php/t-153624.html.
22 Bretth2, "Game Analysis: Final Fantasy VII," ENG 380, University at Buffalo, June 21, 2013, http://eng380newmedia.wordpress.com/2013/06/21/game-analysis-final-fantasy-vii.
23 Loveless, "What Are You Happy That Video Games Taught You?," *Nerd Fitness Rebellion*, November 26–27, 2013, http://rebellion.nerdfitness.com/index.php?/topic/40752-what-are-you-happy-that-video-games-taught-you; Seraf, KoShiatar, Ravendale, and MercenX, in response to Jojee, "How Has FFVII Changed Your Perception of Life?," *Eyes on Final Fantasy*, April 30, 2005, http://home.eyesonff.com/archive/t-61134.html.

24 See James Newman, *Best Before: Videogames, Supersession and Obsolescence* (New York, NY: Routledge, 2012); Jonathan Sterne, "Out with the Trash: On the Future of New Media," and Lisa Parks, "Falling Apart: Electronics Salvaging and the Global Media Economy," in *Residual Media*, edited by Charles R. Acland (Minneapolis: University of Minnesota Press, 2007), 16–47.

25 The Sentient Meat, in response to Jason Schreier, "This Minecraft Recreation of Final Fantasy VII's Midgar Is Absurdly Detailed," *Kotaku*, April 17, 2012, http://kotaku.com/5902834/this-minecraft-recreation-of-final-fantasy-viis-midgar-is-absurdly-detailed.

PART II

Social Ecologies, Mediating Environments

5

MEDIATING INFRASTRUCTURES

(Im)Mobile Toxicity and Cell Antenna Publics

Rahul Mukherjee

> So when it all started . . . when she got diagnosed . . . I was too traumatized . . . all of us were and I blamed it on many things. I thought maybe I did something . . . you go through the motions . . . this happened, that happened . . . and one day while sitting in my husband's office space, I looked out of the window, a giant tower, and I don't know . . . like when you started the show you mentioned, we don't even notice it . . . its true.
>
> —Excerpt from *NDTV's* "We the People"[1]

In her televised testimony on September 16, 2012, Rabani Garg recalls the moment she realized that cell towers might be the reason for her seven-year-old daughter Risa's cancer. She is responding to a question posed by Barkha Dutt, the anchor of the talk show *We the People*. Barkha Dutt is a household name among India's urban middle-class television audience, and *We the People*, which began in 2001, is the longest running discussion-oriented show in the Indian television landscape. Although stories about people who lived close to mobile towers getting cancer had been reported for more than two and a half years, the fact that the cell tower radiation issue made it to *We the People* attests to its national importance.

We are at the halfway mark of the show. Dutt has been interrogating the key actors and experts involved in the radiation debate who are seated in the first and lowest tier of the studio, shaped in the form of an arena. Garg is sitting in the second tier among the audience, and Dutt is there by her side. As Garg is speaking, the camera frames her face in a close-up and, at other times, we see Dutt and Garg framed together with Dutt's hand over Garg's shoulder—a gesture of empathy and solidarity.

Let us stay with Rabani Garg's narrative for a while. Having recently found out that cell towers potentially cause cancers, Garg began seeing towers in every direction: "I went out, I saw the tower, I was really really upset . . . then I go out, I turn left, I see another one. I look the other way and there is another one." And yet, she confesses that she had not noticed the towers for the last year and a half. Garg's experience suggests that even though cell towers were, and are, everywhere in urban India, they blended together with other urban infrastructures, and city dwellers did not care—they did not "notice." The mobile tower was inconspicuous, considered a part of the urban environment alongside street light poles, radio and TV towers, and telephone cables. However, when the discourse about the disruptive potential of cell towers began to be circulated from early 2010 onwards, citizens started to pay attention.

Toward the end of her testimony, Garg points to a key uncertainty at the heart of the cell tower radiation debate: could cell towers be a primary cause of cancer? A number of experts (some present on the show) affirmed the position of cell tower companies: there was no proof that cell towers could be isolated as the factor causing cancer. Garg seems to be anticipating their interrogation when she says: "is that what caused it? I don't know." Another group of experts and anti-radiation activists argued that the burden of proof lay with the proponents of unregulated towers: could they prove that the towers did not cause cancer? Garg seems to be echoing this other group when she refuses to foreclose the possibility that cell towers are dangerous: "But it could be. Can you say for sure [the cell tower] did not cause it?"

Mediating Cell Towers/Cell Antennas

Urban life is sustained by infrastructures. As they move around the city, urbanites communicate via cell phones that depend on cell towers for relaying, amplifying, and directing signals. A cell tower has a certain radius within which it supports cellular communication, and when the mobile phone leaves that radius, it establishes connection with another cell tower. As a result, mobile towers are spread throughout cities, and are very much a part of the Indian urban environment (see Figure 5.1). With 116,454 towers in fifteen regions across the country, the company Indus Towers has the widest coverage in India offering its services to a variety of cellular operators.[2] Yet the infra-ness of infrastructures suggests that while they make life livable, they often remain banal, invisible, and inconspicuous.

As Stephen Graham has argued, it is often only when infrastructures fail or cause disruption that people take notice.[3] Before the ill-health effects of cell towers became news in India, if ever people were reminded of their presence, it was when faced with difficulty in conversing on the phone, that is, in the case of a disruption in the telecom service and in cellular mobility. The problem was attributed to the distance of the tower, or its location out of sight, or the obstruction of the signal by buildings. To sustain the signal and the architecture of mobility, more cell towers were established. Yet when cell towers were associated with cancer, they became disruptive for the Indian public in a different way.

FIGURE 5.1 Dense cluster of cell antennas at Haji Ali in Mumbai.

Source: author.

During the radiation controversy, it was clarified that it was not the whole tower, but the antennas that it housed, which were emitting potentially harmful radiation. More specifically, it was the rectangular/vertical sector antennas that were deemed harmful, rather than the dish antennas, which were involved in benign point-to-point communication. This made apparent the ways that, though cell tower antennas are media infrastructures that support mobile media, they themselves are immobile, standing obdurately on top of an apartment or water reservoir. The power density of signals emitted by a sector antenna is inversely proportional to the square of the distance from the mobile tower. Thus, the environmental impact of a specific mobile tower antenna, its immobile toxicity, is contained within a certain region, even as that toxicity supports the mobility of mobile phones. While cell antennas remain productive in terms of providing network coverage, their fixedness and carcinogenic properties now cause anxiety among the public. In early 2015, news coverage associated the cell tower radiation issue with the issue of dropped calls: with stricter regulation on building cell towers because of the radiation controversy, the cellular infrastructure did not keep up with rising mobile phone subscriptions, and thus could not provide adequate service, which led to frequent call drops.

When an infrastructure or technology turns disruptive, it affects communities, and a wide range of stakeholders gathers around it. The pragmatist philosopher, John Dewey, described how social actors break from the habitual ways of their everyday lives and attempt to forge a public when they find themselves affected by issues or problems beyond their control.[4] Cancer from cell antennas is one such issue, and the cell antenna attracted a public of affected actors: radio-frequency scientists, oncologists, cellular operators, regulators, tower builders, cancer patients, and journalists, among others.

The inability to measure radiation properly and the uncertainty about the health effects of the cell tower signals exacerbated debate between, on the one hand, cancer patients and others living close to mobile towers who wanted the towers removed and, on the other hand, cellular operators whose interests were served by the towers remaining in place. The imperceptibility to human senses of cell tower radiation further complicated matters, and cell tower officials, telecommunication regulators, and anti-radiation activists went about the work of making radiation visible by measuring it, and then arguing with each other about what that visibility actually meant. Radio-frequency scientists looked for ways to measure the signals from the antennas reliably. Oncologists tried to determine whether the signals were indeed carcinogenic. Telecom operators worried because if cell tower antennas were removed, it could affect their capacity to provide cellular service. Tower builders tried to convince apartment dwellers that cell towers were not harmful. Cancer patients felt frustrated that cell towers were potentially dangerous, and yet the ones near their houses were not being removed. Regulators kept trying to come up with a safety standard for the emission of signals that could maintain network coverage and alleviate public apprehension. Journalists covering the controversy started going around the city with radiation detectors and then creating and publishing maps of radiation hotspots.

Even before the controversy erupted publicly, many of these actors had been involved with the infrastructure of cell towers. The radio-frequency scientists I talked with described how they made cell tower components like filters in their laboratories, the tower builders rented space on rooftops of houses and apartments from building owners, while cellular operators and tower builders worked together to ensure smooth cellular network coverage. Infrastructures, therefore, do not only instigate new public actors but also serve as "crystallizations of institutional relations"[5] and manifestations of everyday cultural practices.[6] At the same time, disruptions bring new actors into contact with earlier players, generating new social arrangements.

To understand the environmental and health effects of media infrastructures like cell towers, we may examine the practices and interactions of these different stakeholders, who together comprise the public that gathers around them. These interactions, I demonstrate, take place not only in zones of immobile toxicity, but also in the media coverage of the radiation controversy in local newspapers, talk shows, and lifestyle shows as forms of "mediating infrastructure." Then, to extend this conceptualization of mediating infrastructures, I explore beyond the interactions among stakeholders carried out in person and through media representations such as talk shows and lifestyle shows devoted to cell towers, to the dynamic *intermedial* relations set up between, on the one hand, radiation detectors, microwave ovens, aluminum foil, and cell phones and, on the other hand, cell antennas. These diverse technologies have become part of the knowledge system that makes radiation visible and palpable in popular public demonstrations.

These mediating infrastructures are produced not only through the agencies of human stakeholders, but also via nonhuman actors and dynamic materialities. In the introduction to their edited volume, *New Materialisms*, Diana Coole and Samantha Frost contend that "materiality is always something more than 'mere' matter: an excess, force, vitality, relationality, or difference that renders matter active, self-creative, productive, unpredictable."[7] The mobile tower has properties of excessiveness, eventfulness and unpredictability, which emerge in relation to other components/actors of the telecom infrastructure. It houses antennas with the ability to emit radio waves in particular directions, the power density of which cannot always be measured reliably and the biological effects of exposure to which remain uncertain. Emphasizing the eventful material properties of technological objects, geographers Bruce Braun and Sarah Whatmore argue that such objects "cannot be reduced to things on which decisions are made in the political realm because they are part and parcel of that realm from the outset."[8] These writers call for studying the *performances of things* in relation to the *actions of humans*.

In the field of media studies, there seems presently to be an urgency to study infrastructures that make mediated texts possible, with a greater focus on materiality than discourse (or textual analysis): this is pointed out as a necessary course correction to make up for so much attention having been devoted to the screen, genres, and texts for so long. Moreover, studies of the environmental impact of media technologies and infrastructures engage in just this project. Allison Carruth has interrogated the politics of energy in cloud-computing practices by studying data servers that use nonrenewable sources of energy to maintain optimally cool temperatures.[9] Environment journalist Elizabeth Grossman has written about toxic materials in electronic devices, and has shown how the manufacturing and disposal of such devices involve exposure to such wastes.[10] Jussi Parikka has posited studying the geology of media to ascertain how minerals, metals, and chemicals are mobile and move from inside the earth to become part of batteries and chips of digital machines.[11] Jennifer Gabrys's work seeks to track and reroute the material ecologies and energy flows that connect minerals like coal to savvy digital media.[12] Like Parikka and Gabrys, I am committed to extending media beyond individual objects in order to assess complex media ecologies so as to understand, in my case, mobile media's environmental effects. That said, my intervention is somewhat different than the earlier mentioned scholars. This is because my fieldwork experience suggests that I focus on the radical entanglement of discourse and materiality. I submit that to study environmental effects of cell towers (a media infrastructure), one would have to trace the circulation/mediation of both signs and signals, both representations and resonances, both discourse and materiality. It does not make sense to emphasize one over the other: for example, the material properties of the tower matter, but the public perception of the tower's properties matters, too. Hence, with the mediating infrastructure approach, the focus is not only on the infrastructure but also on the mediation of that infrastructure, where *mediation* is not mere media coverage but a complex sociomaterial process of understanding the infrastructure in its varied relationalities.

In the next three sections, as a way of weaving together various modes of "mediating infrastructure," I foreground three different concepts/practices: "technostruggle" (involving interactions between experts and lay activists in measuring radiation from cell antennas),[13] "intermediality" (interactions between media objects/technologies assembled to demonstrate radiation), and "affective resonances" (interactions between electromagnetic signals of cell towers and bodily intensities at the molecular level). The penultimate section on affect contributes to discussions of phenomenological encounters between human bodies and infrastructural sites, and argues that mediation of infrastructures operates not only at the molar level of discrete bodies or objects, but also at the molecular level where electromagnetic waves and bodily sensations intermingle. In short, these three modes of mediating infrastructure become a means of describing the ways that the human and nonhuman actors of the dynamically reconfiguring public formed around the infrastructure, that is, the cell antenna public, interacted with one another throughout the radiation controversy.

Technostruggle: Measuring Antenna Signals

In 2010, a series of events triggered the radiation controversy and created the ground for the activation of the cell antenna public. On January 3, 2010, Mumbai's afternoon daily, *Mid-Day*, reported that residents of a number of buildings and housing societies in the plush Carmichael Road area of South Mumbai had appealed to the owners of Vijay Apartments to remove the cell tower on top of their building. The tower was considered to be the source of cancer cases reported from the Usha Kiran building that faced it.

Girish Kumar, a radio-frequency scientist and professor at the Indian Institute of Technology, Powai, subsequently sent his report to the Department of Telecommunications (DOT) in December 2010. In that technical report, Kumar not only included this controversy as a case study, but also, in collaboration with *Mid-Day*, used the same photograph that *Mid-Day* photojournalist Bipin Kokate had taken—an overhead shot of the base station tower on the terrace of Vijay Apartments from (most probably) a higher building.[14]

In Kumar's report, another apartment standing in for the Usha Kiran building was added, and the sector antenna's horizontal main beam (marked in red in the original) was shown to cover the three consecutive floors—sixth, seventh, and eighth—in which four cancer cases were reported. The concentration of cancer cases in the zone falling within the main beam of the antenna suggested a strong correlation, Kumar argued.

Like the *Mid-Day* report, other news organizations also incorporated similar semi-epidemiological studies. The news magazine *Tehelka*, known for its investigative journalism, was the first media organization to conduct a citywide radiation survey in partnership with a radiation measurement company, Cogent EMR Solutions Pvt Limited. The magazine carried three consecutive issues—Radiation City 1, 2, and 3—with reports of vulnerable hotspots from radiation surveys in the cities of Delhi, Mumbai, Chennai, and Bangalore during the months of June and July 2010. Here again, the coverage was dominated by anecdotes of people suffering from cancer living in the same building with, or facing, cell towers: the Panchatantra Apartments, a building with seven floors and two wings, had two cases reported from its seventh floor and one each from the first and fifth floor.[15]

The news about cell tower radiation was a moment of infrastructural education for those who were proximate to cell towers. The reports made clear that it was not only the distance from the tower (as a factor related to the strength of the signal) that affected radiation exposure, but that the direction of the antenna and, particularly, the main beam of the signal also mattered. Still other factors affecting the signal strength were highlighted, including buildings on the signal path and the number of cell phones that the mobile tower was supporting at any given moment.

These initial stories triggered further research. Cogent EMR solutions paired up with *Tehelka* to do radiation surveys, and discovered power density levels in two places in New Delhi—the INA market and the Nehru place footbridge—to be 1,500 mW/m^2 and above 4,000 mW/m^2, respectively. The Cellular Operators Association of India (COAI) carried out its own survey in association with several independent technical bodies, such as the Centre of Excellence in Wireless Technology, and came up with different readings at the two declared hotspots. The radiation reading for INA market was 0.000001 mW/m^2 and for the Nehru place footbridge 0.000063 mW/m^2. The critics of COAI studies gave various reasons for this significant disparity in readings. Girish Kumar noted that "the readings (by service providers) must have been taken at spots away from the main beam of the transmitter"[16] Zafar Haq, former chief executive officer of Cogent, offered an explanation:

> Radiation levels do not remain static. High fluctuations happen due to traffic movement and usage patterns. It is less prudent to compare or claim any two readings contradicting if they are not taken at the same time, location and under same conditions.[17]

The mobile media assemblage was so dynamic that, with a change in mobile traffic, direction of antenna, and cell phone usage patterns, the electromagnetic radiation emitted by a particular tower changed. And, therefore, two studies done at different times could not be compared and, thus, could not satisfactorily corroborate or dispute each other.

Cell antenna signals take the form of waves and this radiation is often called an electromagnetic field (EMF). Cellular operators maintained that they were following the International Commission on Non-Ionizing Radiation Protection (ICNIRP) guidelines, which set the permissible limit to be 4.5W/m^2. As a matter of abundant precaution considering public apprehension, the Department of Telecommunications (DoT) reduced the EMF emission levels to 0.45 W/m^2, that is, to one-tenth of the prevailing ICNIRP norms. This did not pacify anti-radiation activists, who highlighted cancer cases close to towers where the recorded power density of electromagnetic radiation was found to be merely 0.001W/m^2.

As part of my research fieldwork, I spoke extensively with Kumar. As an antenna specialist, Kumar is often quoted in newspapers saying that international norms set by ICNIRP (at 4.5W/m^2 as a threshold EMF level) were meant for short-term exposure and mentioned in the guidelines of the ICNIRP written document itself. However, according to Kumar, people living close to cell towers were exposed to antenna signals over a prolonged period of time and hence experienced chronic exposure. Kumar thus was arguing that accounting for the "slow violence" of chronic exposure should lead to setting the threshold at a lower level.[18]

Kumar developed radiometers that showed the radiation levels in numbers measured in decibels per milliwatt (dBm). However, the devices confused people, who felt that it was counterintuitive that decreasing dBm levels meant increasing radiation levels. Too many denominations like watt, milliwatt, and microwatt to convey signal strength emerged and, in the absence of any standard to follow, people using different radiation detectors were either hassled to do calculations or just confused. In order to facilitate easier use of the radiometer, Kumar came up with radiation detectors that conveyed measured radiation through beeping lights in red, yellow, and green, with red showing a high radiation level and green communicating a low radiation level. The shift from green to red through yellow was also accompanied by increasing sound levels.

A number of science and technology studies (STS) scholars have reconceptualized science as a social activity where there can be boundary crossings between science experts and nonexperts to broaden the possibilities of collective decision-making on issues with a high level of uncertainty.[19] Experts like Girish Kumar created detectors so that ordinary citizens could also be vigilant about whether towers were emitting high radiation. In some ways, the citizens now did not need the experts to help them monitor EMF levels. Yet depending on the expert, the design of the detector was different, and hence, laypersons remained dependent on the experts. Kath Weston in her forthcoming book, writing in the context of the post-Fukushima crisis, explains that a wide variety of dosimeters and Geiger counters were being used but the standards each of these devices adhered to were different. Weston uses the term "technostruggle" to characterize this effort of ordinary citizens to use technology as a means of ascertaining environmental and health effects, an effort that yet falls short of any real certainty as to the outcomes.[20] A similar technostruggle ensued in Indian cities as people went about carrying detectors and creating radiation maps, the veracity of which were contested by others.

When the controversy began, measuring signal strength was the primary preserve of radio-frequency experts. However, after a while, concerned citizens also joined the experts in measurement exercises, equipped now as they were with radiation detectors. Technostruggle becomes a way of mediating infrastructures, where there are twin mediations: (1) the technological mediation of the radiation detector that reshapes the relationship between invisible radiation and human subject; and (2) the remediated relationship between expert and laypersons, where laypersons have greater sovereignty over their lives (than before when they did not possess detectors) but still are dependent on experts. Technostruggle highlights that cell antenna publics involve interaction of actors such as regulators, experts, and laypersons and, furthermore, that experts as a group within themselves are far from homogeneous. Experts are themselves divided on what radiation levels are safe, and end up confusing laypersons/concerned citizens who now have technological access to observing radiations but still struggle to discern whether they are safe from radiation. As we will see in the next section, apart from detectors, aluminum foil and microwave ovens also entered the cell antenna publics and, through their intermedial properties, helped to create new mediations of cell towers as infrastructures and reconfigured the existing arrangements within the cell antenna public. I will begin the next section with an anecdote about Prakash Munshi who used aluminum foil and microwave ovens in his presentations about cell tower radiations.

Intermediality: Demonstrating Radiation through Detectors, Aluminum Foil, and Ovens

Munshi is a key anti-radiation activist campaigning for strict regulation of cell tower radiation. He lives in Raj Niketan along the B.G. Kher Marg, a posh area in

Mumbai. His friend and neighbor is the Bollywood actress Juhi Chawla, who has also been fighting against cell tower radiation. Despite their social status, Munshi and Chawla could not get the government or the municipality to pay heed to their requests (made through letters) to remove a cluster of cell tower antennas from the terrace of the Sahyadri Guest House, located on the other side of the road facing their apartment. What finally worked for them was the privilege of having their home in the place that they did: a lot of state government dignitaries stay in the Sahyadri Guest House and ministers tend to hold press conferences there. Munshi organized the residents of the apartments facing the guest house to put up banners, which were addressed to the Sahyadri Guest House and listed the ailments caused by cell tower radiation. One banner asked the question: "Who will be responsible for the parting of our loved and dear ones?" The banners were signed, "From Your Friendly Neighbors." The reporters arriving to cover a press conference organized by the government practically walked right into these banners. The next day, cell tower radiation was news. This is the story Munshi was telling me at his apartment on July 27, 2013, as he got a call about the presentation on cell tower radiation he was preparing to deliver at the Meherabad building on Warden road the next day. On the phone, he said he would definitely require a microwave oven.

The next day, I accompany Munshi to the presentation. We are met by our host Pravet Javeri, who lives in the building, and we make our way through the elevator to his apartment to collect the microwave oven before heading to the apartment's rooftop for the presentation. On the wall of the elevator, I see a notice asking the residents to attend the meeting. The Indus Towers officials had also been invited to present their side of the story and they had come with a blackboard on which to draw figures. The Indus Towers officials wanted to retain their towers in these apartment buildings and did not want apartment residents to be persuaded by anti-radiation activists. Maintaining these towers required lawyers to strike deals with landlords, technical officials to monitor telecom and air-conditioning equipment, and other workers to transport diesel to run the backup generators so as to maintain continuous power supply if the electricity grid failed.[21] Public persuasion in these meetings was also part of maintaining towers.

During the presentation, Munshi switched the microwave on and stood in the front of the oven with the radiometer that Girish Kumar's company, NESA Solutions, manufactured. The LED lights of the radiation detector glowed red. As soon as he moved away from the microwave oven, the detector lights shifted to green (see Figure 5.2). "Ladies should not stand in front of the microwave oven," Munshi cautioned.

Radiation detectors, through their technological mediation, help us perceive the cell tower radiation that we cannot see, smell, or taste in the form of glowing LED lights. Here media shift modes of perception. That the microwave oven has similar heating effects as cell tower signals can be perceived only by bringing radiation detectors into close proximity with (in specific relation to) such ovens. People begin to comprehend that microwave ovens might have similar properties

FIGURE 5.2 Munshi demonstrating with microwave oven.

Source: author.

as cell towers when they became part of a medial configuration generated by the knowledge practice of using radiation detectors. My fieldwork suggests that in order to understand how media work and how people use them, it is important to attend to such intermedial relations, where "intermediality is a concept that brings forth relations that cannot be defined in media as fixed forms."[22] The concept of intermediality helps to ascertain the ways in which media as technological objects are related to one another and to human actors.

I continue listening to Munshi's presentation to this apartment association, which is contemplating removing the cell antennas that are present on the rooftop of the apartment. This apartment complex has received complaint letters from a number of its neighbors living in houses or flats of other apartments, asking them to remove the towers. The residents of the apartment with cell antennas receive considerable money from the cell tower companies for renting out their terrace and, in the face of complaints, are weighing the pros and cons of removing them. Munshi demonstrates how aluminum shielding can block cell tower radiation (and radio waves) completely by wrapping a sheet of aluminum foil around a mobile phone and asking one of the flat residents who is attending the presentation to call that wrapped up cell phone. While earlier the mobile phone could be reached from any other phone, now the phone is unreachable. After the demonstration, Munshi adds that aluminum shields can also reflect back the radiation.

Though there is no conclusive evidence that aluminum can completely block all kinds of non-ionizing electromagnetic radiation, there are a number of demonstrations by citizens worldwide, some of which are on YouTube, that exhibit how mobile phones wrapped in aluminum foil lose their connectivity. The other promised action of reflecting back the EMF signals is not that easy to demonstrate, but again, it has its backers. Certainly Munshi's statement could make his audiences more nervous than usual. Prior to attending this rooftop meeting, they had been evaluating the ethics and morals of benefitting financially from a technology potentially harmful to their neighbors in the path of the sector antenna beams. Now they were learning that their own lives could be threatened from the same radiation, if the residents of adjoining apartments started installing aluminum shields in their respective houses. The EMF signals emitted from cell antennas would then be reflected back, and the residents of the host buildings affected. Media here are not fixed objects but are part of medial events and processes dynamically constituted within emerging systems.

AbdouMaliq Simone writes that infrastructure concerns itself with the "in-between": the things that lie in between humans that are able to both draw humans together and, at other times, make them withdraw from one other.[23] Infrastructures thus not only exert a material force but a socializing force that is closely connected to their materiality. The telecom infrastructure would undergo a change with the addition of aluminum shields to its assemblage. The residents of apartments hosting cell towers, who had presumed themselves immune from the dangers of radiation, find themselves having to rethink the situation. The effect of aluminum shields is not certain, but then neither is the effect of non-ionizing EMF emissions. Munshi, through his presentation, has shared some information that he himself qualifies as indeterminate, but which has generated new ripples of uncertainty about the future of cell towers. Simone's remarks on infrastructures are both pithy and evocative here: "People work on things to work on each other, as these things work on them."[24] It is through the implicit hint of the possibility of aluminum shield deployment that Munshi stands a chance of influencing flat residents to consider dismantling the antennas. Munshi works on the flat residents. The flat residents are worked on by aluminum foil.

Like the radiation detector and microwave oven, the aluminum foil is also a technology that becomes part of the extended media ecology of the cell towers—or, put another way, the aluminum foil is yet another nonhuman addition to the (ontologically) heterogeneous cell antenna public. If technostruggle was one modality to think about mediating infrastructures, this section suggests intermediality to be another one. Mapping the medial relationships among antennas, detectors, ovens, and aluminum shields places cell antennas within an ecology of other technological elements. Mediating infrastructures here enables thinking of infrastructures relationally with media technologies exhibiting new properties when brought in relation to one another. In the next section, affect provides a conceptual language to describe interactions within the cell antenna public that happen not at the level of molar bodies and towers, but at the level of waves and intensities.

Affective Encounters with Proximate Towers

In this environmental controversy, the predominant form of affect was a fear of proximate towers, whether it was directly articulated or, in many cases, expressed as an inability to articulate felt sensations. Affect arose as helplessness at having to continue to see the tower every day without being able to get it removed.

As evidenced in the pages of the local (vernacular/Hindi) newspaper *Rajasthan Patrika* circulating in Jaipur, mobile towers became sources of depression, sleeplessness, heart problems, memory loss, and paralytic attacks. This was not only due to the radiation. Mobile towers are run on generators that cause noise and air pollution, thus leading to sleeplessness. Elderly people are disturbed. Towers are also unsightly—a letter writer to *Patrika*'s editor compares mobile towers to demons (*Pishachas*). In one of *Patrika*'s news stories published on May 15, 2012, Anju Sharma, a high school teacher of English, remembers that when the towers sprung up in her neighborhood, she was scared by their shape. If there was a strong wind or a sandstorm, tears rolled down from her eyes because she felt that the mobile tower would drop on her head.

The lifestyle show *Living It Up*, broadcast on CNN-IBN, took up the cell tower scare issue on October 6, 2012. On the show, gynecologist Geeta Chadha explains that brain-to-body functionality is maintained by electrical impulses going through neurotransmitters and neural junctions. There is no way, she argues, that electromagnetic radiations emitted by cell towers would fail to interact with neural impulses. One might argue that there are no deleterious health effects from such an interaction, but it would not be legitimate to deny that there is indeed an interaction occurring at this level.

How does one write about impulses, intensities, and interactions that cannot be seen and that do not manifest readily into subjective emotions? Affective resonances provide a generative direction of thinking about the materialities of cell antenna publics as they highlight the experiential aspects of human-technology encounters. Affective intensities can take the form of visceral bodily reactions (in relation) to towers, reactions that cannot be equated to—or appropriated as—individual (or social) consciousness.[25]

Living It Up often relies on people to reenact their own stories about fighting illnesses. Karmel Nair is one of those social actors who moved to her new apartment in Mumbai when she was five months pregnant. Nair complains of severe headaches, hair loss, nausea, and sleeplessness, and there is restlessness and frustration in her voice:

> A pregnant woman taking an X-Ray shot, that itself is considered so harmful that doctors do not prescribe an X-Ray to a pregnant woman. Now, I, I have been living under this X-ray shot 24 by 7, under these [?] radiation shots, which is as good as taking an X-ray every second. So how comfortable, how safe is it going to be for my child . . .
>
> (Karmel Nair, *Living It Up*, October, 6, 2012)

Nair is framed sitting next to a window in her house from where the cell towers can be seen outside. She is reading a book placed on her lap and then looks outside, conveying an absent-minded sadness and a feeling of helplessness (see Figure 5.3). The slow and chronic violence of this fear is apparent here: the cell tower signal does not suddenly create an acute pressure point on Nair's body; rather,

FIGURE 5.3 Karmel Nair.

Source: Living It Up (CNN-IBN).

there are small eruptions of fear every day about the fact that the tower remains in the same location, obdurate and still. This fear is part of Karmel's living condition. Nair feels she is being scanned by X-rays all day, all night, for weeks together. Here the body as media and technology interacts with electrical impulses and the electromagnetic environment, an environment of which the body is very much a part. Her body itself mediates infrastructures, and this mediation is as much about signals and resonances as it is about signs and representations.

If the antennas cannot talk, can they at least murmur? If they cannot even murmur, can they generate an intensity, a murmuring sensation in the body of a patient who finds the antenna right in his or her line of sight? Such organizations of social experiences and quotidian feelings, which often remain unrecognized in public discourse, are critical to the political critique of infrastructures and their environmental effects.[26] Along with technostruggle and intermediality, affective resonances provide another modality of mediating infrastructures, where the experience of engaging with cell towers can be described in terms of affective bodily states and behavior of electromagnetic fields. These molecular-level mediations of infrastructures help us to engage with cell antennas within a regime of sensation that is felt by human beings but evades their conscious perception.

Conclusion: Sustaining Health and/or Sustaining Calls

I have espoused the framework of "mediating infrastructures" and "cell antenna publics" with a stress on interactions between different human and nonhuman actors at various levels as a way to argue for the inseparability of discourse and materiality, and to champion a materialism, that is emphatically relational.[27] Each new addition to the cell antenna public, whether of aluminum foil or cancer patient, is not a simple addition but an addition that changes the configuration of the public. Apart from the radiation issue, another issue that further brought in new actors and expanded the cell antenna public was the call-drop issue.

While mobile towers may not be environmentally sustainable, they help to sustain calls. Recently, there has been fresh controversy over call drops. Mobile phone users who subscribe to a number of different telecom companies had to face incomplete conversations. Other problems related to connectivity, weak mobile signals, patchy mobile Internet data services, and busy networks have also been reported.[28] When asked for explanation, cellular companies pointed to insufficient spectrum and mobile towers as the causes. The Indian telecom minister stated that a larger portion of the spectrum could not be allotted and the companies should undertake radio-frequency optimization of their networks in order to improve quality of service for their consumers. He declared that there was no conclusive evidence that mobile tower radiation was harmful, thereby suggesting, implicitly, that more mobile towers could be constructed. Since the cell tower health scare, with the attendant restrictions put in place by civic authorities and state governments, cellular operators believe that the number of mobile towers has not grown sufficiently to keep pace with increasing demand for cell phone service, which includes not only supporting calls but also mobile Internet. In the last eighteen months, sixteen thousand towers were added, but the tower companies felt that to support the 978 mobile phone connections in India, fifty thousand to sixty thousand towers would be required.[29]

The call drops issue became another reason to focus attention on the cell towers. Here again, an infrastructure gains visibility precisely when it does not seem to work. The problem of call drops in turn highlighted the cell tower radiation issue, with Barkha Dutt devoting to it an episode of her talk show *We the People* on June 28, 2015.[30] During the show, Dutt noted that, in their episode on cell phone radiation, Priti Kapoor's husband had asked whether he would need to die to prove that cell towers cause cancer. Now about three years later, he had indeed died of cancer. In this new episode, the old episode where Kapoor's husband utters the sentence about death as proof is replayed, and the image with Kapoor's husband, Kapoor, and Dutt in the frame is stilled and then painted in black and white; the camera zooms in. The replay from the earlier episode and the play of colors points to the present being uncannily invoked/prophesied by the past. Dutt has been arguing with experts, demanding accountability, but when she begins this discussion with Kapoor, she transitions to a different subject-position, a softer empathetic one (similar to the disposition she had in the earlier episode toward Garg as discussed in the beginning of the chapter). The show again hosted a number of different experts and activists who vigorously debated with each other. And again, by the end of the show, it remained unclear whether there was any definite link between cell towers and cancer.

Policy-wise, the government would ideally accommodate both the anti-radiation activists crying foul about cell towers' health effects and the mobile phone consumers enraged about call drops. With the call-drop issue now intertwined with the cell antenna issue, mobile phone users have become key constituents of cell antenna publics. Quite possibly, then, people facing call drops and connectivity problems might also have anti-radiation activists among them. Should we regulate mobile towers to ensure a cleaner environment or should we allow towers to mushroom to meet the need of ever-growing mobile subscribers? Is this really an either/or question? Are these two sustainability issues opposed to one another or can be they resolved together? The questions remain.

Throughout this chapter, by studying practices of mediating infrastructures, I have attempted to understand the field of environmental media as consisting of both the media/journalists representing/covering the environmental debates and, at the same time, the media technologies such as cell antennas and radiation detectors as part of the environment. When environmental media is studied in its radical entanglements, the dynamic medial/relational properties of media become more apparent, and by addressing the tensions and frictions within such entanglements, sustainable environmental discourse can emerge.

Cell towers as infrastructures are neither just radiation-generating technologies nor mere machines that support mobile phone calls and mobile Internet connectivity. Jumping to either of these conclusions would only prompt reactionary moves of evicting towers or disproportionately constructing new ones. I have argued that a mediating infrastructures approach, with its stress on studying varied relations in which the cell towers find themselves, will help us better to comprehend their environmental effects. Put differently, thinking of the media infrastructure ecologically or environmentally, that is, as part of larger media systems and/or as connected with human and nonhuman actors of the public spawned by it, will help us address their environmental impacts in a more holistic way.

Acknowledgments

I am indebted to Charles Wolfe, Bhaskar Sarkar, Rita Raley, Lisa Parks, Colin Milburn, Bishnu Ghosh, and Karen Beckman for their insightful suggestions. Parts of this chapter were presented at International Communication Association and Society for Social Studies of Science conferences. Maria Corrigan, Athena Tan, and Hannah Goodwin read drafts at various stages and provided useful advice. Special thanks to Janet Walker and Nicole Starosielski for encouraging me to write an ambitious chapter. Long conversations with Girish Kumar, Prakash Munshi, and Shipra Mathur were helpful during my research fieldwork.

Notes

1 "Cell Phone Towers: India's Safety Check," *NDTV's We the People*, episode aired September 16, 2012, http://www.ndtv.com/video/player/we-the-people/cell-phone-towers-india-s-safety-check/247080 (accessed July 1, 2014).
2 The number keeps increasing, and for a more up-to-date figure, consult the Indus Towers website, http://www.industowers.com/who_we_are.php (accessed July 12, 2015).

3 Stephen Graham, "When Infrastructures Fail," in *Disrupted Cities: When Infrastructures Fail*, edited by Stephen Graham (New York, NY: Routledge, 2010), 1–27.
4 See Noortje Marres, "Frontstaging Nonhumans: Publicity as a Constraint on the Political Activity of Things," in *Political Matter: Technoscience, Democracy, and Public Life*, edited by Bruce Braun and Sarah J. Whatmore (Minneapolis: University of Minnesota Press, 2010), 177–210.
5 Paul Dourish and Genevieve Bell, "The Infrastructure of Experience and the Experience of Infrastructure," *Environment and Planning B: Planning and Design*, 34, no. 3 (2001): 414–430.
6 Susan L. Star, "The Ethnography of Infrastructure," *American Behavioral Scientist*, 43 (1997): 377–391.
7 Diana Coole and Samantha Frost, "Introducing the New Materialisms," in *New Materialisms: Ontology, Agency, and Politics*, edited by Diana Coole and Samantha Frost (Durham, NC: Duke University Press, 2010), 9.
8 Bruce Braun and Sarah J. Whatmore, "The Stuff of Politics: An Introduction," in Braun and Whatmore, *Political Matter*, xxii.
9 Allison Carruth, "The Digital Cloud and the Micropolitics of Energy," *Public Culture*, 26, no. 2 (2014): 339–364.
10 Elizabeth Grossman, *High Tech Trash: Digital Devices, Hidden Toxics, and Human Health* (Washington, DC: Island Press, 2007).
11 Jussi Parikka, *A Geology of Media* (Minneapolis: University of Minnesota Press, 2015). While Marshall McLuhan's work and recent advances in ubiquitous computing and the Internet of Things demonstrate that media are our environment as they envelop us, Parikka's work suggests that the environment is media too, as elements from environment are embedded within media technologies. Along these lines, John Durham Peters's recent book historically, philosophically, poetically, and playfully conceptualizes the earth, sky, fire, ocean, and time as media. See John Durham Peters, *The Marvelous Clouds: Toward a Philosophy of Elemental Media* (Chicago, IL: University of Chicago Press, 2015).
12 Jennifer Gabrys, "Powering the Digital: From Energy Ecologies to Electronic Environmentalism," in *Media and Ecological Crisis*, edited by Richard Maxwell (New York, NY: Routledge, 2014), 3–18. Also see Gabrys in this volume.
13 Kath Weston, *The Intimacy of Resources: Technology and Embodiment in the Synthesis of Nature* (Durham, NC: Duke University Press, forthcoming).
14 Hemal Ashar, "Towers Sending Tumour Signals," *Mid-Day*, January 3, 2010, http://www.mid-day.com/news/2010/jan/030110-mobile-tower-cancer-cases-carmichael-roadposh-areas.htm (accessed September 12, 2012).
15 Rishi Majumder, "Radiation City," *Tehelka*, June 5, 2010, http://archive.tehelka.com/story_main45.asp?filename=Ne050610coverstory.asp (accessed September 12, 2012).
16 Cited in Vibha Varshney, "Warning Signal," *Down To Earth*, January 31, 2011, http://www.downtoearth.org.in/content/warning-signal (accessed September 12, 2012).
17 Ibid.
18 See Rob Nixon, *Slow Violence and the Environmentalism of the Poor* (Cambridge, MA: Harvard University Press, 2011).
19 For example, see the work of Michel Callon, Pierre Lascoumes, and Yannick Barthe, *Acting in an Uncertain World: An Essay on Technical Democracy* (Cambridge, MA: MIT Press, 2001). Also see Arie Rip, "Constructing Expertise: In a Third Wave of Science Studies?," *Social Studies of Science*, 33, no. 3 (2003): 419–434. Scholarship in this area has also interrogated whether a complete erasure of boundaries between expert knowledge and lay knowledge is possible and, if so, with what consequences. Harry Collins and Robert Evans, "The Third Wave of Science Studies: Studies of Expertise and Experience," *Social Studies of Science*, 33, no. 2 (2001): 235–296.
20 Weston, *The Intimacy of Resources*.

21 Robin Jeffrey and Assa Doron, *Cell Phone Nation: How Mobile Phones Have Revolutionized Business, Politics, and Ordinary Life in India* (Gurgaon: Hachette Book Publishing India, 2013).
22 Katerina Krtilova, "Intermediality in Media Philosophy," in *Travels in Intermedia[lity]: Reblurring the Boundaries*, edited by Bernd Herzogenrath (Hanover, NH: Dartmouth College Press, 2012), 40.
23 AbdouMaliq Simone, "Infrastructure: Introductory Commentary by AbdouMaliq Simone," *Curated Collections, Cultural Anthropology Online*, November 26, 2012, http://www.culanth.org/curated_collections/11-infrastructure/discussions/12-infrastructure- introductory-commentary-by-abdoumaliq-simone (accessed May 24, 2014).
24 Ibid.
25 Brian Massumi, *Movement, Affect, Sensation: Parables for the Virtual* (Durham, NC: Duke University Press, 2002).
26 Kathleen Stewart, *Ordinary Affect* (Durham, NC: Duke University Press, 2007). Also see Lisa Parks, "Media Infrastructures and Affect," *Flow TV*, 19, no. 12, May 19, 2014, http://flowtv.org/2014/05/media-infrastructures-and-affect/ (accessed July 12, 2014).
27 Here my philosophical debt is to Karen Barad. For Barad, matter plays an active role in discursive practices, and those practices are themselves "material (re)configurings of the world." A strict delineation between materiality and discursivity in discussions of practices and phenomena can only be fictive because practices are "material-discursive." Barad's work on relational materialism is exemplary, including her concept of "intra-actions." For more, see Karen Barad, *Meeting the Universe Halfway: Quantum Physics and the Entanglement of Matter and Meaning* (Durham, NC: Duke University Press, 2007).
28 Prasanto K. Roy, "Why India's Mobile Network Is Broken," *BBC News*, December 3, 2014, http://www.bbc.com/news/world-asia-india-30290029 (accessed July 9, 2015).
29 Jayati Ghosh, "Wake-up Call on Call Drops—Glare on Tower Campaign, Spectrum," *Telegraph*, July 8, 2015, http://www.telegraphindia.com/1150708/jsp/frontpage/story_30340.jsp#.VZ_Z93rZhrJ (accessed July 9, 2015).
30 "Mobile Tower: Danger Signals?," *NDTV's We the People*, episode aired June 28, 2015, http://www.ndtv.com/video/player/we-the-people/mobile-towers-danger-signals/373046 (accessed July 9, 2015).

6

THE LACK OF MEDIA

The Invisible Domain post 3.11

Minori Ishida

The Great East Japan earthquake occurred on March 11, 2011. Following the massive quake, an unusually large tsunami swallowed up people and towns on the Pacific coast from Tōhoku down to the North Kantō region. That day, all I could do from my own town located on the other side of Japan was to watch the terrible scenes transmitted on television. After the tsunami subsided, images of completely devastated landscapes stunned me again. Television news reports powerfully and visually mediated for the public the piteousness, sadness, and anger of the people affected by the tremendous power of nature.

However, the news concerning the accident at the Fukushima Daiichi Nuclear Power Plant (FDNPP) run by the Tokyo Electric Power Company (TEPCO) was a completely different story. Television coverage did not successfully convey the details of this crisis to the Japanese public. When news of the incident first broke, although I initially turned on the TV, I soon began to search for information on the Internet.

Due to the tsunami, all power to the Fukushima Daiichi plant was lost, and therefore the ability to cool its four reactors. Three reactors went into meltdown, and a large quantity of radioactive material was released into the atmosphere and the Pacific Ocean. One month later, the Japanese government characterized the situation as a "Level 7" accident. According to the International Nuclear Event Scale, the Fukushima accident is equal to that of the Chernobyl disaster of 1986.[1] However, as will be discussed in this chapter, prior to the government evaluation Japanese television news failed to inform the public of the facts of the meltdown as it was developing. This failure made the accident less visible. Coincidentally, the radioactive material scattered over the environment is equally invisible, even without the failure of television news. Hence a truly ironic situation occurred within this audiovisual medium.

How did this happen? Answering this naïve question requires an understanding of institutional relationships within the nuclear energy policy in postwar Japan, including politics, the mass media, and national identity. The first section of this chapter analyzes the television news coverage of the accident, for the purpose of clarifying and historically contextualizing the sociopolitical factors that made the nuclear disaster invisible.

The second section of this chapter will argue in favor of seeing the nuclear disaster within a new media environment, where labor-intensive individual and collective efforts undertaken by members of the public are striving to make radioactivity visible. This spontaneous effort by those who want to know the effects of radioactivity may be regarded as a kind of media activism for surviving the daily effects of this nuclear catastrophe.

The Failure of Television News Coverage during "The Decisive Moment"

Scarcity

At least four serious explosions occurred at the FDNPP. However, only two explosions were photographed at the precise moment of disaster: the explosion of Unit 1 at 3:36 p.m. on March 12 and the explosion of Unit 3 at 11:01 a.m. on March 14. The scarcity of images is distinctive, compared to the abundance of images that came from the 9/11 attacks. The explosion and the collapse of the World Trade Center towers were photographed by all sorts of optical devices, from personal cell phones with digital cameras to television cameras. By contrast, at Fukushima only one fixed observation camera, owned by Fukushima Central Television (FCT), filmed the explosions from a distance of 17 kilometers from the FDNPP. If this camera, installed more than ten years ago, had malfunctioned, we would have lacked all visual evidence of the explosion.[2]

Delay

Compounding the scarcity, the video of the Unit 1 explosion is the first-ever visual recording in human history capturing the precise moment of a nuclear power plant explosion. It was filmed under difficult circumstances in the aftermath of the massive earthquake. The uniqueness of these circumstances may be construed as a "decisive moment" in Henri Cartier-Bresson's terms.

This concept, renowned in photojournalism, is drawn from Cartier-Bresson's 1952 book. Its title, *The Decisive Moment*, was coined for the English version. The original French title is *Image à la sauvette*, which may be translated as "images taken on the sly." Regarding the original title, photography historian Eiko Imahashi claims that, by naming his book as he did, the photographer emphasized the underlying nature of his photography.[3] For Cartier-Bresson, photography was a way of instantly or hastily catching a moment—snatching an image on the sly—in spite of all the difficulties.

The explosion image met with further difficulty in its circulation. Even though the video of the Unit 1 explosion captures a genuinely decisive moment, its television broadcast was belated. Although FCT promptly reported the explosion of Unit 1 as breaking news, the national network Nippon Television Network Corporation (NNN), which is FCT's parent station, did not broadcast the video until 4:50 p.m. This was one hour and fourteen minutes after the initial explosion. As for the delay, an NNN spokesman explained that they had feared to create panic by transmitting the explosion video without an analysis of its context. Moreover, they also required time to consult with experts in the field.[4] However, as will be argued in this chapter, the network's commentary was full of deception, having ignored the advantage of immediacy that TV provides as well as the possibility of extended, in-depth coverage. Therefore, this delay became catastrophic.

Monopoly

Regarding the explosion image data, there is another serious problem that needs to be explored. NNN, considering the video a "scoop," monopolized it, and did not provide it gratis to the other national networks. Only after the BBC released the video of the explosion on the Internet at about 7:00 p.m. did more Japanese people realize the seriousness of the matter.[5] Media scholars Mamoru Ito and Hiroaki Mizushima criticize NNN for withholding this video to protect their copyright because the information it contains seriously affected people's security at that time. For this reason, they argue, the explosion video belongs in "the commons."[6] NNN proved unable to respond to a crisis in which a large quantity of radioactive material was dispersed. Therefore, NNN's response was not only lacking in haste, but it also betrayed the public's interest that it should seek to protect. The incident clearly reveals that the TV network—ostensibly an important source of public information—is also a profit-focused company.

A further problem stemmed from the need for the pursuit of profit. For decades, TEPCO and other electric power companies have been significant sponsors of commercial broadcasting companies. Ryū Homma demonstrates how the electric power industry poured resources into advertising costs. Since the 1970s, in the mass media including TV, whenever a nuclear plant accident happens in the world, the industry has eagerly promoted "nuclear power is safe" campaigns.[7] In other words, the monopoly over information regarding the explosion of Unit 1 was a natural consequence of commercial broadcasting companies being dominated by the electric power industry, and may also have led to a delay in releasing information.

Interpretation

The collapse in the credibility of television news, however, is not only a matter of the domination of commercial broadcasting stations by the electric power industry. The Japan Broadcasting Corporation (NHK), a public broadcaster supported by fees received from television viewers, also faltered in transmitting the explosion image. NHK reported the explosion at 4:52 p.m., one hour and sixteen minutes after the initial explosion. By this time, Unit 1 had already lost its containment housing. This image was "the decisive moment" for NHK, for it was photographed with a high-efficiency camera, developed by NHK at the expense of profitability. Nonetheless, at this moment, I once again turned to the computer and began to look for information on the Internet, as I could not easily understand the studio conversation between the reporter and the expert on atomic energy engineering. They explained that what appeared to be an explosion might be a series of operations to open the blasting valves in order to control the pressure in the nuclear reactor. But that commentary was factually irrelevant to the image of Unit 1. As Ito notes, the commentators' explanation was rather strange.[8] It even felt mysterious that while they were watching it with their own eyes, they could not take in this image as the onset of the meltdown.

Of course, a single image does not articulate meaning by itself. For example, an image of a gun shows only that there is a gun,[9] while an empty image cannot prove there is no gun. That is why additional manipulation is necessary to extrapolate clear meanings from a vague image. One of these potential manipulations is commentary on an image by language, and television news reports adopt this form in essence.[10] In a studio, comments are added to an image in an attempt to articulate its meaning. On the afternoon of the explosion, NHK commented on the image as usual—but the meaning they articulated was far from the real situation at the FDNPP.

Robert M. Entman's concept of "framing" is an effective way to understand this disconnected commentary. According to Entman, framing is an interpretative act in which we "select some aspects of a perceived reality and make them more salient in a communicating text, in such a way as to promote a particular problem definition, causal interpretation, moral evaluation, and/or treatment recommendation."[11] If the framing of an image is variable, meanings will also differ.

A nuclear energy engineer, functioning as an in-studio expert, erroneously reported that the nuclear reactor remained under control. This is why both the host and the expert came to interpret the explosive image of Unit 1 as a consequence of the control process itself and not as an indicator of a meltdown. The same thing happened on NNN. Responding to the image capturing the moment of the Unit 1 explosion, another nuclear energy engineer told a worried newscaster that the image indicated an "intentional" explosion in the control process.[12] As noted earlier, NNN explained that the need for this analysis of the explosion video was one of the reasons for the delayed broadcast. Nonetheless, the commentary did not take sufficient account of what the expert actually saw. Instead, he bent it to his own existing frame of reference: a nuclear reactor could not be out of control.

Postwar Japan and Nuclear Energy

Before the disaster occurred at the FDNPP, the framing that came to be adopted in the television news coverage of the explosion of Unit 1 prevailed and, in fact, became the dominant version for Japanese society. The notion that nuclear energy could be controlled had long been accepted as self-evident, part of the complicated and inseparable relationship cultivated between postwar Japan and nuclear energy.

Radiation exposure experience in Japan was not limited to the atomic bombings of Hiroshima and Nagasaki in 1945. In 1954, the Daigo Fukuryū Maru, a tuna fishing boat, was involved in the US thermonuclear test on Bikini Atoll; the crew died from the nuclear fallout. In short, the Japanese were tormented by nuclear power. And yet, as many were surprised to discover, at the time of the FDNPP accident, 54 nuclear reactors existed across Japan.

According to Akihiro Yamamoto, in a detailed analysis of the discourse on nuclear energy from 1945 to the 1960s, public opinion on nuclear energy was composed of two seemingly contradictory approaches. One is "the memory of radiation exposure" held by victims, and the other is "the nuclear dream" of a new technology expected to solve energy resource problems.[13] Under the occupation of the Allied powers led by Douglas MacArthur from September 1945 to April 1952, information about the atomic bomb was strictly censored by the authorities. In the summer of 1952, when *Asahigraph* magazine (issued on August 6) published the ghastly photographs of the A-bomb aftermath, the public reacted fiercely to the cruelty manifested in images of the victims. Two years later, on March 1, 1954, the Daigo Fukuryū Maru accident occurred. A small group of housewives in Tokyo immediately began collecting signatures against atomic and hydrogen weapons, and many people, spontaneously and irrespective of political affiliation, participated in this campaign. This grassroots movement spread in a flash throughout the country, and the First World Conference against Atomic and Hydrogen Bombs was held in Hiroshima on August 6, 1955. In this way, "the memory of radiation exposure" was shared as a national problem, driven by the nuclear threat from the US-USSR Cold War. In 1954 the king of monsters, Godzilla, was born from "the memory of radiation exposure" actualized by the Daigo Fukuryū Maru accident.

At the same time, a plan calling for "the peaceful use of atomic energy" was steadily advancing. On the day following the Daigo Fukuryū Maru accident, the conservative parties submitted an atomic energy budget to the National Diet. In November 1955, three months after the First World Conference against Atomic and Hydrogen Bombs, the Japanese government signed the Agreement for Cooperation between the Government of the United States of America and the Government of Japan Concerning Peaceful Uses of Nuclear Energy. This agreement promised a loan of enriched uranium from the United States, and the development of a nuclear energy research reactor was started. In December, the Genshiryoku Kihon Hō [Atomic Energy Basic Law] was promulgated. Its purpose, declared in Article 1, is

> to secure energy resources in the future and achieve the progress of science and technology and the promotion of industries by encouraging the research, development and utilization of nuclear energy, and thereby contributing to the improvement of the welfare of human society and of the national living standard.[14]

With this law, the Japanese government attempted a practical use of nuclear power generation. As a result, the success of the Tokai village experimental reactor in 1957 led to the start of commercial nuclear power generation in 1966.

Regarding the coexistence of "the memory of radiation exposure" and "the nuclear dream" in postwar Japan, Yamamoto explains that Japan promoted nuclear power generation in order to keep up with advanced countries. But at the same time, as the only nation that had experienced nuclear devastation, it explicitly rejected the use of nuclear weapons.[15] In addition, we cannot overlook the ambition and pride of the scientists who focused on taming the destructive power that once attacked the nation. In 1949, four years after the end of the war, theoretical physicist Hideki Yukawa received the Nobel Prize in Physics for the prediction of the existence of mesons. In light of such events, it is not strange that scientists worked hard on nuclear energy research. Japanese society welcomed this nuclear energy research as a sign of scientific progress as well as national advancement. For example, in the popular movie *High Teen* [Late Teens] (dir. Kazuo Inoue, 1959), high school students visit the Japan Atomic Energy Research Institute at Tokai village on a school excursion and learn about nuclear energy.

In postwar Japan, all components of society, including the government, industry, science, and culture, were closely united in propelling the development of nuclear energy. This nuclear dream gradually turned into the so-called safety myth, the belief that nuclear energy could be controlled by high level technology, as Japan experienced rapid economic growth through technology and industrial power.

The End of the Safety Myth

If we return to the TV news reports of March 12, 2011, and consider the relationship with nuclear energy in postwar Japan, it is not difficult to understand how the strange commentary on the decisive moment was made and why confusion would ensue at the studio: for a Japanese society satisfied with this safety myth, the striking image of Unit 1 was too difficult to look at.

At around 5:00 p.m., NHK divided the TV screen into two, juxtaposing the two images of Unit 1. In the upper portion of the screen, an image from 9:00 a.m. showed the external wall intact; in the lower half of the screen, an image from 4:30 p.m. showed that the wall had been destroyed. At that moment, by putting the two images one above the other, the horrible situation was revealed. The reporter at the studio, mentioning the possibility of an explosion, addressed the further possibility of a large release of radioactive material, and warned that people should remain indoors in order to avoid any radioactive fallout. Finally, he accused the government, the Nuclear and Industrial Safety Agency (NISA), and TEPCO for dictating how to provide the public with information, and called for the provision of factual and honest information regarding the present situation to the public and the media.[16]

The Loss of TV News Coverage

In spite of this reporter's appeal for more precise information, the news outlets that only barely managed to convey the scene of the accident to the public then became largely absent from the night of March 12 on. The reason for this was that most of the mainstream media, including television stations, withdrew their staff from proximity to the power plant. For example, NHK restricted its employees from entering the area within thirty kilometers of the nuclear power plant for one month, until April 11.[17] Therefore, instead of investigative reporting, the mainstream media were devoted to interpreting announcements from the government, NISA, and TEPCO. This decision meant the abandonment of their own independence, because it was these supposedly competent authorities that were the main axis of the safety myth. Chief Cabinet Secretary Yukio Edano euphemistically described "some kind of explosive phenomena" at the announcement of the explosion of Unit 1.[18] Judging by the name, NISA, a jurisdiction section of nuclear power generation, had also participated in the construction of the safety myth. Of course, TEPCO wanted to underestimate the accident. Afterwards, the situation went from bad to worse. Explosions occurred at Unit 3, then Unit 2, and finally Unit 4. Monitors in the power station showed abnormal levels of radiation. But even in such situations, TV news coverage repeatedly stated either that there were no serious problems or that the situation was not as bad as Chernobyl.[19]

When TEPCO provided a photograph of the nuclear power plant to the public on March 16, I could hardly believe my eyes. Units 3 and 4 were squashed, still emitting white smoke. What television audiences saw from March 12–16 was the paralysis of television stations. Here follows a brief summary of the role Japanese television has played historically, as a means of highlighting the social import of the loss of television news coverage.

During World War II, the media were inseparable from the national power, and the nation pushed forward to its own destruction. In 1953, when NHK started full transmission, television became part of society, and has since become a main medium in postwar Japan for news, entertainment, and education.[20] And above all, this new medium has represented democracy: people have expected television to convey information transparently and, thus, to constitute a foundation for a democratic society. As a public broadcasting station, NHK operates, in principle, independently of all political power. Witnessing the failure of television and, specifically, NHK regarding the accident at the FDNNP, most of us must have noticed that Japanese television lost its position as a leading pillar of the public sphere.

The Invisible Domain

A Fear of Invisible Things

On March 16, milk polluted with radioactive material was found in Fukushima prefecture and, on March 18, polluted spinach was discovered in neighboring Ibaraki prefecture.[21] At this point, the extent of radioactive material release became clear, and the nuclear plant accident thereby advanced to the next stage. Yet, according to government statements, people would only experience around one-fifth of the radiation exposure of a CT scan, even if polluted foods were eaten for a year.[22] Television news coverage agreed with this government statement.

On March 22, radioactive iodine-131 levels at more than the then government standards for infants (100 Bq/kg) were detected in the tap water in Tokyo, about 220 kilometers away from the power plant.[23] The danger from radioactive pollution was no longer restricted to the power plant in Fukushima and its neighboring prefectures. Although measurement with a Geiger counter or a scintillator is possible, one cannot directly see and taste radioactive material. This additional example of a lack of visibility, made even more obscure by the obfuscations of the media, has increased the fear of radiation exposure. In short, we have faced a double lack of media.

Overcoming Invisibility

Responses to the double lack of media have confined themselves, by and large, to fear and anxiety. Several voluntary steps, however, have been taken to overcome these difficulties. The radiation hygiene scholar Shinzo Kimura resigned from a research institute related to the Ministry of Health, Labour and Welfare. On March 15, he entered the twenty-kilometer exclusion range from the FNDPP to measure radiation. He ran the whole course of Fukushima prefecture by car, collecting soil samples. With other scientists, Kimura analyzed these samples and created a pollution map.[24] Their activity and results should be regarded as a sincere reply from within the realm of science, one that is completely contrary to the expressed views of the atomic energy engineering experts in the television studios, who repeated their claims that the nuclear reactors were healthy.

I turn now to the investigations by amateurs who do not have any special scientific knowledge. Until the end of March 2011, many people used Geiger counters to measure fallout in their own personal environment, and reported the numerical values on the Internet along with photographs and videos. Although at first a large number of obvious mistakes occurred in measurement and evaluation, many consider the independent actions taken by ordinary people to control the presence of radioactive material in their daily lives to be even more important than the scientific measurements.[25] From such innumerable attempts, I will discuss two cases in particular, both of which involve looking "directly" at the radioactive material.

The first case concerns a photograph of Unit 2 released by TEPCO on April 11. On April 18, this photograph became the topic on 2channel, the largest Japanese message board on the Internet (Figure 6.1a).[26] TEPCO's release of the photo was for the purpose of reporting on the high tsunami and inundation situations depicted therein. The argument on the message board, however, started from an entirely different viewpoint. Contributor ID:bLphu8Ak0 uploaded the TEPCO photograph and related it to the photographs taken at the Chernobyl power plant in which points generated by radiation appear, as the TEPCO photo shows. Because others replied to this contributor that they could not see any points in the photo, he or she posted the image again, enlarged. In the enlarged photo, it is easy to see the innumerable multicolored points (Figure 6.1b). The contributor insisted that these innumerable points must be the radiation noise that passed through the imaging pixels of the digital camera and destroyed them.

FIGURE 6.1A Photograph of Unit 2 released on April 11, 2011.

Source: © TEPCO.

FIGURE 6.1B Because TEPCO did not permit the author to publish an enlargement of the photograph in Figure 6.1a, this photograph, rendered by Mizuho Maki, is substituted for it. The circles are added for the purpose of emphasizing the colored points that became the focus of an online discussion of the effects of radioactivity.

Of course, there were several objections to this opinion. Contributor ID:O8ILGFyC0 stated that "the noise appears with such an old digital camera, so we cannot assert it was generated by radiation." Despite such objections, the discussion advanced considerably. Contributor ID:9oz/wrz20, taking into account the mechanism of the digital camera, claimed that it is much more difficult to catch radiation with a digital camera since, in that case, radiation passes through the sensor only at the moment of recording an image, instead of at the moment of exposure, as is the case with film. In spite of this difficulty, the participants concluded that the occurrence of noise was evidence of high dose radioactivity.

Furthermore, another contributor, ID:fFB+8j5n0, introduced TEPCO's operation video of the Unit 4 pool, and also pointed out the appearance of white noise.[27]

This argument is important. All the evidence in the discussion, such as the photographs and videos about the FDNPP, was announced on TEPCO's homepage. In other words, ordinary people obtained the documents directly, without the mediation of mainstream media. Further, they freely and frankly exchanged their opinions about the evidence they were seeing. Of course, in order to rule out other reasons for the camera noise it would be necessary to investigate the TEPCO camera. But more important to my mind is the fact that this discussion of the relationship between radiation and optical devices did not occur on any TV stations cooperating with the government and the sponsors; the open and free discourse was the purview of amateurs representative of no specific interests.

The second case is that of the video entitled "Radiation noise appearing in the camera," contributed to YouTube on May 5, 2011.[28] The video begins with a black screen, which remains consistent throughout. The videographer's voice-over describes the method, place, and purpose of the video. According to his explanation, he is in a zone within Fukushima prefecture where the radiation dose is more than 100 μSv per hour. His intention is to confirm whether the digital camera is affected by radioactivity. His photography apparatus is a common digital video camera made by Sanyo (six million pixels of CCD) used in a normal mode called "scenery photography." With the lens closed, he slowly lowers the digital camera to the ground. Blinking lights begin to flash on and off, thinly visible in many spots on the black screen. Despite this easy description of the appearance of the blinks, in reality they are so minute as to be easily overlooked if one does not watch the screen carefully. In addition, the viewing environment determines the visibility of the blinks: while they are perceptible on a computer screen, they are not easily seen on a projector.

This video is certainly an unstable and fragile way to prove radiation noise. Some consider it unscientific. However, the black screen not only demonstrates the difficulty of seeing radioactivity; it also drives us to stare at the screen. Gradually, our desire to see radioactivity is more and more strongly roused, all the more so because the substance is not visible to the human eye. The largest positive effect of this video would be the permeation of our consciousness by the awareness of radioactivity that many people have had since the FDNPP accident. Through this desire to perceive the existence of invisible radiation, we see all the more clearly the cover-up of the accident effected interdependently by government and mainstream mass media.

Grasping the Unusual in Daily Life: The Practice of Shuji Akagi

On December 16, 2011, the government declared an end to the meltdown with the cold shutdown of the reactors, and announced that the process of the decommissioning work had advanced.[29] However, while environmental surveys of the inside of the reactors' buildings are not yet possible because of high radiation, there is no choice but to release contaminated water into the ocean. Under these circumstances, how is it possible to maintain any visible track of radioactivity? It is not only its invisibility that makes radioactive material dangerous to human beings. These materials have an extremely long life span: cesium-137 has a half-life of about thirty years, plutonium-238 has a half-life of eighty-eight years, and uranium-235 has a half-life of 704 million years. In other words, we will need to endure their existence for a long time.

This last subsection of the chapter argues for the importance of Shuji Akagi's photographs. Akagi is an art teacher and resident of Fukushima City, who continues to post photographs on Twitter (@akagishuji). His series of photographs are extremely suggestive, because they make us conscious of the passage of time.

Akagi began to post his photographs on Twitter on March 12, 2011. His time line shows the confusion that occurred, even after the nuclear plant accident commotion gradually settled down, and encourages us to avoid thinking that daily life has been restored in Fukushima. As time goes on, scenes that did not exist before the accident appear.

Let me focus on a series of pictures taken in Fukushima City on May 21, 2013. In the photograph published here as Figure 6.2, we see that all the grass has been unnaturally stripped from the ground. On June 25, Akagi posted the following conversation alongside his photographs:

> "Are you going to tear off that slope, too?" "Yeah, I'm putting on turf instead of grass." I [Akagi] was preoccupied with the junior high students behind.

This tweet provides context for the photographs. People were removing the radioactive layer when he visited the park. Akagi intentionally included in the frame junior high students in their school uniforms. They go to school, passing near the undressed slope, and Akagi's photograph powerfully captures the contrast between normal everyday school life and this stark new reality.

Because of the process required to remove radioactive material, these strange scenes have appeared suddenly within ordinary sight. Akagi has been comprehensively photographing decontamination operations to reveal the unusual aspect

FIGURE 6.2 Photographed on May 21, 2013.

Source: © Shuji Akagi (image courtesy of the photographer).

that daily life now possesses: the pollution from radioactive material. In Fukushima City, where many people continue to live their daily lives, radiation doses are not high enough to generate noise on a digital camera. While the radioactivity itself is difficult to visualize, the existence of radioactive material can nevertheless be shown, even if only indirectly, through the exhaustiveness and unnaturalness of the decontamination work. The work reminds us of the accident and the radioactive material scattered over the environment. In other words, the "trace" that decontamination work leaves behind appears as a kind of monument, approaching even the emotional and memorial level of the Atomic Bomb Dome (Genbaku Dōmu) standing at ground zero in Hiroshima.

In 1945, those who were at the center of the explosion were robbed of life immediately, due to the heat, the blast, the light, and the high dose of radiation. The radiation exposure experience that the atomic bomb brought is explained well by Akira Mizuta Lippit's concept of "avisuality," namely, the extreme optical experience that brings us to a point between the visible and the invisible, the exteriority and the interiority, the living and the dead.[30] Therefore, the Atomic Bomb Dome that endured avisuality has functioned as a great symbol of pacifism and democracy. In other words, it has been the major support of the collective memory for seventy years.

On the other hand, the city of Fukushima is considered an environment of low-dose radiation, where people can and do live during decontamination work. Currently, the inside of the reactors at the FDNPP—where a human being would die within several minutes and all electronic equipment would break down—is the only place of avisuality.[31] That is why the passage of time is crucial to the retention of our sense that the environment is always in danger.

Akagi's photographs illustrate this fact. Look at the picture taken on May 3 and posted on Twitter on June 1, 2014 (Figure 6.3). This photo depicts the same slope as the 2013 park photo. Here, the lawn grows, and even the weeds grow thick. One year before, the slope had provided an opportunity to recall the invisible radioactive material. But, as Akagi comments, "there has been no trace of work anymore"; the trace of the decontamination has disappeared over time. It is too vain to insist that the grassy slope is a monument. But these days, the oblivion in our daily lives is our biggest menace. The practice of Shuji Akagi conveys, calmly, that this is still a new catastrophe.

The impact of the accident at the FDNPP on Japanese society is as yet immeasurable. It is no exaggeration to say that the whole system built in postwar Japan is being shaken from the bottom up. The loss of trust in television as a major pillar of the public sphere is a definite symptom of this change. In the middle of such a change, the individual practices that have been expressed may seem too personal and too ephemeral. But such practices are required to visualize the nuclear disaster. Now, with the benefit of contemporary information technologies, shared images and discussion among individuals can become an alternative

FIGURE 6.3 Photographed on May 3, 2014.

Source: © Shuji Akagi (image courtesy of the photographer).

hub of information and action. The continued attempt to see with both our eyes and bodies is necessary if we are to confront and survive the invisible catastrophe that has become our daily lives.

Notes

1 "News Release," Ministry of Economy, Trade and Industry, April 12, 2011, http://www.meti.go.jp/press/2011/04/20110412001/20110412001–1.pdf (accessed November 2014).
2 Mamoru Ito, *Dokyumento: Terebi wa Genpatsujiko o Dou Tsutaetaka* [*Document: How Did TV Media Broadcast The Nuclear Power Plant Accident?*] (Tokyo: Heibonsha, 2012), 90.
3 Eiko Imahashi, *Foto Riterasī: Hōdō Shashin to Yomu Rinri* [*Photo Literacy: Photojournalism and Moral for Reading*] (Tokyo: Chūōkōron Shinsha, 2004), 17–24.
4 Asahi Shimbun Tokubetsu Hōdōbu, *Prometheus no Wana: Akasarenakatta Fukushima Genpatsu Jiko no Shinjitsu* [*Prometheus's Trap: The Hidden Truth of the Fukushima Nuclear Power Plant Accident*] (Tokyo: Gakken, 2012), 247.
5 See "BBC ga Fukushima Daiichi Genpatu Ichigōki no Bakuhatsu no Shunkan no Mūbī o Kōkai" ["BBC Shows a Video of the Fukushima Nuclear Power Plant First Unit Explosion"], *Gigazin*, http://gigazine.net/news/20110312_nuclear_fukushima_movie/ (accessed November 2014).
6 Ito, *Dokyumento*, 91–93; Hiroaki Mizushima, "Fukushima Daiichi Genpatsu no Bakuhatsu Eizou: 'Kokyozai' toshite Shakai de Kyoyu o" ["Explosion Images of the FDNPP: Our Society Needs to Share them as 'Commons'"], *Journalism*, July 10, 2012, http://www.asahi.com/digital/mediareport/TKY201207060365.html (accessed November 2014).
7 Ryū Homma, *Genpatsu Kōkoku* [*The Advertisement of the Nuclear Power Plants*] (Tokyo: Aki Shobō, 2013), 14–44.
8 Ito, *Dokyumento*, 99.

9 Christian Metz, "Le cinéma: langue ou langage?," *Communications*, 4 (1964): 76.
10 Joseph Hillis Miller, *Illustration* (Cambridge, MA: Harvard University Press, 1992), 61–66.
11 Robert Entman, "Framing: Towards Clarification of a Fractured Paradigm," *Journal of Communication*, 43, no. 4 (1993): 52.
12 Ito, *Dokyumento*, 94–96.
13 Akihiro Yamamoto, *Kaku Energī Gensetsu no Sengoshi 1945–1960: 'Hibaku no Kioku' to 'Genshiryoku no Yume'* [*The Postwar History on Nuclear Energy Discourse 1945–1960: 'The Memory of the Radiation Exposure' and 'the Nuclear Dream'*] (Kyoto: Jimbun Shoin, 2012).
14 Genshiryoku Kihon Hō [Atomic Energy Basic Law], Law number 186, December 19, 1955, http://law.e-gov.go.jp/htmldata/S30/S30HO186.html (accessed November 2014). An English translation is available at http://en.wikipedia.org/wiki/Atomic_Energy_Basic_Law (accessed November 2014).
15 Yamamoto, *Kaku Energī Gensetsu no Sengoshi 1945–1960*, 153–164.
16 Ito, *Dokyumento*, 100–102.
17 Asahi Shimbun Tokubetsu Hōdōbu, *Prometheus no Wana*, 47–50.
18 Yukio Edano, Chief Cabinet Secretary Press Conference, March 12, 2011, http://www.kantei.go.jp/jp/tyoukanpress/201103/12_p.html (accessed August 2014).
19 Ito, *Dokyumento*, 116–159.
20 Regarding the history of TV broadcasting in Japan, see Takumi Sato, *Terebi Teki Kyoyo: Ichioku Sōhakuchika eno Keifu* [*Television Education: A Genealogy of the Idea That All Japanese Are Intellectuals*] (Tokyo: NTT Shuppan, 2008).
21 "Levels of Radioactive Contaminants in Foods," Ministry of Health, Labour and Welfare, March 19, 2011, http://www.mhlw.go.jp/stf/houdou/2r98520000015iif-att/2r98520000017ee2.pdf (accessed August 2014).
22 Yukio Edano, Chief Cabinet Secretary Press Conference, March 19, 2011, http://www.kantei.go.jp/jp/tyoukanpress/201103/19_p.html (accessed August 2014).
23 "Result of Radioactivity Measurement of Tap Water: The 17th Report," Bureau of Waterworks Tokyo Metropolitan Government, March 23, 2011, https://www.waterworks.metro.tokyo.jp/press/h22/press110323–01.html (accessed August 2014).
24 Shinzo Kimura, *Hōshanō Osen Chizu no Ima* [*The Present of the Radioactive Contamination Map*] (Tokyo: Kōdansya, 2014).
25 Regarding amateur measurements and their radiation monitoring maps, see Atsuro Morita, Anders Blok, and Shuhei Kimura, "Environmental Infrastructures of Emergency: The Formation of a Civic Radiation Monitoring Map during the Fukushima Disaster," in *Nuclear Disaster at Fukushima Daiichi: Social, Political and Environmental Issues*, edited by Richard Hindmarsh (New York, NY: Routledge, 2013), 78–96.
26 Although this discussion is not on 2channel in 2014, it is archived at http://2r.ldblog.jp/archives/4525278.html (accessed August 2014). A web archive is also available at http://megalodon.jp/2014–0809–0827–59/2r.ldblog.jp/archives/4525278.html (accessed August 2014).
27 "Sampling in Spent Fuel Pool of Unit 4, Fukushima Daiichi NPP," YouTube video, 1:48, posted by "Sibuta98," April 16, 2011, http://www.youtube.com/watch?v=y0WcnZnG4wg (accessed August 2014).
28 "Kamera ni Utsurikomu Hōshasen no Noizu" ["Radiation Noise Appearing in the Camera"], YouTube video, 2:00, posted by "Extremefct," May 5, 2011, https://www.youtube.com/watch?v=YscBQI_mUpg (accessed August 2014).
29 Yoshihiko Noda, "Prime Minister Press Conference," December 16, 2011, http://japan.kantei.go.jp/noda/statement/201112/16kaiken_e.html (accessed August 2014).
30 Akira Mizuta Lippit, *Atomic Light: (Shadow Optics)* (Minneapolis: University of Minnesota Press, 2005), 81–103.
31 An investigation video of the Unit 2 released by TEPCO on January 20, 2012, shows innumerable noises generated by radiation: http://www.tepco.co.jp/tepconews/library/archive-j.html?video_uuid=l3gn6trv&catid=61703 (accessed August 2014).

7

PING AND THE MATERIAL MEANINGS OF OCEAN SOUND

John Shiga

Emerging from early twentieth-century transportation catastrophes and military, political, and economic crises precipitated by submarine and nuclear warfare, sonar embodies a model of underwater sound conducive to military and commercial efforts to control movement in ocean space. As Matthew Axtell writes, while military officials and scientists involved in sonar research and development have tended to conceptualize the ocean environment as an expansive body of water containing various marine organisms and other "natural" entities and structures, the ocean has also been "mediated by transducers, battleships, hydrophones, and human science, making it a highly technological space . . . the world that Navy officers and marine scientists made . . . was both a natural habitat and a technical workshop."[1] The tension between the ocean as natural or wild space and the ocean as technical workshop in military, industrial, and commercial practices is particularly legible in the case of sonar.

This chapter suggests that acoustic knowledges, techniques, and discourses were key to rendering the ocean as highly technological space, and this process of intensive mediation was oriented toward institutional interests in long-distance communication and human-machine integration in underwater environments. The chapter sketches two stages in the production of the ocean as technical workshop. The first centers on the listening devices such as underwater trumpets, acoustic lenses, and hydrophones (underwater microphones), which initially linked the ocean environment to human ears by means of a nonelectric and largely human-powered assemblage. The second stage begins with the turn in anti-submarine research and development to "active acoustics" or pinging sonar to detect objects in the water. Active acoustics coincides with the broader process of the electrification of media and continues through computerization.

By contrasting hydrophone-based and ping-based technologies, I hope to bring into focus what Jacob Smith (2015) calls the "eco-sonic" dimensions of media, including the ways in which media enable aspects of the material environment to be captured, altered, and controlled and, just as important, the "strange agencies of the nonhuman world in modern media."[2] Through an eco-critical lens, the history of underwater acoustic techniques can be reappraised in light of their contributions to the contemporary eco-crisis. Long-forgotten technologies, practices, and discourses might also be revisited and reconsidered as a model for what might be called, borrowing from Smith, a more "convivial" underwater media infrastructure that aims for sustainability rather than the objectives that have guided sonar infrastructure development such as fidelity, clarity, accuracy, and ubiquity.

Toward an Eco-centric Analysis of Sonar History

Sonar contributes to the production of ocean space through its mobilization as a "logistical medium," which Judd Case defines as those media that

> intrude, almost imperceptibly, on our experiences of space and time, even as they represent them . . . they are devices of cognitive, social and political coordination . . . Lighthouses, clocks, global positioning systems, temples, maps, calendars, telescopes, and highways are just a few of them.[3]

As a logistical medium, sonar represents space and time by capturing and processing portions of the ocean's acoustic field and rendering them audible and/or visible to human operators. But as Case notes, logistical media do not produce neutral descriptions of the world; they also reshape experiences of, and coordinate thought and action in, space and time. Just as airspace and aeromobility collapse without the continuous operation of radar, sonar enables ocean-space to be "continually 'beckoned' into being through the generative relationship of technology and human practice."[4] Sonar is key to the production of the ocean not only as an abstract space in maps and simulations but also as spaces of lived experience "whose embodied, emotional and practiced geographies remain to be adequately charted."[5] Those spaces include mobile spaces such as submarines, container ships, and research vessels as well as relatively fixed spaces and infrastructures such as undersea cables and the Clarion-Clipperton Fracture Zone (an area of the seafloor rich in minerals and thus attractive to deep sea mining firms); these spaces would cease to be functional or would be inaccessible if sonar's network of technologies, practices, and discourses were to fail. Attending to the logistical dimensions of sonar foregrounds the many military, commercial, industrial, and recreational activities in the ocean facilitated by underwater "pinging," which in turn have both short-term and long-term impacts on the material constitution and embodied experience of subsurface spaces.

In addition to the experiential, representational, and coordinative dimensions of logistical media, an eco-critical analysis might also explore the ways in which sonar incorporates, mobilizes, and transforms various environmental elements through the processes of research and design, manufacturing, distribution/

deployment, use, and disposal. Indeed, focusing exclusively on the way sonar impacts ocean space through its use in military, governance, industrial, and other activities may detract from an analysis of the spaces "behind" sonar (or the spaces on which the existence of sonar depends) that enable devices and techniques to be produced in the first place; an "eco-centric" materialism would direct attention precisely to the circulation of "both material and sonic goods through infrastructural systems and the ways in which those human-made systems depended on natural systems to provide raw materials."[6] Ranging from the use of explosives to test the sonic properties of the ocean to large-scale mining of quartz for transducer arrays in anti-submarine sonars during the Second World War to the intended and unintended ways in which the sonar ping acts as a radiant in the ocean environment, the "long tail" of sonar stretches far beyond the aquatic spaces it which it is used.[7] Understanding sonar as an eco-sonic assemblage is key to tracing these imbrications between medium and environment and developing a critique of the broader political and ethical implications of sonar.

I conceptualize the relationship between sonar and ocean environment in terms of what Lisa Gitelman calls the "material meanings" of media, that is, the "nexus of cultural practices, economic structures, and perceptual and semiotic habits that make tangible things meaningful."[8] Like the conceptualizations of mediation in the work of Katherine Hayles and Bruno Latour, Gitelman's concept of material meanings emphasizes the manner in which media are products of ongoing alignments and interactions between widely dispersed social, cultural, and technological processes.[9] The concept of material meanings focuses attention on what is so often taken for granted in discussions of media transitions: the apparent "thingness" of a given medium—the sense that a medium is a self-contained instrument, device, or system rather than a disparate group of processes and practices—needs to be continually reproduced through the alignment of cultural, economic, semiotic, and perceptual elements. While media often survive despite changes in practices, economies, or habits, in some cases, those changes disrupt the meanings attached to the medium and to its material substrates, which in this case include water and sound as well as the various resources that are used in sonar's life cycle. Gitelman's concept of material meanings suggests that the "content" of media might be productively understood as consisting of transformations of material resources (vibration, electricity, labor, etc.) through (temporarily) stabilized networks of people, things, concepts, and institutions. Media transitions are key sites for eco-centric analysis because they tend to foreground a wide array of elements that are constitutive of a medium as well as the manner in which those resources may at times resist the roles assigned to them by designers and users.

The Hydrophonic Ocean

The material meanings of sonar were shaped by new forms of underwater mobility and new types of catastrophe produced at the intersection of modern transportation and communication technologies; the increasing speed of transportation led to a demand for early-warning systems that could detect distant threats. Initial designs for acoustic early-warning systems relied primarily on the sound-conducting properties of air or water. While over-the-air systems that used steam-powered sirens to project a warning signal to ships had been installed near lighthouses in the United States and Great Britain in the late nineteenth century, engineers soon turned their attention to underwater acoustic systems because of the efficiency of sound transmission in water as compared with air. In an early experiment in 1826 to measure the speed of sound in water, Swiss physicist Daniel Colladon designed an underwater trumpet that could pick up the sound of an underwater bell struck with a hammer ten miles away. Colladon calculated the speed of sound in water to be 1,435 meters per second (about three meters less than the current standard for fresh water) and "marveled that so little energy at the source could be transmitted so great a distance through the water medium and could still be detected by the trumpet receiver."[10]

Colladon's enthusiasm for water-based sound transmission may seem incidental to the development of contemporary global sonar networks and the debates surrounding them. Indeed, the cumbersome trumpet, which had to be lowered into the water by the operator who simultaneously pressed his or her ear to the trumpet's end, straining to hear the signal against a cacophony of waterborne noise, may seem to be little more than a proto-sonar curiosity. But the issue of energy transmission is key to the eco-sonic dimension of sonar, and the underwater trumpet design enfolded human, hydrological, and acoustical elements in what can be recognized today as a remarkably low-impact assemblage, in part because the trumpet system used the acoustical properties of the water itself rather than electrical amplifiers to transmit sound.

The design of the submerged trumpet and early hydrophones shared a number of key characteristics with acoustic early-warning systems designed to detect airborne threats, such as Alfred Mayer's topophone (or "sound placer"), which worked like a binaural stethoscope for detecting the grumbling of icebergs at sea and, later, the distant engine sounds of approaching war planes; Japanese "war tubas" that enabled the Japanese military to detect planes approaching the coast during the Second World War; and the massive sound mirrors constructed by the British military, which focused sound into a mobile microphone to detect and determine the bearing of aircraft engine noise.[11] All of these acoustic locators or "macrophones," as Case calls them, were motivated by the increase in the speed of attack and the consequent reduction in warning time brought about by the use of airplanes in the First World War.

The intensification and acceleration of marine transportation similarly made conventional maps and optical sighting untrustworthy. Coal- and oil-powered motors led not only to increased speed, but as Willem Hackmann notes,

> Ships were also growing in size—the outcome of improvements in steel production and steam-engine design. The increased draughts of these ships

> made it imperative to know the contours of the sea bed, the position of wrecks and of other underwater hazards. This information was also vital to those laying the numerous submarine telegraph cables.[12]

The sinking of the *Titanic* in 1912, which is "the UK's deadliest peacetime disaster," signaled a broader breakdown of the economic structures, perceptual habits, routines, and signifying systems that supported transoceanic transportation and communication.[13] At that time, acoustic warning systems consisted of underwater bells installed on or near rocks and other hazards and rung electrically, pneumatically or by wave power (see Figure 7.1). Ships could hear the warning bells through a battery-powered underwater microphone, or "hydrophone." The Boston-based Submarine Signal Company cornered the market for these devices in the early twentieth century, and by 1912 had established over one hundred shore stations to operate warning bells marking underwater hazards in Europe, Asia, and North America.[14] Despite the scale of the warning bell network, systems based on hydrophonic listening were of limited use for detecting silent and mobile threats at sea, such as the iceberg that sunk the Titanic.

The deployment of submarines in twentieth-century warfare intensified the need for instruments that could probe the depths for threats, and the solution in first two decades of the century was found largely in nonelectric systems such as the Walser apparatus (see Figure 7.2). Developed by Lieutenant G. Walser for the French Navy in 1917, the Walser gear captured sound through a pair of nonelectric, acoustic lenses, each three to four feet in diameter, mounted on the underside of ships, with horn-tubes transmitting the sound to the operator's headphones. In nonelectric systems like the Walser gear, large lenses or horns (or a combination thereof) captured, focused, and amplified sound picked up from the water. Many of these systems depended upon the use of the human binaural sense to determine bearing through differences in amplitude between the two ears, or by adjusting the length of the air channels in a grooved plate or rubber tubing between the lenses and the headphones until the sound transmitted to each ear had the same intensity.

While nonelectric, acoustic locators were used to detect aircraft well into the Second World War, several factors led to the shift away from this technology in underwater early-warning systems by the end of the First World War. While the development of Colladon's trumpet and other devices based on nonelectric focusing of sound were motivated by scientific and commercial interests, the military concern with detecting submarines moving through the water became the primary driver of underwater acoustics research and technological development from 1914 through to the Cold War. The material meanings of underwater acoustic media began to shift away from civilian applications and toward undersea warfare. As a consequence, the relatively large sound lenses required to focus acoustic energy in the water were perceived to be a major drawback.

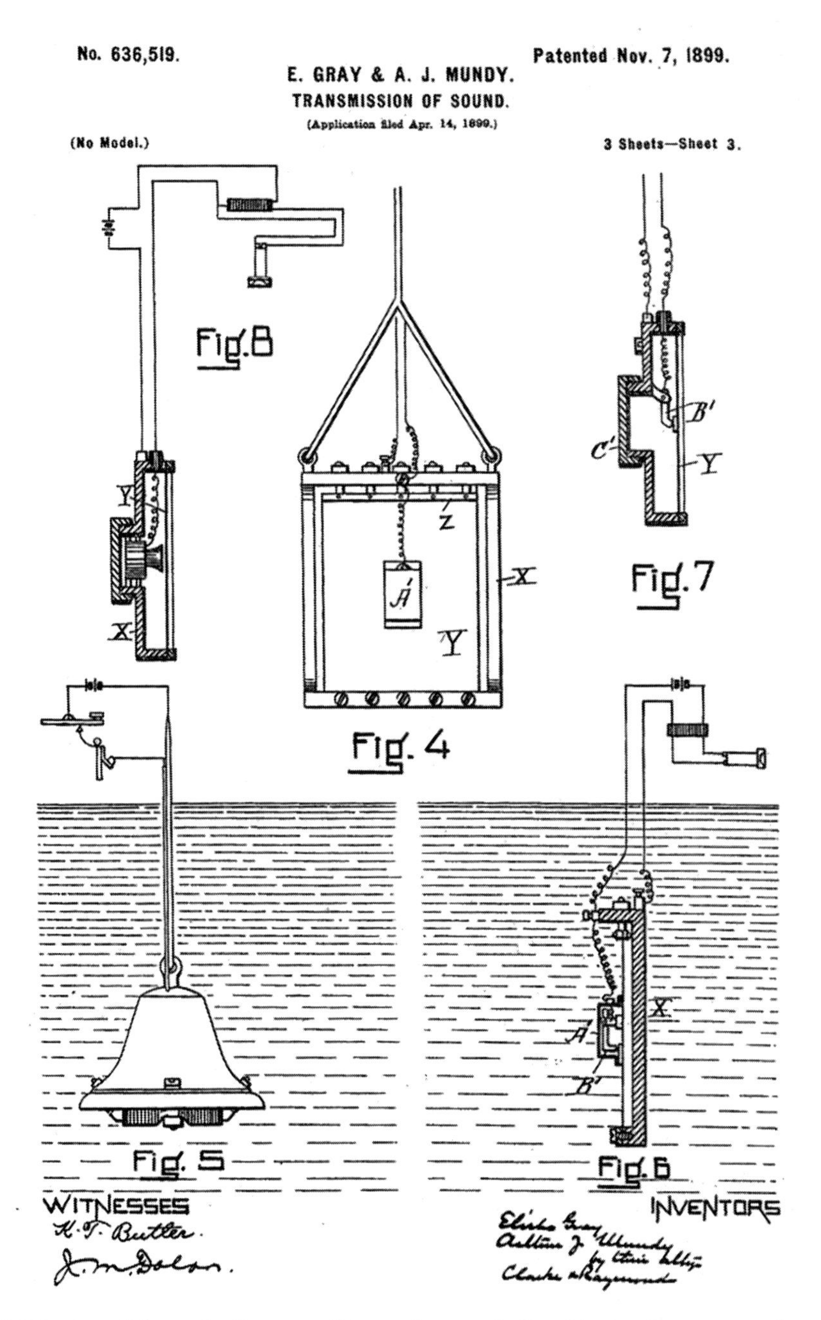

FIGURE 7.1 Schematic diagrams of the underwater bell and hydrophone system from Gray and Mundy's patent for "a new and useful improvement in the transmission of sound." While the patent notes that the bell could be rung pneumatically, the illustrations show an electromagnetically rung bell (Fig. 5) and battery-powered receiver in various configurations (Figs. 4, 6, 7, and 8).[15]

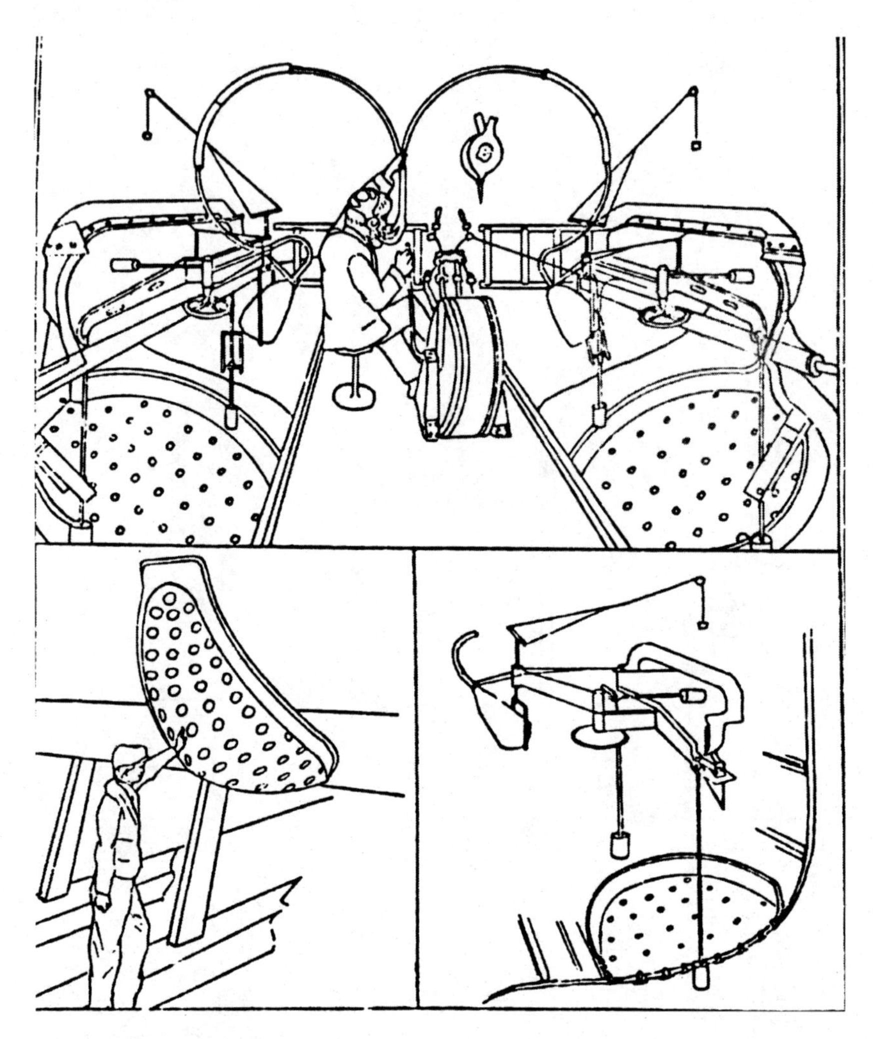

FIGURE 7.2 Installation of the Walser apparatus, with hull-mounted sound lenses and horn-tubes placed above each lens, carrying sound through adjustable tubing to the headphones.

Source: From Hayes, 1920, p. 20.[16]

As Hackmann points out, the British Admiralty refused to install Walser gear on Royal Navy ships because

> it was considered too fragile for sea use, repairs could only be effected in dry dock, and it required a great deal of space. Moreover, Royal Navy ship architects had a great reluctance to cut holes in the hulls for any form of extraneous apparatus.[17]

Nonelectric, binaural locators were capable of filtering out noise even while the search ship moved at high speeds, but during the First World War the American and British navies nevertheless turned to relatively noisy hydrophones, which were distributed to anti-submarine flotillas in the thousands.[18]

While signal-to-noise ratios remained a central problem in Allied anti-submarine operations in the First World War, the industrialization of warfare led to a new concern with increasing the scale or mass of military deployments while simultaneously organizing militaries into "small, homogenous, and comparatively informal" units.[19] Military forces were not only larger than ever before (soldiers now numbered in the millions) but also consisted largely of civilian reservists.[20] American and British navies enlisted merchant fishing ships and organized them into submarine-hunting flotillas. While remarkably precise, sensitive, and capable of continuous monitoring of the ocean's sonic environment, nonelectric listening systems were simply too large to be installed on most ships, and the context of industrialized warfare demanded that equipment be produced quickly and "in large quantities."[21] Simple hydrophones based on Elisha Gray's design (inspired by the telephone transmitter) that could either be lowered into the water on the end of the pole or built into the ship through a relatively minor alteration of the hull solved the problem of mass production and rapid and large-scale deployment much more easily than the Walser gear and similar systems with their bulky sound lenses and intricate systems of acoustic "plumbing."

Although it required relatively little electricity, the hydrophone inaugurated the age of electrified underwater listening since, like the telephone transmitter on which it was based, it required a continuous current to transform sound as variations in water pressure into modulated electrical signals. While research and design continued to attend to the problem of noise reduction, engineers pursued solutions involving electrical delay lines, compensators, and amplifiers rather than mechanical and binaural designs through the interwar period and the Second World War. From an eco-centric perspective, the merit of earlier technologies based on human muscle power and listening labor is that they "distributed the energy costs" of acoustic detection and localization "between human exertion and apparatus."[22] But in the context of the "first machine war," when "it was possible to grasp the relative warmaking potential of most countries simply by glancing at a diagram depicting annual coal and steel production," traditional forms of sonic knowledge associated with nonelectric systems became increasingly devalued.[23] The war of 1914–18 led to a decidedly industrial logic of listening: with the hydrophone and, later, the "pinging" sonar oscillator, the "burden" of listening was increasingly "borne by electric power created by the burning of fossil fuels."[24]

The electrification of underwater listening was driven by naval concerns with the extraction of locational information from underwater sound (for the purpose of targeting) and with reducing noise that interfered with the detection and identification of submarines and other vessels. Chief among the sources of noise in the sea was the "self noise" of the search ship itself, the engines and propeller of which were so loud that anti-submarine flotillas had to turn off ship engines to hear submarines through hydrophones. This slow, methodical search strategy was soon displaced by systems that would permit relatively clear signals to be heard through equipment even while the search ship was in motion. The concern with speed, with more rapid and continuous acoustic detection and tracking, and with faster responses to suspicious sounds with patterns of depth charges, drove the development of devices containing multiple hydrophones (over twenty in some systems) that could be towed behind the search ship, the incoming signals focused and amplified with electrical delay lines and compensators.[25]

Curiously, this new category of noise-suppressing, electrified, towed hydrophones were called "fish," and individual devices were given the names of animals, such as the Eel and the Rat—devices deployed by the Royal Navy in the First World War. Sonar incorporated a familiar and comforting image of "nature" at the same time as underwater listening became increasingly distant and insulated from the motor noise generated by fleets of search vessels and increasingly dependent upon electricity generated by coal-powered engines. Naturalized in name, electro-acoustical fish make legible the broader sonic division of the ocean whereby signals or sounds that are considered to be useful are captured, processed, and analyzed in spaces that are sealed off from the roar of engine emissions, propeller noise, explosions, and the sound of sonar itself.

Active Acoustics

If, in the First World War, the war-making potential of a nation could be estimated based on its production of coal and steel (key constraints for the production and operation of weapons, ships, and railroads), by the Second World War another set of figures became more significant in terms of war-making potential: "the number of automobiles made, the quantity of aluminum (for aircraft) extracted, and the quantity and quality of the electronic products assembled."[26] Between the two world wars, speed became paramount in decision-making about the development and organization of both transportation and communication infrastructures. The shift in military command to "speed rather than sheer mass" had significant implications for the eco-sonic dimensions of undersea warfare. As vessels became faster and capable of communicating by means of radio with each other via relays through headquarters, decision-making could be decentralized to smaller units. The notorious "wolf pack" tactic developed by German U-boat commanders was a particularly devastating form of decentralization, in which U-boats would disperse in the North Atlantic searching for Allied convoys. Upon spotting a convoy, the U-boat would communicate with others and coordinate an attack. Moreover, the invention of the snorkel (permitting the diesel engines to run with most of the ship submerged) meant that the U-boat could remain fully or partially submerged for extended periods and dive to greater depths and sail at higher speeds below water to avoid retaliation. In response to the tremendous increase in the mobility, invisibility, and flexibility of U-boat fleets, Allied navies invested in the development of electrified listening systems. While "unplugged" hydrophonic devices with complex systems of sound lenses and adjustable air-filled channels to focus and amplify distant sounds continued to be deployed in anti-submarine flotillas in the First World War, the desire for detection at greater ranges and for the extraction of more accurate locational information from underwater sound led to an arms race in submarine design and acoustic countermeasures. Nonelectrical systems were not necessarily simpler or easier to build and use than their electrified successors, but the growing demand for more accurate targeting information at greater ranges from the target led to the application of electrical techniques for amplifying signals picked up by hydrophones—and, by the interwar period, to expansive research and development projects on active acoustics in the United States and in Great Britain.

The shift to active acoustics—what the British Admiralty called "asdic" and the US Navy called "echo-ranging"—occurred during the interwar period and continued through the Second World War. Whereas horn-, lens-, and hydrophone-based systems listen "directly" to the sounds emitted by objects in the water, active acoustic systems project a sonic or ultrasonic pulse ("ping") into the water, listen for echoes, and use the delay between ping and echo to calculate the distance between the search vessel and the target. Although nonelectric "pings" can be generated by nonelectric means (e.g., via explosive charges), virtually all active acoustic systems developed during and between the two world wars relied on relatively high-powered, ultrasonic (i.e., above the range of human hearing) transducers. The principal reason for this shift was the fear that U-boats would soon be able to travel more efficiently in three dimensions; that is, that they could operate at greater depths and for prolonged periods on relatively quiet, electric motors. At the end of the First World War, British scientists advised their American counterparts that "submarines could be silenced to such a degree that passive sonar systems would not be efficient against them . . . Echo location gear, active sonar, was recommended."[27] Even with electrification, it was anticipated that hydrophones would soon become obsolete with the development of U-boats that could remain submerged for extended periods on battery power. The proposed solution to the silencing of submarines and their ability to drop out of sight and out of the range of hydrophones was to ensonify or acoustically "light up" the ocean with ultrasonic pings.

While the passive/active dualism obscures the way hydrophones and other listening techniques act upon ocean sound, the notion of "active" acoustics is useful to the extent that it highlights the manner in which acoustic force is applied to ocean space to extract information rather than collecting and repurposing already existing sound waves in the ocean. Initial forays in the use of active acoustics for underwater detection were made by the Canadian inventor Reginald Fessenden, who in 1913 developed a device he called an "oscillator"—a magnetostrictive transducer that radiated a continuous tone into the water and could then oscillate or switch to a "passive" mode to receive echoes reflected off objects in the water.[28] Around the same time, French physicist Paul Langevin and Russian émigré Constantin Chilowsky experimented with thin sheets of mica to generate ultrasonic frequencies, echoes of which would be picked up by a hydrophone. By 1918, Langevin developed a piezoelectric device in which a quartz mosaic sandwiched between two steel plates would act as both the transmitter of ultrasonic pulses as well as the receiver of returning echoes.[29] In the United States, following the First World War, and in anticipation of another undersea war, the National Defense Research Committee and the Office of Naval Research helped enlist scientists and engineers in sonar research from institutions including Columbia University, Harvard University, and the University of California, as well as telecommunications and electronics firms such as AT&T, General Electric, RCA, and Westinghouse.[30] This new alignment of military, industrial, and academic institutions once again altered the economic structures of underwater sound, which were now oriented not only to problems of navigation and threat detection but also to the problem of targeting increasingly fast, quiet, and deep-diving submarines. Military funded researchers redefined the problem of how to destroy a submarine from a question of building better sonar gear to "fundamental research" in oceanography and hydroacoustics and the drilling of such knowledge into sonar operators, or "ping men" (or simply "pings") as they were informally referred to in the Navy, to increase the efficiency of sonar in diverse water conditions and to enable echoes from targets to be distinguished from background noise. "Pinging" was not only a defensive measure to enable threat detection and avoidance but also a method of attack based on the extraction of locational information from underwater sound that would then facilitate more accurate barrages of depth charges.

According to the *Oxford English Dictionary*, the term "ping" appears in literature in the early nineteenth century, when it was used interchangeably with "ring" to denote "a short, resonant, high-pitched (usually metallic) sound, as that made by the firing of a bullet, the ringing of a small bell." The popularization of the term in the Second World War coincides with the incorporation of hydrophone and echo-ranging techniques into the broader category of "sonar," which referred to a range of passive and active techniques for manipulating underwater sound. The electro-acoustical discourse of ping produced a new experience of underwater acoustic transmissions as force and more specifically as firepower, an association that appears early on in the development of echo-ranging, and centers around a set of analogies between the transmission of electrical power and the mechanical acceleration of projectiles. In his patent for the echo-ranging oscillator, filed in 1916, Fessenden boasted that his invention could reach full power (1,600 watts) in under one-thousandth of a second, which, if translated into mechanical acceleration, "would give a 12 inch shell three times the velocity it has when fired from a 12 inch gun."[31] The association of pings with precisely controlled, destructive force is also evident in the discourse of sonar operators in the Second World War. Frank Curry, who served as sonar operator in the Royal Canadian Navy during the Second World War, described ping in terms of projectiles in an entry in his war diary in 1943: "Quite an interesting time pinging transmissions off all the neighboring ships and to follow a convoy down the harbor, ship by ship."[32] Decades later, Curry's short book describing his experience as a sonar operator continues to articulate ping as weaponized sound, cutting through the water: "We [took] up continuous anti-submarine search, our piercing asdic operating continuously, night and day . . . we beamed out transmissions, 3,000 yards every few seconds, listening carefully to echoes in our earphones as we trained the oscillator."[33]

The figuration of ping in terms of firepower was motivated in part by the close relationship between echo-ranging and targeting. The deployment of depth charges was to a large extent controlled by the interpretation of echoes in the sonar room; sonar was no longer a means of detecting and evading submarines but a system for targeting them. Ping was used as weaponized sound in diverse ways during and after the war. Curry notes that, as a result of depth charges with shallow settings, the crew frequently

> found the ocean surface covered with thousands of fish, floating dead. Those caught in the centre of the explosions were mangled and torn, but those further out were simply killed by shock . . . The fish were cleaned and turned over to the cooks.[34]

After the war, whalers devised new uses for ping as a sonic weapon. Initially, surplus naval sonar devices were sold to European whalers for use in whale hunting in the Southern Ocean but these were soon replaced by sonar equipment specifically designed for whale hunting. By the 1950s, whalers in Norway, Great Britain, Denmark, and Japan were using commercial sonar equipment to track whales while they dove, as ships would position themselves to kill the whale with explosive harpoons once it surfaced. In baleen whale hunting, pings were used not only to track and locate the whale but also to frighten them,

> resulting in escape behavior in which the animals swam at high speed near the surface in a straight line away from the sound source. This caused them to tire more quickly and made it easier to follow the whale and kill it.[35]

In both military and commercial whaling applications, the discourse and embodied experience of ping as a piercing beam or projectile rested on certain assumptions about ocean water. If mechanized underwater bells and nonelectric hydrophones were being displaced by electroacoustic pings, this meant that the model for underwater acoustic space was no longer tied to the concentric emanations of biological, geological, or mechanical sounds but rather to the linear transmissions of electric media. "Good water" was pliable, seemed to have no mediating effects on the range, direction, or intensity of pings, permitted pings to move in predictable trajectories, and returned clear echoes to the listening gear. Yet the ocean rarely conformed to this ideal. In addition to breakdowns of equipment, echo-ranging was frequently disrupted by vast "scattering layers" of marine organisms, which reflected sonar pings, and variations in water conditions, which bent transmissions downward and produced "shadow zones" beyond the reach of sonar. One postwar report by the US Navy noted that despite the efforts of engineers to build more powerful and discriminating sonar systems, there had been only modest gains in terms of the range at which submarines could be detected. According to the report, "This is because the limitations are generally not in the apparatus itself but in the sea water which carries the sound wave. And the sea has its full quota of perversities to trouble the sound man."[36] Although the report describes itself as a "record of achievement" by the National Defense Research Council (NDRC) and the academic and industrial institutions it supported during the war, more efficient and powerful sonar gear only led to marginal increases in the effectiveness of anti-submarine detection. The key part of the medium that had yet to be fully understood was the ocean water itself, which could propagate transmissions over long distances in ideal conditions but more frequently scattered, bent, absorbed, and otherwise distorted those transmissions.

The notion of ocean "space," then, was complicated by the growing awareness that this space was not empty but crowded with the action of waves, interactions with organisms, the atmosphere, and structures in the water formed by heat and pressure gradients that could facilitate or resist the transmission of pings. Consequently, Allied navies initiated large-scale research programs in underwater acoustics, which sought to identify and map the various hydrological, geological, and biological structures that interfered with the smooth flow of pings and echoes through the ocean. In 1941, NDRC, Division 6 was established for anti-submarine research and development, and by 1945 there were approximately three thousand scientists and other staff working on this project.[37]

But not all interactions between sonar and the ocean environment were considered to be worth studying; what is noteworthy about the postwar frenzy of activity in ocean acoustics is the manner in which concern with enhancing military command and control systems influenced the way scientists problematized the sonar-ocean relationship. Of particular interest to the US Navy from about 1937 through the Cold War was a layer of water approximately 4,000 feet below the surface called the "deep sound channel," which Maurice Ewing demonstrated in an experiment in 1945 could carry low-frequency sound over 1,000 miles.[38] The deep sound channel would later be instrumentalized or perhaps "infrastructuralized" as the hydroacoustic backbone of the US Navy's Sound Surveillance System (SOSUS), a global network of hydrophones and signal processing stations developed in the 1950s to track the movements of Soviet submarines.[39] More recently, the deep sound channel has been exploited in US Navy's Surveillance Towed Array Sensor System (SURTASS), a high-powered, low-frequency, active sonar system, which has generated considerable controversy due to the lethality of its pings demonstrated in mass stranding events, and also due to the Navy's declaration that it plans to deploy SURTASS in 75 percent of the world's oceans.[40]

Despite the ongoing efforts of environmental groups to litigate the navy's use of SURTASS, US courts have ruled that the value of high-powered sonar for national security outweighs the public interest in biodiversity. Rather than merely representing or coordinating movement through a preconstituted space, ping reshapes the ocean according to the motivation of dromological institutions to eliminate obstacles to the projection of bioacoustic force. Global sonar networks, which once consisted of hydrophones anchored to the seabed and other fixed infrastructure, now float freely according to what Paul Virilio called the "dromocratic" ideal of "displacement without destination in time and space," perhaps most clearly articulated in the right of the state to transform the ocean into a "vast logistical camp" of military preparedness, even if this entails the continuous ensonification of international waters and bodies therein.[41] The contemporary ocean is now an acoustic system, the mandate of which is to operate as a platform for the free flow of naval vessels and their bioacoustic projections.

While the interplay between sonar and the ocean environment has been scrutinized by generations of scientists working within and outside of military institutions, many interactions between ping and the ocean environment have been neglected until recently. Because the predominant use of sonar in naval operations is to act as a navigational, surveillance, and targeting mechanism, sonar's capacity to act more directly on bodies as a sonic weapon was practically ignored until the development of high-powered, low-frequency, active sonar in the 1980s. As Nate Cihlar notes, "the [United States] Navy has never addressed the possible effects of direct transmission of acoustic energy into bodily tissue and resonant cavities occurring when bodies are submerged in water."[42] Sound transmits more efficiently through a uniform medium; changes in the medium (e.g., moving from water to air) disrupt the transmission of sound. Whereas the body reflects 99.7 percent of acoustic energy in air, 100 percent of that energy will penetrate rather than reflect off a body in water if that body is composed mostly of water. The relatively uniform sound-conducting medium formed by bodies of water and bodies composed of water means that, beyond locational information, ping also produces "tissue rupture and hemorrhaging in various organs; yet, the Navy has failed to satisfactorily address the issue."[43]

Conclusion

In his short story, "Ping," ([1966] 1995) Samuel Beckett draws attention to what is so often obscured in postwar military discourses of sonar—that is, that ping does not only operate through prostheticization of vision and hearing; ping also exerts acoustic force on bodies in the water. While "Ping" seems to describe something like echo-ranging from the standpoint of a scanning, locating, targeting subject, it also incorporates fragments of an experience of *being pinged* in a breathless sequence of words with machinic regularity: "Ping perhaps a nature one image same time a little less blue and white in the wind. White ceiling shining white one square yard never seen ping perhaps away out there one second ping silence."[44] The story's repeated references to the storyteller's body parts (legs, toes, hands, etc.) are suggestive of the sonar scan, rotating back and forth along an arc searching and monitoring subtle changes in environment, which in "Ping" blurs into the body. As Elisabeth Bregman Segrè notes, the short, repeated words and phrases "may appear easy to understand" but in the end "our sense of logical, linear progression is thwarted, particularly by the interminable repetitions. Our habitual processes of understanding the written word prove to a large extent inoperable."[45]

Through its repetitions and refusal to cohere or progress as a linear narrative, "Ping" is one of the many cultural texts in this period that engage with post-war society's increasing occupation with the potential for catastrophe in the vast and complexly interconnected sociotechnical systems. Catastrophe, as Mary Ann Doane argues, emerges through the conjuncture of technological failure and confrontation with death.[46] "Ping" may recall or anticipate a number of potential and actual catastrophes, but for Western readers during the Cold War, the catastrophe mostly closely associated with sonar may have been nuclear war. Kathryn Schulz reinterprets "Ping" in relation to humanitarian uses of military sonar to search for missing Malaysian Airlines flight MH370 in March 2014:

> We ping in search of connection, and we ping to indicate our presence. It is the latter sound we hope to hear this week, as a device called a Towed Pinger Locator searches the Indian Ocean for a missing plane, like a child playing Marco Polo, in an unimaginably large pool, alone.[47]

Reflecting the tone of news media coverage of the sonar-based search, Schulz extracts sonar from the political history of relations between media and environment and celebrates technological and scientific breakthroughs enabled by the collaboration of military and commercial organizations in the search for the missing plane. Beckett's "Ping" becomes an homage to the irrepressible human desire for connection, diverting attention from the catastrophic potential of logistical media demonstrated by MH370's disappearance.

But in light of sonar's long history of violent manipulations of the ocean environment, it is also possible to read the nonlinear, cyclical, repetitious "Ping" as an allegory for, or perhaps an anticipation of, contemporary anxieties at the intersection of logistical media technology and *ecological* catastrophe; in this sense, we might understand the circularity of "Ping" as a gesture toward eco-sonic ruptures in the illusion of technological progress. Ping has played a major role in eco-catastrophe by facilitating the instrumental mapping of ocean depth for the expansion of military and commercial shipping, the deep sea extractive industries, whaling, and the seafood industry; by producing mass stranding events as legally acceptable and even necessary collateral damage through the application of low-frequency, acoustic force to nonhuman bodies in sonar testing, training, war games, and simulations; by reconfiguring the subsurface ocean as a vast platform for the maneuvering of nuclear arsenals; by insulating military operations through layer upon layer of electronic mediations from the anthropogenic noise such operations produce; and by consuming ever-greater quantities of nuclear or fossil fuels in the name of increasing the range, accuracy, and informational payload of sonar pings.

While it may seem tempting to embrace passive acoustics as an alternative to the projection of acoustic force in active systems, it is worth remembering that SOSUS and other hydrophone-based systems sustained military fantasies of perfect surveillance and control over the flow of nuclear weapons in the oceans for nearly half a century, and that this precarious circulation of nuclear missiles on submarines has the potential to literally end the world. Rather than arranging our options in stark terms such as passive versus active acoustics, we can instead develop a "green-media archeology" that will "uncover alternative media histories that are models for emergent practices."[48] We might look past the hydrophone to the nonelectric underwater trumpets, grooved plates, and labyrinthine air tubes, not as archaic curiosities but as sources for a new model of media development that strives toward the minimization rather than the expansion of media infrastructures.

Notes

1 Matthew Axtell, "Bioacoustical Warfare: *Winter v. NRDC* and False Choices between Wildlife and Technology in U.S. Waters," *Minnesota Review*, 72/73, no. 3 (2009/2010): 206.
2 Jacob Smith, *Eco-Sonic Media* (Oakland: University of California Press, 2015), 13.
3 Judd Case, "Logistical Media: Fragments from Radar's Prehistory," *Canadian Journal of Communication*, 38, no. 3 (2013): 380.
4 Peter Adey, Lucy Budd, and Philip Hubbard, "Flying Lessons: Exploring the Social and Cultural Geographies of Global Air Travel," *Progress in Human Geography*, 31, no. 6 (2007): 776.
5 Ibid., 775.
6 Smith, *Eco-Sonic Media*, 14.
7 John Shiga, "Sonar: Empire, Media and the Politics of Underwater Sound," *Canadian Journal of Communication*, 38, no. 3 (2013): 368; John Shiga, "Sonar and the Channelization of the Ocean," in *Living Stereo: Histories and Cultures of Multichannel Sound*, edited by Paul Théberge, Kyle Devine, and Tom Everrett (New York, NY: Bloomsbury, 2015), 86.
8 Lisa Gitelman, "Media, Materiality, and the Measure of the Digital; or, The Case of Sheet Music and the Problem of Piano Rolls," in *Memory Bytes: History, Technology and Digital Culture*, edited by Lauren Rabinovitz and Abraham Geil (Durham, NC: Duke University Press, 2004), 203.
9 See N. Katherine Hayles, *How We Became Posthuman: Virtual Bodies in Cybernetics, Literature and Informatics* (Chicago, IL: University of Chicago Press, 1999) and Bruno Latour, "Drawing Things Together," in *Representation in Scientific Practice*, edited by Michael E. Lynch and Steve Woolgar (Cambridge, MA: MIT Press, 1990), 19–68.
10 Marvin Lasky, "Review of Undersea Acoustics to 1950," *Journal of the Acoustical Society of America*, 61, no. 2 (1977): 283.
11 Case, "Logistical Media," 386–389.
12 Willem Hackman, *Seek and Strike: Sonar, Anti-Submarine Warfare and the Royal Navy, 1914–1954* (London: Her Majesty's Stationery Office, 1984), 4.

13 Iain McLean and Martin Johnes, "'Regulation Run Mad': The Board of Trade and the Loss of the Titanic," *Public Administration*, 78, no. 4 (2000): 729.
14 Gary Frost, "Inventing Schemes and Strategies: The Making and Selling of the Fessenden Oscillator," *Technology and Culture*, 42, no. 3 (2001): 466.
15 E. Gray and A.J. Mundy, *U.S. Patent No. 636519* (Washington, DC: US Patent and Trademark Office, 1899). Retrieved from https://www.google.com/patents/US636519 (accessed August 17, 2015).
16 H.C. Hayes, "Detection of Submarines," *Proceedings of the American Philosophical Society*, 59 (1920): 20.
17 Hackmann, *Seek and Strike*, 56.
18 Gary Weir, *An Ocean in Common: American Naval Officers, Scientists, and the Ocean Environment* (College Station: Texas A&M University Press, 2001), 9.
19 Martin Van Creveld, *Command in War* (Cambridge, MA: Harvard University Press, 1985), 149.
20 Ibid., 150.
21 Lasky, "Review of Undersea Acoustics to 1950," 287.
22 Smith, *Eco-Sonic Media*, 25.
23 Van Creveld, *Command in War*, 189.
24 Smith, *Eco-Sonic Media*, 31.
25 See Hackmann, *Seek and Strike*, 58, and Lasky, "Review of Undersea Acoustics to 1950," 287.
26 Van Creveld, *Command in War*, 189.
27 Lasky, "Review of Undersea Acoustics to 1950," 288.
28 Frost, "Inventing Schemes and Strategies," 471–472.
29 Hackmann, *Seek and Strike*, 77–82.
30 Lasky, "Review of Undersea Acoustics to 1950," 292–293.
31 As cited in Frost, "Inventing Schemes and Strategies," 471.
32 Frank Curry, War Diary. Frank Curry fonds, R2552–0–1-E, box 1. National Archives of Canada, Ottawa, 1940–45, 85.
33 Frank Curry, *War at Sea: A Canadian Seaman on the North Atlantic* (Toronto: Lugus Productions, 1990), 34–36.
34 Ibid., 91–92.
35 John Fornshell and Alessandra Tesei, "The Development of SONAR as a Tool in Marine Biological Research in the Twentieth Century," *International Journal of Oceanography* (2013), 2.
36 John Herrick, *Subsurface Warfare: The History of Division 6, NDRC [National Defense Research Council]*. Department of Defense, Research and Development Board Report for the Office of Scientific Research and Development, https://archive.org/stream/subsurfacewarfar00herr#page/n1/mode/2up, 8–9 (accessed November 23, 2015).
37 Lasky, "Review of Undersea Acoustics to 1950," 293.
38 Ibid., 294.
39 Weir, *An Ocean in Common*, 172.
40 Nate Cihlar, "Navy and Low Frequency Active Sonar: Stripping the Endangered Species Act of Its Authority," *William and Mary Environmental Law and Policy Review*, 28, no. 3 (2004): 913–950.
41 Paul Virilio, *Speed and Politics* (Los Angeles, CA: Semiotext(e), 1977/2006), 64.
42 Cihlar, "Navy and Low Frequency Active Sonar," 934.
43 Ibid.
44 Samuel Beckett, "Ping," in *Samuel Beckett: The Complete Short Prose, 1929–1989*, edited by Stanley Gontarski (New York, NY: Grove Press, 1995), 194.
45 Elisabeth Bregman Segrè, "Style and Structure in Beckett's *Ping*: 'That Something Itself,'" *Journal of Modern Literature*, 6, no. 1 (1977): 127.
46 See Mary Ann Doane, "Information, Crisis, Catastrophe," in *New Media, Old Media: A History and Theory Reader*, edited by H.K. Chun, W. Kennan, and T. Keenan (London: Routledge, 2005), 251–264.

47 Kathryn Schulz, "The Meaning of Ping: Electric Signals and Our Search for Connection," *Daily Intelligencer*, April 11, 2014, http://nymag.com/daily/intelligencer/2014/04/meaning-of-ping-flight-370.html (accessed July 2015).
48 Smith, *Eco-Sonic Media*, 144.

8

"GOING THE DISTANCE"

Steadicam's Ecological Aesthetic

Amy Rust

> *The Rocky Steps. . . . Photo Op: Once you reach the top and mimic Rocky's triumphant celebration, turn around for a breathtaking view of the scenic Benjamin Franklin Parkway and the Philadelphia skyline. And be sure to tag your photos with the hashtag* #visitphilly.[1]

An odd start for a chapter about Steadicam, I admit. Still, this excerpt from VisitPhilly.com exhibits the economic and ecological concerns that, I contend, condition the device's historical emergence in the mid-1970s. One of three films to inaugurate Garrett Brown's camera stabilizer in 1976, *Rocky* (dir. John Avildsen) forever transformed phenomenal and psychological relationships to Philadelphia, its urban environment, and the steps of the Philadelphia Museum of Art. "This [is] the place, that, after Rocky [Sylvester Stallone] was running for so long, he ran up," a tourist from Puerto Rico tells CBS News some thirty years after the film's release. Retracing his steps, she adds, "You get the feeling of 'I did it. I made it.'"[2] Here, identifications with character and city mitigate geographic distance and the passage of time. One finds, in fact, that contradictions between location and dislocation, perdurance and change, characterize not only spectatorial experiences and commercial appropriations of *Rocky*, but also, I argue, its narrative preoccupations and Steadicam's very aesthetic.

The device is, after all, deeply ambiguous. Worn by the cameraperson, though at a distance from his or her body, it grants unprecedented access to physical spaces, even as it eliminates the corporeal traces of this newfound mobility. Its abstract and free-floating presence, meanwhile, suggests both visual omnipotence and the movement of comparatively grounded and concrete characters. In this sense, VisitPhilly's "photo op" gains fresh significance. Encouraging visitors to "mimic Rocky" before "turn[ing] around for a breathtaking view," the website

FIGURE 8.1 Invented by Garrett Brown, Steadicam is a hands-free camera technology first adopted by Hollywood in 1976.

Source: Photo by Steadicam operator John E. Fry, 2008.

repeats Steadicam's approach to the scene. In *Rocky*, that is, Brown and his device run alongside the boxer and then hover behind him, locating and dislocating spectators with character and camera, as first one then the other spins toward the place where parkway meets skyline. As a result, Rocky and Steadicam, operator and spectator, each "go the distance," a phrase used by the boxer to express his desires for perseverance. Rocky does not aim at victory, in other words, during the film's climactic bout; he merely wishes to endure the beating he receives for the appointed fifteen rounds. "Going the distance" thus implies a temporal as well as spatial achievement. Animating the boxer's trajectory from "bum" to contender, the phrase at the same time replaces change with endurance.

The same may be said of Steadicam's many locations and dislocations, which trace Rocky's movement with long takes through Philadelphia's postindustrial milieus. In these shots, the device glides through abandoned shipyards and trash-filled streets that differ dramatically from VisitPhilly's gentrified cityscapes. Navigating this environment alongside the film's hero-in-training, spectators thrill at Rocky's tenacity, which the film ties to economic and ecological renewal. Still, because they survey as well as traverse Rocky's milieu, operator, Steadicam, and spectator alike potentially register the impoverished and contaminated landscapes that individual triumphs cannot alter. It matters, for this reason, that Steadicam's long takes ground and unground spectatorial identifications, revealing meetings and departures among character and camera, human and world, by way of their ambiguously embodied and extended durations.

In what follows, I approach Steadicam as an ecological aesthetic, by which I mean a phenomenal and psychological form of relation that organizes—perhaps even reorganizes—encounters between living and nonliving milieus. To approach Steadicam in this way means thinking how *technologies* (literally, *logics of making*) not only mediate, but also shape *ecologies*, or *logics of dwelling*, for people and things. It means considering how the device's aesthetic expresses and yet also exceeds the diegetic worlds that it depicts. For these reasons, then, I situate Steadicam in its historical moment, when the rise of neoliberalism and environmentalism forge uncertain encounters with and between city and country, capitalism and activism, persistence and transformation. Looking at two films that mark the device's bicentennial appearance—*Rocky* and the Woody Guthrie biopic, *Bound for Glory* (dir. Hal Ashby, 1976)—I trace how each locates and dislocates spectatorial identifications with unemployed men who "go the distance" amid economic and ecological devastation.[3] For Rocky, this means navigating the poverty and pollution of an urban East Coast environment; for Woody, a trek West through Depression-era rural America. Regardless, each film allies individual perseverance with collective change in ways that corroborate historical reroutings of the New Deal toward neoconservatism, even as modern environmentalists organize against the abuses of industrial and neoliberal capitalism alike.

Steadicam, I submit, registers, repeats, and resists these actions. It does so, moreover, through the decidedly ambiguous relationships it organizes between operators, spectators, and on- and off-screen environments. Indeed, to the extent that it removes and returns each from and to the other, the device joins private interest to public imperative, figuring both the limits and possibilities of the era's economic and ecological narratives and politics.

Operator

I want to begin with the relationships Steadicam establishes between operator and environment. Invented to isolate cameras from the men and women who wear them, Garrett Brown's stabilizer obscures the intimate contact with people and things its development served to facilitate. One might suggest, therefore, that Steadicam separates individuals from the milieus with which it likewise connects them. Or, to borrow terms from Marshall McLuhan, it amputates the very bodies it extends into the world. For McLuhan, humans deploy technologies to respond to phenomenal and psychological needs, but in so doing often deny the responsibility such devices imply between users and their surroundings.[4] As a result, people regard technology less as the place where human and nonhuman meet and more as a mechanism for separating individuals from the world and its material limits.

In the case of Steadicam, these limits include the spatiotemporal restrictions of its closest progenitors—handheld camera work and the dolly, or tracking, shot. Indeed, the tracks upon which dolly shots depend evoke density, inflexibility, and finitude, when compared to Steadicam's unrestrained mobility. "The dolly shot," writes Jean-Pierre Geuens, "is bound by its massive physicality. . . . It is yoked to the ground."[5] Steadicam, by contrast, brings free-floating autonomy to its predecessor's fluidity. This liberty extends, moreover, to financial considerations. While dollies demand expensive collaborations among operators and actors, focus pullers and grips, Steadicam—at least at first glance—requires but one individual working in relative isolation.[6] In this sense, the device resembles the handheld camera, which also supplies independence and portability in the face of nontraditional settings. Unlike handheld, however, Steadicam hides the bodily traces, the corporeal immediacy, for which the former is known. Pivotable and rotatable connections displace the operator's movement, while the camera's own disarticulated construction uproots its gravitational center.[7] From this point of view, Steadicam dislocates operator and apparatus alike, decentering the grounded, locational force with which handheld and dolly shots are mutually, even if diversely, linked.

Still, for all its apparent bodilessness and dislocated abstractions, Steadicam nonetheless hinges on concretely locatable and sensuous perceptions. "Steadicam [can] go anywhere," notes Geuens, but there are limits, including "the director's imagination" and—more importantly—"the operator's physical stamina."[8] Weighing forty-six pounds by 1976, the device's bulk makes it exceedingly taxing to carry.[9] Beyond its heft, moreover, Steadicam proves sensitive and unwieldy to operate. Springs and cantilevers may supply hands-free operation, but camerapersons must navigate tight spaces and convoluted paths—often backwards—with but a poorly lit video feed to guide their movements. Famously difficult to work as a consequence, Steadicam "implies," its inventor writes, "a thorough and frustrating relearning of gestures, shooting postures, and ways of monitoring the shot."[10] Indeed, such painstaking labor may account for the detailed descriptions of arduous shoots with which operators regale interlocutors.[11] These tales, which affirm Steadicam's links to specialized training in the midst of Hollywood's post-studio reorganization, subtly subvert its purported flights from materiality and self-effacing classicism, respectively.

After all, though Steadicam foregoes the handheld's amateur shakiness, it does not simply revert to the dolly's stolid professionalism. Instead, its operation proves far more paradoxical. Ungrounded yet unwavering, the device's unencumbered stability lends images limited, even unstable, force. One thinks, for instance, of the well-known introduction to John Carpenter's *Halloween* (1978), which uses Panaglide (a short-lived Steadicam competitor) to situate viewers inside the killer's body. The result, notes Carol J. Clover, certainly produces an assaultive, omnipresent view, but it likewise produces a look that is resolutely partial and unsteady. "The credibility of the first-person killer-camera's omnipotence is," she argues, "undermined from the outset. One could go further," in fact, "and say that . . . Steadicam sequences [draw attention] to the very item the filmmaker ostensibly seeks to efface: the camera."[12] The device does not, in other words, simply cleave operator from environment or concreteness from abstraction. Rather, despite its construction, Steadicam entangles location with dislocation and persistence with change.

Environment

Near the end of this chapter, I trace the aforementioned entanglements by way of Steadicam's ambiguous identifications, which, I submit, at once ground and unground omniscience and subjectivity. For now, however, I want to investigate the broader environment these identifications register and repeat as well as to which they respond. To examine economy and ecology during this period is, after all, to discover how they—like Steadicam—divide what they at the same time draw together. As movements, that is, neoliberalism and environmentalism also "go the distance," recommending endurance or, in what often amounts to the same, freedom in the face of material limitations.

Among these limitations are the economic and ecological crises for which the 1970s are known: unemployment, stagflation, oil shortages, and inner-city decline, not to mention white flight, air and water pollution, and the rise of the Sun and Rust Belts. For the era's principal "operators"—its workers—increased automation, union decertification, and the emergence of service and clerical sectors only exacerbate these crises. The result, according to historian Jefferson Cowie, is a "wholesale meltdown in working-class identity."[13] Displaced New Deal Democrats find ground in Chicago School Republicanism that resists, among other things, labor reform, collective bargaining, and unionized work in the South. However paradoxical, this drift from 1930s collectivism to the individualism of the soon-to-be 1980s means laborers retreat from public engagement to preserve private gains. The 1970s may begin, therefore, with shop-floor strikes and proposals for full employment, labor reform, and national health insurance, but the decade concludes with "hardhats" and Southerners who incongruously identify with Northern, middle-class privilege.

Mediating this turn, meanwhile, are the 1960s, the celebrated freedoms of which unexpectedly link self-expression to self-reliance and activism to capitalism. On one hand, student protesters join labor halls increasingly composed of ethnic white men, women, and African Americans. Together, these groups challenge bureaucratic authorities, including both industry management and union leadership. On the other hand, however, these alliances open a gulf between labor's old and new constituencies that hampers their mutual goals. As the economy worsens, moreover, the inclusiveness of post–Civil Rights liberalism stagnates its efforts at change. Workers unite around labor reform, but split at affirmative action, while proposals for full employment strain political affiliations among blue- and white-collar citizens.

Again and again, fractures between and within working and middle classes help route public imperatives toward private interests by way of the freedoms they share. Neoliberals, as David Harvey notes, marry noninterventionist economic theory to a "practical strategy" of "liberty . . . with respect to lifestyles, modes of expression, and a wide range of cultural practices."[14] They dismantle the welfare state without alienating those who pursue social justice. A related tactic pits groundless, intellectualized cosmopolitanism against rural, anti-elitist roots, thereby encouraging industrial and agrarian laborers to endure rather than transform inequitable economic arrangements.[15]

When it comes to ecology, similar strategies take hold, as meetings between liberal and conservative agendas split shared interests and convene divergent realities. At issue, in large part, are two tracks that typify modern environmentalism. The first, or what historian Benjamin Kline calls the *mainstream* trajectory, looks to "legislation, administrative and regulatory action, the courts, and the electoral sphere" to effect environmental change.[16] In the 1970s, examples include the National Wildlife Federation and Environmental Defense Fund, which, like their peers, advocate private licenses and industrial limits for corporations through established economic and political systems. The second, or *alternative*, path rejects such structures of power in favor of grassroots, direct-action approaches.[17] Friends of the Earth and Greenpeace, for instance, rely upon public demonstrations to pursue comparatively radical goals, including those of "deep ecology," which pursues—by reputation, at least—nonanthropocentric encounters with nonhuman worlds.[18]

Though hardly identical to historical precedents, such groups nonetheless evoke the conservationism and preservationism of Theodore Roosevelt and John Muir, respectively. Approaching environment as *resource*, Roosevelt and adviser Gifford Pinchot sought to safeguard nature in perpetuity with efficient use policies and the scientific management of state and national parks, wildlife areas, national monuments, and the like. Muir, by contrast, regarded nature as *refuge*, not resource, and aimed to protect it from the interests of modern industry, even as he garnered funds from Henry Ford, John D. Rockefeller, and J.P. Morgan.

However inexactly, mainstream and alternative groups retain philosophies of resource and refuge for the 1970s. They also repeat the uneasy encounters with capitalism their forebears fostered. As with Roosevelt and Pinchot, mainstream organizations often adopt corporate practices, relying upon the market's abstract instruments to control natural reserves from apparently unimplicated positions. Following Muir, alternative activists defend the environment against industrial interests, even as this protection frequently cleaves human from nonhuman and complicates the "total-field image" of nature toward which deep ecology turns its nonanthropocentrism.[19]

For these reasons, I argue, each group—despite commitments to regulative limits or collective action—not only splits resource from refuge, but also inadvertently contributes to neoliberal efforts to cleave and conflate private advantage and public engagement. What is more, because these organizations draw their support from divergent constituencies—one working, the other middle class—they frequently strengthen oppositions between rootless and affluent radicals who call for change and grounded and gritty realists who know perseverance.[20] By the 1980s, in fact, "no attack on environmentalism, no proposal to loot the country's natural assets, is complete," according to historian Jim O'Brien, "without a phony-populist sneer at those who try to defend the environment."[21] By this logic, activists, at best, combat urban pollution and promote public health by conserving resources for and through institutions that sustain working-class people. At worst, they are liberal elites who, according to conservatives, can literally afford to constrain the market and put nature before human beings.

Spectator

With the foregoing in mind, I want to return now to Steadicam's ecological aesthetic and the ways that it forms—and potentially transforms—the relationships to labor and capital, human and nonhuman, that neoliberalism and environmentalism express. The device does so, I have suggested, by mingling omniscience and subjectivity within narratives that equate "going the distance" with economic and ecological freedom. To wit, *Rocky* and *Bound for Glory* feature characters who, each in his own way, embrace persistence as transformation. Spectators, too, "go the distance," thanks to Steadicam's long takes, but as I shall demonstrate, the device's many (dis)locations offer change in addition to perdurance.

Take *Rocky*, which opposes the street-smart "Italian Stallion" to the black Apollo Creed (Carl Weathers), who absurdly embodies both the effete middle class and the narrative's own exploitation of the American Dream. On one hand, Creed supplies the beating by which Rocky overcomes his environment. On the other, mere survival hardly seems adequate to the film's triumphant ending.[22] In going fifteen rounds, Rocky may offer, as the match's promoter declares, "the greatest exhibition of guts and stamina in the history of the ring." Yet his performance does little (at least in this first installment of the franchise) to alter the boxer's material circumstances. Most of the film, in fact, bears witness to Rocky's impoverished and debased surroundings: "seedy, Philadelphia neighborhoods," as Vincent Canby describes them, which—save for the conspicuous absence of African Americans—evoke photographs from the Environmental Protection Agency's DOCUMERICA project of the 1970s.[23] Indeed, because the film imagines success only inside the ring, it cannot help but register the failures that precede and exceed it.[24]

Steadicam, I argue, heightens these traces, corroborating yet challenging the film's—not to mention the era's—economic and ecological cleavages and conflations. When, for instance, Rocky climbs the steps of the Museum of Art, Garrett Brown and Steadicam, as I previously described, first run alongside the boxer then leave his side. As they do so, the image unsteadily bobs down and then up, exhibiting both the stabilizer's famous ungroundedness and its uncannily embodied style of movement. The effect constitutes Steadicam's ambiguous identifications. Mediating Rocky's literal and figurative ascent, the device invites audiences to share his experience without, at the same time, replicating his exact position in time and space. As a result, spectators do not move from omniscience to subjectivity and back again, but rather identify at once with camera, character, and perhaps even the operator.

To get at this hybrid experience, French film theorist Christian Metz provides a good start, particularly since his notions of primary and secondary identification emerged at roughly the same time as Brown's device. With primary identification, writes Metz, the spectator identifies with the camera, and in so doing, also "*identifies with himself* . . . as a pure act of perception."[25] The result is the "all-perceiving . . . transcendental subject" for which Metz and others condemn cinema and to which Steadicam, at least in part, refers.[26] Metz even describes primary identification as a "diffuse, geographically undifferentiated *hovering*" that remains "ready to catch on preferentially to some motif in the film . . . without the cinematic code itself intervening to govern this anchorage."[27] Literalizing this hovering, Steadicam expresses its floating omnipotence. Still, as with primary identification, it, too, touches down inside films.

Indeed, according to Metz, secondary identification temporarily extends omniscience into more subjective positions by way of on-screen characters, "uncommon" angles and framings, or, one might add, unusual movements.[28] In this way, the "hovering" that Metz describes serves two simultaneous functions: first, it grants abstract and unrestrained access to on-screen environments; second, it tethers this freedom to comparatively concrete and restricted positions. Either way—and contra Metz's own suspicions regarding spectatorial identifications—films continuously locate *and* dislocate viewers, removing *and* returning them from and to what remain linked but distinct possibilities.[29]

As I have described it, Steadicam gives this process discernible shape, extending omniscience into subjectivity and back again without dividing or flattening their terms. One might suggest, in fact, following scholar Richard Kirkland, that the device creates an "almost but not quite" relationship to character and, I would add, an "almost but not quite" identification with camera.[30] "The freedom of . . . Steadicam," writes Kirkland, "is manifest but at the same time limited": it "gestures toward the implication of the subject . . . while hinting that this interpellation is ultimately conditional, that we can, and will, range beyond our own perspectives as necessary."[31] The device thus proves crucial for the 1970s, when ambiguities among individual and collective, city and country, capitalism and activism, are increasingly leveraged toward persistence rather than transformation. Fluidly locating and dislocating spectatorial identities, the apparatus underscores *and* undermines the stability and freedom for which it was widely touted.

The first production to adopt Steadicam, *Bound for Glory*, like *Rocky*, turns these (dis)locations toward explicitly economic and ecological matters. Set (with no small significance) in the 1930s, the film follows Woody Guthrie from Oklahoma to California, depicting his encounters with migrant workers who suffer by and sustain the landscapes of the Great Depression. The result, according to period reviews, presents "a portrait of an uncomfortably ambiguous man."[32] Woody exhibits concern for poverty and social injustice while at the same time "evad[ing] . . . family responsibilities and party politics for [a] rendezvous with freedom atop a moving train."[33] Whether financial or geographic, in other words, the singer's mobility places him both inside and outside the communities for whom he organizes and about which he writes music. In "going the distance," he mingles public engagement with private advantage and thereby expresses the film's own historical moment. Indeed, when Woody rejects wife and children for an itinerant life on the road, *Bound for Glory* not only ties New Deal to New Left, but also anticipates neoconservatism, which itself joins self-expression to self-reliance so as to answer—as the film's title suggests—latent restriction with manifest destiny.[34]

Steadicam, for its part, repeats and reorganizes these encounters with its mix of omniscience and subjectivity, which, I have demonstrated, removes and returns operators and spectators from and to the environments with which it mediates their contact. In *Bound for Glory*'s most famous use of the device, Brown and Steadicam descend in a crane to follow Guthrie, as he navigates a labor camp. Together, operator, camera, character, and spectator pick through this milieu with its clotheslines and tent flaps, makeshift furniture, drifting dust, and smoke. After nearly a minute, the shot culminates, as all four push into a crowd of migrants who travel in, around, and between them. At this moment, viewers partake of Woody's experience. Having moved from air to ground, they are "transformed," according to Brown, "from a third-party voyeuristic perspective to a participatory point of view."[35] And yet, as workers press near character and camera, another look—what Metz might call the "out-of-frame"—also presents itself to spectators. An invisible intermediary, the out-of-frame refers to an off-screen character that, according to Metz, spectators "go through" —"as one goes through a town"—before "dispersing all over the surface of the screen" into the looks of on-screen entities.[36]

Viewers pass through the gaze of someone they cannot see before settling into the looks of characters they perceive. The out-of-frame thus names the place where primary hovering and secondary groundedness meet. Spectators do not simply leap from voyeurism to participation, as Brown imagines, but rather enjoy omniscience and subjectivity as simultaneous yet nonidentical experiences.

This is what I mean, then, by Steadicam's *ecological aesthetic*, which locates and dislocates the ambiguous identities that characterize meetings with labor and landscape during the 1970s. In fact, to the extent that the device mingles primary with secondary identifications, it also supplies an *ecological ethic*, one that draws freedom and responsibility together without the rifts and replacements that *Rocky* and *Bound for Glory*—as much as their era—repeatedly exhibit. What is more, because Steadicam's out-of-frame does not belong, as in Metz's formulation, to an actual character, it potentially extends omniscience and subjectivity to both living and nonliving entities. In (dis)locating character and camera at the same time, the device mediates neither fully point-of-view positions nor, as Jean-Pierre Geuens suggests, thoroughly "synthetic [and] once-removed" perspectives. Instead, to the extent that it remains "almost but not quite" first and third person, Steadicam evokes disembodied as well as embodied points of view.[37] One thinks, for instance, of the scene from *Rocky* in which raw meat hangs from an overhead conveyer. As the shot tracks backward, sides of beef swing in and out of the frame, conjuring both the film's unseen operator and views from the carcasses themselves. Coupled

FIGURE 8.2 Implying first and third person at once, Steadicam frequently evokes the perspective of unseen people and things, as in this scene from *Rocky* (dir. John Avildsen).

Source: United Artists, 1976.

with Rocky's sense that he, too, is a "piece of meat," this look demonstrates coincident, though not interchangeable, demands for care on the parts of humans, animals, and nonliving things.

Compounding this gesture is the shot's duration, which, like most long takes, invites viewers to attend to environments in addition to narrative agents. The result, which troubles distinctions between foreground and background, joins omniscience with subjectivity, or, to paraphrase André Bazin, presents "facts" that are at the same time "hallucinations."[38] Indeed, with its facility for the long take, Steadicam expresses Bazin's "faith in reality," whereby spectators proceed *in and through* concrete perceptions to arrive at abstract representational meanings.[39] Revealing "the *freedom* that," according to John David Rhodes, "Bazin relishes ... [as] *coercion*," the device grants spectators liberties to which identificatory points of view reciprocally commit them.[40]

In this sense, Steadicam recommends something more than merely "going the distance." Demanding that spectators tarry with worlds that, in turn, restrict them, the device joins private interest to public imperative and opens perdurance to possibilities for change. In the context of the 1970s, these possibilities include relationships to economy and ecology that neither collapse self-expression and self-reliance nor divide cosmopolitan privilege from rugged anti-elitism. Rather, Steadicam suggests coincident but nonidentical encounters between labor and capital, human and nonhuman, that refuse both resource and refuge as models of freedom and responsibility among living and nonliving entities.

Notes

1 See "The Rocky Statue and Rocky Steps," at Visit Philadelphia: http://www.visitphilly.com/museums-attractions/philadelphia/the-rocky-statue-and-the-rocky-steps/ (accessed June 10, 2014).

2 See Jason Straziuso, "Nostalgic Jaunt to 'Rocky' Steps," July 1, 2004, http://www.cbsnews.com/news/nostalgic-jaunt-to-rocky-steps/ (accessed June 11, 2014). According to Straziuso, "Scores of tourists daily seek out the Art Museum to emulate [Rocky's] iconic run."

3 The third film to employ Steadicam in 1976 is *Marathon Man* (dir. John Schlesinger), which I have elected to forego for brevity's sake. That said, the film certainly celebrates "going the distance" with its protagonist, Babe (Dustin Hoffman), a naïve academic and aspiring marathon runner, who escapes the past by enduring present-day tortures, all in the context of environmental protests and labor strikes.

4 See Marshall McLuhan, "The Gadget Lover: Narcissus as Narcosis," in *Understanding Media: The Extensions of Man* (Cambridge, MA: MIT Press, 1994), 41–47.

5 Jean-Pierre Geuens, "Visuality and Power: The Work of the Steadicam," *Film Quarterly*, 47, no. 2 (Winter 1993–94): 11.

6 I write "at least at first glance" because Steadicam operators require assistants to pull focus remotely, and grips and other technicians to remove obstacles from their paths or adjust elements such as lights.

7 In his efforts to provide balance to Steadicam, Garrett Brown increased the camera's moment of inertia by spreading out its components along a central pole.

8 Geuens, "Visuality and Power," 13.

9 Brown's final prototype—which featured a fiber-optic viewfinder rather than detached video monitor—weighed twenty-three pounds. See Garrett Brown, "The Iron Age," *Steadicam Letter*, 1, no. 4 (March 1989): 6. Heavier versions were developed by Cinema Products Corporation, to which he first sold Steadicam's rights. Today, the device weighs as much as seventy pounds or more. For the roughly ninety-minute long take in Aleksandr Sokurov's *Russian Ark* (2002), Steadicam operator Tilman Büttner carried an eighty-pound rig. "I had a special dolly built," he told one interviewer. "The Steadicam was always on my body, but occasionally, for thirty seconds at a time, I would have to rest and stretch by half sitting on a barstool placed on the dolly and wheeled over by a grip." See Louis Menashe, "Filming Sokurov's *Russian Ark:* An Interview with Tilman Büttner," *Cinéaste*, 28, no. 3 (Summer 2003): 23.

10 See Garrett Brown, "The Moving Camera, Part II," http://www.garrettcam.com/movingcamera/article2.htm (accessed June 18, 2014).

11 See, for instance, Ted Churchill's characterization of his work in "Steadicam: An Operator's Perspective, Part II," *Stabilizer News*, October 29, 2012, http://www.stabilizernews.com/steadicam-an-operators-perspective-part-ii/ (accessed June 18, 2014). "Steadicam is entirely dependent upon the skill of the operator," he writes, "and skill with the machine requires training and practice. It is rare that someone can simply pick it up and make a good shot the first time out."

12 Carol J. Clover, *Men, Women, and Chainsaws: Gender in the Modern Horror Film* (Princeton, NJ: Princeton University Press, 1992), 187. Worth noting in this regard is Carpenter's own estimation of Panaglide, which complements the approach to Steadicam for which I am arguing: "So it doesn't have the rock-steadiness of a dolly, but it also doesn't have the human jerky movements of a handheld—it's somewhere in between" (Clover, 187 n.46).

13 Jefferson Cowie, *Stayin' Alive: The 1970s and the Last Days of the Working Class* (New York, NY: New Press, 2010), 169.

14 David Harvey, *A Brief History of Neoliberalism* (Oxford: Oxford University Press, 2007), 42.

15 There are, Cowie writes, things "southern, western, gritty, masculine, working class, white, and soaked in the reality of putting food on the table" and things "northern, eastern, radical, effete, leisurely, affluent, multicultural, and full of pipe dreams." *Stayin' Alive*, 178.

16 Benjamin Kline, *First along the River: A Brief History of the U.S. Environmental Movement* (Lanham, MD: Rowman & Littlefield, 2007), 96.

17 "Alternative" is, once again, Kline's designation.

18 Norwegian philosopher Arne Naess coins the term "deep ecology" in 1973. Whereas "shallow ecology" struggles to "fight against pollution and resource depletion . . . [for the] health and affluence of people in the developed countries," deep ecology, according to Naess, values nature "independent of the usefulness of the nonhuman world for human purposes." See Bill Devall, "Deep Ecology and Radical Environmentalism," in *American Environmentalism: The U.S. Environmental Movement, 1970–1990*, edited by Riley E. Dunlap and Angela G. Mertig (New York, NY: Taylor & Francis, 1992), 52; and Naess, "The Basics of the Deep Ecology Movement," in *Ecology of Wisdom: Writings by Arne Naess*, edited by Alan Drengson and Bill Devall (Berkeley, CA: Counterpoint Press, 2010), 111. I write "by reputation, at least" in this context to register challenges to deep ecology's supposed nonanthropocentrism. Eric Katz discovers anthropocentric bias, for instance, in Naess's conception of self-realization, whereby human ontological expansion comes by way of identifications with nonhuman worlds. I am struck, for my part, by the centrality of identification in deep ecology, since the term animates my own understanding of Steadicam's many (dis)locations. Moreover, while I concede that deep ecology is not free from anthropomorphic—perhaps even anthropocentric—claims, I nonetheless find its distinctly *nonidentical* forms of identification inhibit too easy divisions or collapses between humans and nonhumans. See Eric Katz, "Against the Inevitability of Anthropocentrism," in *Beneath the Surface: Critical*

Essays in the Philosophy of Deep Ecology, edited by Eric Katz, Andrew Light, and David Rothenberg (Cambridge, MA: MIT Press, 2000), 17–42. For more on the potential value of distinguishing anthropomorphism from anthropocentrism, see Jane Bennett, *Vibrant Matter: A Political Ecology of Things* (Durham, NC: Duke University Press, 2010).

19 "Total-field-image" is Naess's term for the fundamental relationality that, he argues, constitutes humans, nonhumans, and their environments. Though frequently a concern for scholars who worry that the "total field" slips in its conception of mutual, though not identical, interests among entities, Naess's description of "knots in the biospherical net" suggests an attention to "the precarious balance of sameness and difference" that Val Plumwood, among others, rightly demands. See Arne Naess, "The Shallow and the Deep, Long-Range Ecology Movement. A Summary," *Inquiry*, 16 (1973): 95; and Val Plumwood, "Deep Ecology, Deep Pockets, and Deep Problems: A Feminist Ecosocialist Analysis," in *Beneath the Surface*, 63. It is precisely Naess's coincident, though not interchangeable, meeting of "subject, object, [and] medium" within "one gestalt" that, I contend, Steadicam forges for operators, spectators, and environments without, at the same time, cleaving or conflating such registers, as conservationism and preservationism, however unwittingly, do. See Arne Naess, "Self-Realization: An Ecological Approach to Being in the World," in *Ecology of Wisdom*, 96.

20 Because they welcome nonutilitarian approaches to wilderness spaces, alternative organizations tend to gather support from middle and upper classes. Mainstream groups, by comparison, are more closely allied with laborers and nonrural minorities, since the former's work in licenses and limits suits them to issues of urban pollution and public health.

21 Jim O'Brien, "Environmentalism as a Mass Movement: Historical Notes," *Radical America*, 17, nos. 2–3 (1983): 7.

22 Even Stallone, who declares he has "had it with anti-this and anti-that," notes of the screenplay's genesis: "I said to myself, 'Let's talk about stifled ambition and broken dreams and people who sit on the curb looking at their dreams go down the drain.'" See Judy Klemesrud, "'Rocky Isn't Based on Me,' Says Stallone, 'But We Both Went the Distance,'" *New York Times*, November 28, 1976.

23 Vincent Canby, "'Rocky,' Pure '30s Make-Believe," *New York Times*, November 22, 1976. Conceived as a photodocumentary project in the spirit of the Farm Security Administration of the 1930s, DOCUMERICA sought to record adverse changes in the American landscape as well as relationships between humans and their environments. In the end, the project produced more than twenty thousand images from nearly one hundred photographers between 1971 and 1978. Given the importance of the New Deal to economic and ecological thought during the 1970s, I am struck by the 1930s influence on both DOCUMERICA and the title of Canby's review.

24 It is worth noting in this context that the production moves from Philadelphia to Los Angeles to shoot the climactic bout, as if fleeing the devastated exteriors that neither art direction nor narrative can fully circumvent. That *Rocky* fails to abandon these exteriors, according to my reading, corroborates the work of Tony Williams, for whom the boxing film historically articulates "the dark repressed underside of capitalist existence." See Tony Williams, "'I Could've Been a Contender': The Boxing Movie's Generic Instability," *Quarterly Review of Film and Video*, 18, no. 3 (June 2009): 315.

25 Christian Metz, *The Imaginary Signifier: Psychoanalysis and the Cinema*, trans. Celia Britton and Annwyl Williams (Bloomington: Indiana University Press, 1982), 49.

26 Ibid., 48, 49.

27 Ibid., 54.

28 Ibid. "In other cases," Metz writes,

> certain articles of the cinematic code . . . are made responsible for suggesting to the spectator the vector along which his permanent identification with his own look should be extended temporarily inside the film. Here we meet . . . subjective images,

out-of-frame space, [and] looks (. . . no longer the look, but the former . . . articulated to the latter.)

29 In his own work, Metz tends to emphasize the locative functions of identification—whether primary or secondary—which connote stability and mastery without the disorienting limits that, as Steadicam makes legible, likewise accompany them.
30 Richard Kirkland, "The Spectacle of Terrorism in Northern Irish Culture," *Critical Survey*, 15, no. 1 (2003): 86.
31 Ibid., 86–87.
32 Aijean Harmetz, "Gambling on a Film about the Great Depression," *New York Times*, December 5, 1976. See, too, Molly Haskell's "The Land Is Made for Passing Through," *Village Voice*, December 13, 1976, which declares the film "a richly ambiguous character study."
33 Joy Gould Boyum, "This Land Is Your Land," *Wall Street Journal*, December 6, 1976.
34 See Cowie, *Stayin' Alive*, 173–174. Significant in this context are Cowie's observations concerning the "Okie," who is "nationalized," he argues, during the mid-1970s. Whereas Okie "once . . . referred specifically to the uprooted peoples of the Southwest, . . . the ideal had grown," he writes, "to a conglomerate of American identity." Appropriated by neoliberals, Okies evince rebellion without threatening capitalism; they are "people who have known suffering, [but] . . . are tough enough to rise above it."
35 Garrett Brown, as cited by Serena Ferrara, *Steadicam: Techniques and Aesthetics* (Boston, MA: Focal, 2000), 27.
36 Metz, *The Imaginary Signifier*, 55–56.
37 Steadicam thus offers a precursor to more contemporary forms of identification, including those forged by video games and digital animation effects, as described by scholars such as Alexander Galloway and Scott Richmond. See, for instance, Galloway, "Origins of the First-Person Shooter," in *Gaming: Essays on Algorithmic Culture* (Minneapolis: University of Minnesota Press, 2006) and Richmond, "The Exorbitant Lightness of Bodies, or How to Look at Superheroes: Ilinx, Identification, and *Spider-Man*," *Discourse*, 34, no. 1 (Winter 2012): 113–144.
38 André Bazin, "The Ontology of the Photographic Image," in *What Is Cinema? Volume I*, trans. Hugh Gray (Berkeley: University of California Press, 1967), 16.
39 See André Bazin, "The Evolution of the Language of Cinema," in *What Is Cinema? Volume I*, 23–40.
40 John David Rhodes, "Haneke, the Long Take, Realism," *Framework: The Journal of Cinema and Media*, 47, no. 2 (Fall 2006): 19.

PART III

(Un)sustainable Materialities

9

ECOLOGIES OF FABRICATION

Sean Cubitt

Are contemporary media sustainable? To the extent that our dominant technical media bear the stamp of the political-economic regime that gave them birth, they can sustain themselves only as long as capital can sustain itself. The top one percent of the world's population owns over half of its wealth; this is not a sustainable ratio.[1] As the authors of the 2011 UN Human Development Report note, "inequitable development can never be sustainable human development." Their definition of sustainable development bears careful analysis: we define "sustainable human development" as "the expansion of the substantive freedoms of people today while making reasonable efforts to avoid seriously compromising those of future generations."[2] The overall definition of development given by the UN Development Programme (UNDP) is tied to the idea of freedom as choice, a form of words that shows the wrangling that must have occurred to commit the UNDP to a model of consumerism. This definition asks us to consider sustainable media in light of what impacts they may have on future generations. The question is then whether consumerism is a sustainable model for media. This chapter investigates one aspect of this question: whether the number and scale of media technologies that we use in the developed countries can be expanded to the rest of the world, and whether that expansion can be sustained. The development perspective places greater demands on tactics of sustainable design because it asks whether there are enough materials and energy available in the finite system of the planet to provide them, in the forms we are now familiar with in the wealthy world, to the three billion people still living beyond the range of our most fundamental technologies.

The still-current crisis of capital that began in 2008 is technically over at time of writing in 2014 because gross domestic product (GDP) in the core metropolitan countries is once again showing growth, albeit at a reduced level compared to precrisis figures. The GDP is, however, a crude yardstick which notoriously ignores internal difference within nations, as well as the vast and increasing, if occasionally controversial, evidence (see the storm over Piketty's 2014 attempt to use economic accountancy's own tools against it) that that difference has accelerated as a result of the crisis, even compared to the growing gulf between rich and poor that preceded it.[3] Austerity measures of the kind once reserved for developing nations by the International Monetary Fund (IMF) are now employed by European Union and North American polities to ensure the effects of crisis are felt most by those least able to bear them. Sustainable media theses then have to deal also with the impoverishment of metropolitan populations as well as those in developing nations.

Campaigns by labor organizations and environmental groups addressing the sweatshop conditions of workers and the ecological impacts of electronic industries have recently begun to impact on eco-critical media studies (see, for example, the chapter by Gabrys in this volume). The question of sustainability points us directly to the immense use of energy in the manufacture and use of digital media; the immense quantities of, in some cases, rapidly diminishing stocks of minerals for their making; and the immense challenges of dealing with waste electronic goods. This chapter singles out manufacture, specifically of integrated circuits and subassemblies, as a critical node in this environmental cycle. Its main ambition is to argue that environmentalists need to expand their political horizons to include human victims of anti-ecological practices, and to argue that these include not only workers and those living in the immediate vicinity, but everyone involved in the circuits of neoliberal capital. To make media committed to sustainability, to sustain the very media we use, and to make a world where media are sustenance requires a commitment to solidarity and community between different classes of human victims with the nonhuman environment, without which green politics, for lack of global human commitment, cannot sustain itself. The chapter works on the premise that an understanding of the industrial-consumption cycle of electronic commodities is a necessary first step in building such communities.

There are three large-scale mechanisms driving crises of capital. Rosa Luxemburg may have been the first to recognize that accumulation not only named the brutal expropriation of common land at the origin of modern capitalism in Europe but also continued as the equally brutal dispossession of colonized peoples.[4] Accumulation by dispossession, in David Harvey's usage, is the continuing employment of enclosure, through seizures of common goods such as land and geology, seabeds, water and air, and public goods such as health, welfare, and security by capital.[5] The second mechanism is financialization, whose essential characteristic is the trade in risk, intended to reduce the intrinsic insecurity of investment by trading in future values. This trade in futures, itself dependent on computers and network communications, both closes down options for change and simultaneously creates conditions of debt peonage, while increasing the rate of transfer of wealth from poor to rich.[6] The third mechanism is the application of extended reproduction whose theory Marx propounded in *Capital*, volume 2:

the devotion of economic resources to growth rather than to satisfying fundamental needs. It is this feature that underwrites the other two mechanisms, both of them responses to earlier crises produced by excessive or failed growth, crises of overproduction or overaccumulation. It is growth itself, the engine of capital, that opposes sustainability.

Expanded reproduction in the twenty-first century has been characterized as cognitive or immaterial by writers as diverse as Hardt and Negri and André Gorz.[7] Typical accounts of this analysis concentrate on the exchange of symbols (intellectual property, electronic cash flow) but omit or diminish the continuing role of material production and distribution on which this new development rests. Without the infrastructure of processors, memory and outputs, and the network of cables, routers, cellular networks, and satellite communications, there would be no cognitive capitalism, since it would lack the means to create its products and services and the ways to get them to market. Engineering and design, closely allied with software and content, are the high-value industries of the twenty-first century, but their realization depends upon the existence of this infrastructure and its capacity for innovation in waves driven by the synergistic demands of both tiers: new designs demanding new forms of content; new content demanding new forms of software; and new software demanding new hardware designs, in a spiral that promises the level of growth that neoliberalism demands. In this sense the immaterial sector of the economy is as committed to growth, and as equally unsustainable, as the material.

Indeed, separating the two is only an analytical exercise: empirically they act entirely in consort. As other authors in this collection have argued, this material infrastructure and its perpetual innovations incur immense environmental costs in terms of materials, energy use, and waste. Like them, I want to argue that the environmental costs of (digital) media are also human costs, on the ecological principle that human societies are entirely integrated into their environments. In this chapter I want to concentrate on manufacture: the material production of goods and the labor required to produce them, along with the extra-economic consequences of manufacture. Products like semiconductors have been the objects of intense investment from which they emerge as intellectual capital in the form of patents. They require physical production in factories (semiconductor plants prefer the term "fabrication" to "manufacture": facilities for their production are known in the industry as "fabs"). As a consequence of the mechanisms of accumulation, financialization and extended reproduction, these fabs have migrated in two intertwined but distinguishable forms: outsourcing and offshoring. Outsourcing refers to the practice of subcontracting elements of manufacture to smaller companies, often outside the contracting company's country; offshoring to building fab plants and other facilities beyond the home country's borders where wages, health and safety costs, environmental controls, and the tax burden required to educate workers are far lower than those won by working-class movements in the contracting parties' country of origin. Implicit in both outsourcing and offshoring are the environmental costs of transporting semi-finished goods or subassemblies to centralized final assembly plants, along with biopolitical aspects of the policing of intellectual property when subassembly is entrusted to subcontractors.[8] Before these can be addressed, we must first engage with the nature of component fabrication and the dispersed structure of the manufacture of subassemblies.

The labor of producing semiconductors can be divided between high-value design (cognitive labor) and low-value manufacture and assembly (physical labor). The policing of patents operates on the same principle as that ascribed to al-Qaeda cells: each subcontractor operates in ignorance of the central planning within which his/her separate activities alone make sense. As a result, labor in subassembly plants and component manufacture is kept in as great a state of ignorance as is compatible with the efficient production of the units involved. This ignorance is not a native state but one that must be constantly produced, since any passage of the cognitive capital involved to the workers would arm them with the capacity to seize control of the means of production.

It is also important to note that many factory workers, even in sweatshops, prefer the wage labor of factory employment to the even more precarious and brutal conditions of a demeaned agricultural sector, which offers the only alternative for displaced populations such as those of Indonesia, India, and China.[9] Whenever we argue against the subcontractual regimes of outsourcing and offshoring, we need to remember that the alternatives to sweatshop labor need to be better rather than worse than the existing state of affairs, not only from our perspective but from that of the workers themselves. The challenge of sustainability requires us to face an ethical problem should we determine to promote the well-being of the environment over the well-being, real or imagined, of the sweatshop labor force. For ecological utilitarians, our acts are to be judged by their outcomes, and the best outcome is the one that increases well-being for the largest number, not limited to humans. The problem of this consequentialist ethos is that it is prepared to sacrifice the well-being of the minority to the well-being of the majority. For a materialist ethics, there can be no sacrifice of even one entity. In the deontological perspective of eco-philosophers like Paul Taylor, every living thing has its intrinsic worth, compelling us to recognize its claim to live and be happy, so chiming with the materialist ethos, while however concentrating on the individual.[10] Ecology, as the study of the connectedness of everything, and ecomedia, as the study of the intermediation of everything, cannot rest on individuality but must work on the level of community, communication and communion. Thus a political analysis of sustainable media must not restrict itself to the human beneficiaries, like the UNDP, or to the environment at the expense of the human, as in deep ecology, but faces instead the greater intellectual and political challenge of creating an ethos that embraces both the nonhuman and the human. Likewise it must deal with well-being not only as a biopolitical measure of successful rule, but as expressed in the aspirations, desires, and demands of human and nonhuman agents alike. Ultimately this is political to the extent that the ethical concerns what I should do, where the political concerns what we should do. From this it follows that the term "corporate ethics" is an oxymoron, a fact demonstrable in the recent history of improvements to manufacturing conditions.

By no means can all fabs or assembly plants be treated as sweatshops. Many companies have been forced by consumer boycotts and campaigns to ameliorate working conditions in the computer industry as they have in at least some cases in the garment trade. Similarly, even in head offices, there can be deep inequalities between classes of employees. Dell Computer, for example, agreed in a $9 million class action settlement in 2009 that it had failed to offer women employees equal access to training, equal pay, or promotions, and established a Global Diversity Council to monitor its policies thereafter, extending them down its international supply chain. Such companies are to be applauded for their eventual acceptance of community values, but not for the preceding decades of oppression, nor for the lives their previous policies stunted. In a similar vein, while many companies have attempted to clean up their atmospheric emissions, waste material dumping and water pollution policies in the last five to ten years, the legacy of their previous actions is not thereby cleansed. Some perfluorocompounds (PFCs, emitted as gases from chemical vapor deposition and plasma etching procedures in fabrication plants) persist in the atmosphere for thousands if not tens of thousands of years, and have up to twenty thousand times more impact per part than carbon dioxide on the greenhouse effect.[11] Other mineral and solid waste, much of it composed of known carcinogens and other compounds whose long-term effects are unknown, will persist in the vicinity of the plants for equally lengthy periods of time. For the many female employees who bear children, those effects last long after they might terminate their employment, and affect children with otherwise no connection to the plants, present or past. To the extent that today's media restrict the lives of future adults, they are unsustainable in the UNDP's terms; to the extent that they restrict the vitality of regional environments, they inhibit the emergence of human-ecological community. This unsustainability of the computer industry extends geographically to include the connectedness of by-products to aquifers, ocean currents, and atmospheric circulation connecting distant places with the source of pollution.[12] Sustainability points us towards the legacies of long-abandoned factories in close and remote places, near and distant futures.

In a 2011 overview of the industry's environmental and health hazards, Corky Chew notes that PFCs are less frequently used in semiconductor fabrication than previously, but that remaining dangerous chemicals include heavy metals, rare earths, solvents, epoxy, corrosives and caustics, fluorides, ammonia, and lead.[13] Process redesign focuses on treatment of solid, liquid, and gas wastes, which themselves use acids and caustics to neutralize pH levels in wastewater, and include incineration and landfill. Some of the energy required comes from flammable by-products, but even with this saving, the costs of these processes are in general less than those of recycling materials. Other documents, such as the International

Finance Corporation (IFC)/World Bank Guidelines, use a discourse peppered with expressions like "amelioration," "abatement," "improvement," "optimization," and "minimizing," in the context of a detailed set of recommendations for improving the environmental performance of the industry. The IFC Guidelines admit their applicability is greater in new facilities than in retro-engineered existing plants, and note that their application is always subject to "site specific targets and an appropriate timetable for achieving them," adding that site-specific variables include such factors as "host country context, assimilative capacity of the environment and other project factors."[14] While asserting that the industry should, in case of conflicting guidance, apply the more stringent of the options, these final notes on host countries, their relative weakness in enforcing environmental regulation, and their environmental capacity to somehow digest waste are distressing in their gesture towards a rule that can be bent. They exemplify what John Urry refers to as "a kind of regime-shopping [which also] preclude[s] the slowing down of the rate of growth of CO_2 emissions, which presupposes shared and open global agreements between responsible states, corporations and publics."[15] The IFC Guidelines clearly aspire to that kind of transparency, while at the same time indicating ways it can be ignored.

The sad truth is that the increasing imbrication of the Internet in the operation of daily life from trade to traffic signals, the explosion in mobile media use, and the prospects for an increasingly embedded Internet of things heralded by the move to Internet protocol version 6 with its vastly expanded address space, all suggest that we are stepped too far in to go back. The Semiconductor Industry Association reported "that worldwide semiconductor sales for 2013 reached $305.6 billion."[16] It seems impossible to convert that figure into an estimate of the numbers of chips produced, given the mix of mass and specialist products involved: a unique and secure device created for the military will be priced differently to the one in a cheap watch, an RFID (radio frequency identification) tag or a credit card. The numbers, however, are growing, even as the prices tend to drop in line with Moore's law, despite the years of downturn since the global financial crisis and the increasing costs of key minerals including indium, gold, and the lanthanides. A 2002 report suggested that two hundred billion discrete components (diodes, transistors, rectifiers, etc.) were produced annually, with around another billion units of optoelectronics (LEDs, laser diodes, CCD chips), memory, logic, microprocessing, and other devices.[17] Today we could expect that annual production is at least tenfold. Each chip is tiny, but the collective weight of the metals and chemicals required to make them is great.

This is especially true of the water needed to build chips. According to Global Water Intelligence, "creating an integrated circuit on a 300 mm wafer requires approximately 2,200 gallons of water in total, of which 1,500 gallons is ultrapure water."[18] Ultrapure water (UPW), which typically requires 1,400 gallons of ordinary water to produce 1,000 gallons of UPW, is so pure it is considered an industrial solvent. It not only provides washing and lubrication for the polishing processes required between steps in manufacture, but unlike normal water does not carry any dissolved minerals that might interfere with the nanometer scale electronics. However, this requires the safe removal of those minerals, while also demanding the removal of the by-products of the polishing processes. Some of the mineral effluent is valuable, and occurs in large enough quantities to be worth rescuing through flocculation, coagulation, centrifuge, and for nanoscale molecules, hollow-fiber membranes. US plants use a series of these processes, plus various chemical reagents to neutralize acids and caustics, but much of what is produced is defined in federal and state legislation, notably that of California, as toxic waste. In other jurisdictions, wastewater ponds are built to allow dangerous materials to sink, but these are vulnerable to flooding and seismic activity and are illegal in the United States and the European Union. Illegality, however, is no guard against illegal behavior, which becomes increasingly attractive as top-end consumers of semiconductors in computer, phone, and games markets pressure their suppliers to cut costs. In fact KPMG auditors report that "Losing [market] share to lower cost producers is perceived as posing the single greatest threat to their business models by global semiconductor manufacturers" (their second greatest fear is "Political/ regulatory uncertainty," a reference to environmental regulation among other factors including continuing fallout from the global financial crisis).[19]

On December 9, 2013, Taiwanese company Advanced Semiconductor Engineering (ASE) of Kaohsiung City, a municipality approaching 3 million people, was fined the maximum amount of NT$600,000 (just over US$20,000) for dumping wastewater containing acids and metals into the main river used for irrigation in the area. In June 2014, the Taiwanese Environmental Protection Agency upped to NT$20 million what the maximum fine would be for future infringements. In the same month, ASE announced that despite the partial closure of the plant, the company was planning to increase production in the third quarter of the year, and would be raising up to NT$15 billion to support the expansion.[20] Although water can account for up to 1.5 percent of operating costs, including its reuse and recycling, it is clear that given the scale of operations, fines are routinely written off as part of that cost: ASE was reported to have paid seven fines for ongoing pollution dumps between July 2011 and October 2013. The same report quotes activist assertions that the company had enjoyed tax exemptions of NT$3 billion.[21] Both the strategic importance of the industry—ASE is the world's largest supplier of semiconductors and testing services—and its association with the technocratic dogma in development policy tend to ensure that violations of the law are treated leniently, leading to the assumption that the IFC/ World Bank recommendations to conform to host country standards present an opportunity to cut costs in the interests of increasing sales to end-users, in this case manufacturers of consumer electronics. (ASE blamed a one-off employee error and promised an internal investigation).

The water issue is strategic since it involves a common good. In Taiwan as in other countries, companies pay for metered water use, but as in other countries there appears to be more relaxed metering of outflows from semiconductor fabs. This is particularly worrying in China. Fabs can use up to 30–50 megawatts of peak electrical capacity. In China, this power is most likely to come from hydroelectricity. Growth in the sector, which ran at 24 percent per annum for the decade following 2001, thus competes with itself for consumable water for power or UPW. Competition with agriculture and with other industries as well as human consumption is at its highest where the greatest densities of fabs are found: in the Yangtze River delta (Shanghai and Jiangsu) and in the environs of Beijing, regions that are accounted as "Dry" in the standard UNEP/UNDP measure, having less than a thousand cubic meters of water per person, while Zhejiang to the south of Shanghai is reckoned "At Risk." In total, over 80 percent of the country's fabs are based in Dry or At Risk regions. The industry is making steps towards less profligate use of the resource, including reduction, reuse, and recycling projects and migration from intensive use of UPW. According to research by the non-governmental organization (NGO) China Water Risk (2013) into the records of the Institute of Public and Environmental Affairs, there were "over 10,000 environmental violations for key semiconductor companies," the major effluents including arsenic, antimony, hydrogen peroxide, and hydrofluoric acid.[22] Of these, arsenic is a major carcinogen in humans and animals;[23] high levels of antimony are especially toxic to aquatic life;[24] hydrogen peroxide, despite being used extensively in wastewater treatment, is classified as a corrosive and in concentrated or aerosol form has a variety of ill effects on humans and animals;[25] and hydrofluoric acid is corrosive and toxic for both humans and animals. The more water is recycled, the more concentrated the remaining toxins become.

Chip fabrication employs a range of technologies besides the chemical. Chip "burning" is a test process subjecting semiconductors to high levels of heat and voltage; ion implantation is used in doping (the practice of introducing tiny quantities of rare earths into silicon crystals to define their electronic qualities); and X-rays are used to check quality. It is unclear whether these processes contributed to a spate of cancers among workers in Samsung fabs in South Korea in the 2010s.[26] Volatile organics like benzene, trichloroethylene, and methylene chloride are also common in "clean rooms" where chips are handled by human operators and may have contributed to the problem. Some three years after Grossman reported on this for *Yale Environment 360*, noting that the Semiconductor Industry Alliance protested that studies of links between fabs and cancer clusters were "scientifically flawed," Samsung apologized and promised compensation to a group of ex-employees who have suffered from cancer, without, however, accepting a link between chemicals or physical processes and their illnesses.[27] Liability may lie with the South Korean government, to whom companies pay a levy from which claims for industrial injury are paid.

These three stories from Taiwan, China, and Korea are, in some sense, typical of the kinds of tales that we discover in any environmental analysis of industries of all kinds. Even the division of high-risk, low-paid labor in manufacture from low-risk, high-paid labor in research and development parallels similar structures in the garment sector where design is highly paid and respected, unlike the work of sweatshop laborers, or indeed for example automotive, aerospace, and other transport manufacture.[28] These all comprise, in their various ways, aspects of communications; indeed, transport was typically included in the sociology of communications through the 1970s. The case of semiconductors and other electronic components is, by contrast, rather more specific, in that consumption of the end product, by both consumers and businesses, is also a source of high-value innovation, especially when a proportion of that innovation is undertaken by unpaid consumers who pay for their own equipment in order to provide content for corporations like Facebook. We should, however, consider two important distinctions: that between research and development (R&D) or design, on the one side, and content production, on the other; and that between innovation and invention. In the latter, the atmosphere of intense competition over price and new product lines belies the deep standardization of core tools like semiconductors, constrained to work with now entrenched protocols including the use of binary logic, Internet protocol, and shared standards like MPEG, encouraging innovation—new fashions in standard forms—while discouraging real invention. In the former, while both design and content are productive of revenues, only R&D and design are paid. The trend to innovation within standards (as opposed to invention beyond them) and the differential use of paid or unpaid creativity are, however, ultimately linked and equally engaged in answering the question of sustainability.

The developing salience of the produser[29] has become a core feature of consumption in the twenty-first century, offering not merely new ways to innovate[30] but the possibility of a wholesale remaking of the principles of political economy.[31] Produsage blurs the distinction between users and producers in value chains: production is always incomplete, as in the case of computers delivered without software installed, so that the end-user has to participate in the production process. The opportunities for a cashless commons of shared benefits based on principles familiar from open-source software, Wikipedia, and open participation science projects are immense. In the field of semiconductors, however, the entry costs are far higher than those for producing content or code. Contemporary integrated circuit (IC) design faces key challenges in accelerating the performance of processors, now approaching the scale where quantum effects hinder logic design, in improving performance-energy ratios, and ameliorating battery and display designs to reduce power and increase performance. These challenges mean that new chips require US$30–40 million to produce, figures that the peer-to-peer community cannot raise to date. Such sums are even challenging for venture capital seeking start-ups, while large corporations, paradoxically, are reducing their in-house R&D in favor of acquiring start-ups that have passed a threshold of risk (economically a safer bet) or of licensing the intellectual property they require, a sure route towards standardization.[32] Venture capital, since the global financial crisis, has been hard-put to find investors interested in risk, and has become, as a result, increasingly risk-averse itself. The end product of this has been a diminution of invention, a shift towards investing in applications that have a better risk-to-profit ratio and, even more perversely, a shift in the number of patents being secured away from the United States, which has been traditionally the home of innovation in IC design, towards East Asia, where major corporations are increasingly becoming "fabless," like their US counterparts.[33] A "fabless" corporation typically takes on the lucrative design work, then subcontracts the fabrication of its chips. Keshavarzi and Nicol cite Nicky Lu, CEO of Etron, to the effect that China is investing US$14.2 billion in fabless design companies.[34]

The concentration and mobility of intellectual capital has always been characteristic of capital, but this shift to East Asia is a prime indicator of the hypothesis advanced by Arrighi and others that China is in the process of leading a new era of capital centered in East Asia.[35] The newly diversifying concentrations of R&D and IC design on either side of the Pacific are built on the equally mobile but far more precarious labor in fabs that are increasingly migrated further offshore, especially into South East Asia, Indonesia, and the Philippines. The maquiladoras, sprawling subcontracting factories along the US-Mexican border that have become major economic zones since the introduction of the North American Free Trade Agreement (NAFTA), have been extensively documented for their poor workplace health and safety, exploitation of women, and environmental impacts. Summarizing much of the literature, Schatan and Castilleja argue that lax environmental regulation and enforcement, while giving a cost advantage, ultimately imprisons Mexico in the low-value end of the market, excluding it from the high-value, "clean" product of the fabs north of the border.[36] What they do not note is that this depression of the potential of Mexican fabs is typical of NAFTA's one-sidedness. High-value fabrication remains the preserve of the dominant economy in the partnership; while it suits the US-based corporations that low-value and dirty production, from which they also benefit (albeit at a far lower profit per unit), be kept discrete. In the same way, the employment of young and often uneducated rural women makes competition (or even theft) of intellectual property unlikely, while militating against workplace organization.

Ironically, the export of poverty-level employment and environmental recklessness as a result of NAFTA has had the foreseeable result that pollution is now crossing the border. According to the US Environmental Protection Agency (2013), deforestation has increased runoff in the watershed of the Tijuana River whose estuary lies in San Diego County, California. The runoff from storms carries with it fertilizers, pesticides, metals, and polychlorinated biphenyls (PCBs) from the maquiladoras, as well as sewage from the unplanned expansion of slum housing along the river. Two major sewage spills in April of 2012 totaling four million gallons emphasized the lack of adequate infrastructure for the massive population expansion in the factory zones and for the poverty experienced by their inhabitants. The local San Diego paper reports that one result has been concentrations of drug-resistant genes in bacteria in the estuarine wetlands, which deliver genetic material traceable to human waste flowing down the river onto a popular surf beach. Meanwhile in the twin cities of Ciudad Juarez and El Paso, on either side of the Chihuahua/Texas border, and Nogales on either side of the Sonora/Arizona border, air pollution travels without regard to boundaries, carrying ozone and particulate matter less than 10 micrometers (PM10, PM2.5)—dust so fine it penetrates deep into the lungs of air-breathing creatures.[37] On the one hand, this has allowed Nogales to claim exemption from federal air quality controls because the dust originates in Mexico, in the tradition of blaming the poor for pollution; while on the other hand promoting in both cities consumer-oriented campaigns to reduce domestic and automotive emissions while continuing to turn a blind eye to industrial pollution, especially that sourced from US-owned or contracted plants. By no means does all of this waterborne or airborne pollution derive from the electronics industry, but it certainly contributes, and its workers are constrained to drink, wash in, and breathe the results.

Even without tracing the sources of minerals and energy used in fabrication, the processes employed are clearly already deeply embroiled not only in human but in nonhuman atmospheric and aquatic cycles, local and regional, up to at least the scale of the Pacific. The responsibility for the ecological fallout from these processes has frequently fallen on citizens and consumers, whose boycotts of sweated and environmentally dangerous goods and campaigns against industrial practices have been significant. It is clear, however, that corporations resist taking responsibility, spending instead vast sums on legal actions blocking charges against them and on public relations campaigns (including the expensive scientists whose reports they commission). Governments from Mexico to Taiwan and South Korea recognize the importance of their electronics industry to GDP and therefore to inward investment, both markers of development that keep them free of the IMF and World Bank structural readjustment that has historically been a tool for exporting capital from countries afflicted by it. To the extent that taking responsibility is a human action, we must infer that the refusal to recognize and take responsibility is not. While, with Latour, we acknowledge the involvement of the nonhuman in the networks of labor, manufacture, and waste involved in semiconductor fabrication, we must also acknowledge the role of inhuman actors, primarily corporations, and beside them the political elites who deny involvement, and ease the operation of corporate irresponsibility.[38]

Workers in the North suffer from the export of jobs to offshore and outsourced subcontractors. They envy the industrial employment of their circum-Pacific neighbors. On the other hand, workers in offshore fabs, typically kept in the dark about the segregation of US society and the existence of a vast African American and Hispanic underclass, envy the levels of consumption available to their North American and European counterparts. The geography of this new division of labor is complex but can be expressed as the increasing spatial divorce of productive and consumptive work. Consumption becomes work when, under conditions of produsage, it is undertaken not for the fulfillment of needs or the realization of aspirations, but as a disciplined function required by capital to remove the excess product manufactured in the pursuit of expanded accumulation and growth. For capital to continue to grow, the working class of the wealthy nations now has as its chief function not production but the mass consumption of excess product, in cycles that range from overconsumption of junk food and pharmaceuticals, to exercise and diet products to counter the effects of the former.

In the pair, work discipline and consumer discipline, one constant is the passage of responsibility for accidental spillages, toxic waste, and carbon footprints to the productive (laboring) and consumptive classes. This amounts to a migration from political matters—how should we live?—to the ethical level of individual responsibility, a theme frequently on the lips of neoliberal politicians who are otherwise averse to corporate or governmental responsibility. The diminution of global problems like polluted aquifers and airborne toxins to the scale of ethical decisions by citizens, and of ethics to the level of consumer choice (for the constraints on which corporate citizens take no responsibility) is not designed to maximize efficiency in the use of resources. If it were, the problem of waste would not be integral to the financialization response to crises of overproduction, which has been to offer unpayable loans to the poor and open a trade in bad debt. Nor is it a democratic process, since democracy is definitionally concerned with the construction of an "us" capable of acting in consort. It is, rather, a projection of corporate irresponsibility and inhumanity onto the very people who suffer most from it, in the same logic of blaming the poor that drives austerity packages and attacks on public welfare.

A second constant of the division of production from consumption is the migration of aesthetic labor and enjoyment to the elites, and a parallel anaesthesis of the workers. In the productive realm, this is easy to see in the degradation of working and living environments and of the surrounding country; among consumers it is grounded in the depreciation of skills associated with living well, such as home cooking, homemade or crafted clothes and furniture, and vernacular architecture. In their place, "value-added" manufacture provides standardized products with customized additions (T-shirt emblems, a differently colored front door), while comedians vie with one another to deride amateur music or home-knitted garments. This is not to suggest that popular culture has not produced works of great depth and beauty, but that the industrial structure of their production and dissemination scrapes away their intimacy, devalues their capacity for permanence, and, through celebrity cultures and intellectual property regimes, diminishes the possibility of communities taking ownership of the cultural events on offer for themselves. For the producers and consumers of Top 40 radio shows, the object of consumption is not individual works but "music." In this sense, consumption moves to occupy itself no longer purely with use-values but with exchange values. Marx distinguished between living labor, the production of use-values, and objectified labor, its abstract form in which what is produced is not things but exchange value, "undifferentiated, *socially necessary general labour*, utterly indifferent to any particular content."[39] What he could not predict was that this indifference to particular content would become a characteristic of consumption under conditions of neoliberal disciplined consumerism.

Bifo notes of this abstract form of labor that "it means the distribution of value-producing time regardless of its quality, with no relation to the specific and concrete utility that the produced objects might have," adding that in info-production "labor has lost any residual materiality and concreteness."[40] The irony is that consumption of symbols too has lost its materiality, its specific and concrete utility, and is instead entirely devoted to the production of value, first in various forms of payment for the consumed objects and services, and second in generating more value through paying attention to the advertisements and marketing that are so embedded in the flow of media. Thus, the division of labor between those forced to work and those forced to consume, while unjust and divisive, is at least equitable in divorcing both productive and consumptive classes from meaning and pleasure, while at the same time using that division to minimize the possibilities of a common revolt against their condition of abstraction and anaesthesis.

This anaesthesis extends to the absence of truth in media, specifically truth about themselves and their foundation in toxic conditions of work and both local and global pollution. Metaphorically it might be feasible to speak of certain forms of media message as toxic (violent pornography, race hatred), but metaphorical violence is rarely as directly threatening as actual toxicity. In this instance, the metaphor hides the truth of toxic media, the toxicity of production processes integral to the integrated circuit. The same is true of the digital sweatshop, those call-centers and data-processing centers where the semiotic labor of shifting symbols or converting human conversations into data are undertaken in states of high abstraction. Digital labor, the work of translating into symbols and manipulating those symbols, already has a high degree of abstraction in terms of the relation between the worker and the content, in much the same way as the intense division of labor in the subassembly supply chain deprives workers of a relationship to any finally useful product. But that abstraction is driven to a higher level by the mock-Hegelian logic that Bifo identifies:

> Absolute Knowledge is materialized in the universe of intelligent machines. Totality is not History but the virtual assemblage of the interconnections preprogrammed and predetermined by the universe of intelligent machines. Hegelian logic has thus been made true by computers, since today nothing is true if it is not registered by the universe of intelligent machines . . . When History becomes the development of Absolute Computerized Knowledge difference is not vanquished: it becomes residual, ineffectual, unrecognizable.[41]

The universality of these machines, in which are captured and codified the wisdom and skills of previous generations, is overstated in Bifo's analysis, to the extent that this universality appears such only from within the universe of intelligent machines. The externalities retain their reality, among them in first place the environment, an economic externality in the sense that it does not enter into the accounts of corporations, and in second place, and with greater intensity with each passing year, the residual difference of populations. Human labor, in production or consumption, is "given," like the environment, in that capital does not pay for its reproduction or education, and is external in the sense that waste product, or indeed junk product, is regularly dumped into populations as into reservoirs, with no account taken of potentially lethal effects. In the cyborg logic of the corporation—a vast network of interlinked computers with human biochips inserted—the human difference is now an externality: a source of creativity to be exploited and a sump of waste to be discounted. That this course is suicidal does not register in the ethos of the cyborg for which profit alone is calculable, and all other effects are simply left out of account.

Nevertheless, Bifo does retain an iota of hope in the form of the residual and unrecognized difference that, I would argue, is not as ineffectual as he says here. The indifference of capital is twofold: the indifferentiation of the objects of labor and consumption under the regime of exchange value, and the indifference to externalities, both human and ecological. This indifference is premised on the universality of neoliberalism, embodied in the claim that there is no alternative, and enacted in the inertia of political classes faced with such tasks as mitigating climate change or enforcing environmental regulations. This universality is, however, premised on its externalities and incapable of functioning without them. To the extent that both workers and environment are now external to capital, they are thrust out of the universal claimed for neoliberalism. At the same time, the IFC/World Bank note on "assimilative capacity of the environment" places a demand on ecological systems that they assimilate the fallout of semiconductor production, a demand equally borne by fab workers, while at the same time requiring individual workers and consumers shoulder responsibility.[42] To the extent that produsers and workers undertake that responsibility, they typically either collapse into inaction before the scale of the task, or move towards that kind of bitter melancholy that can become the basis for political action. When the ideological weight placed on the family became unsustainable in the 1950s and '60s, there was a rush to divorce; when the weight of ideological individualism is crushing, the individual falls apart. Negatively, this appears as mental illness, a frequent accompaniment to sweated labor made frighteningly public in the Foxconn suicides.[43] More positively it can lead to the realization that the self is no longer the source of action, leading to participation in group formation and political activism and a turn towards a new politics of nature. Sustainable media will demand not only a sustainable community of workers but equally a sustainable commons embracing workers and consumers, and beyond that a community of workers, consumers, and their environments. It is in this sphere that the aesthetics of sustainable media may be capable of counteracting and subsuming the anaesthesis of the contemporary division of labor as well as the division between human and nonhuman environments. It is not only because both economics and politics have failed to create sustainable ways of life, or even to address them, that we need to turn to aesthetics. Traditionally aesthetics has looked towards the organic unity of the artwork. In reality our cultural artifacts are riven by contradiction: to create unity would be a lie. An aesthetic approach must consider both the sustainability of the media themselves as material practices and their role in mediating between phyla and among humans, a movement through communication as a means towards communication as goal.

Marx observed that in the commodity "the relation of the producers to the sum total of their own labor is presented to them as a social relation, existing not between themselves, but between the products of their labour."[44] Eco-critique adds: in the commodity form, the relationship between producers, consumers, and the externalized environment appears in disguise, hiding the true involvement of all three under the sign of exchange value. The commons on which sustainable media might be built involves a migration from the bogus universality of capital to the active integration of the indifferentiated human and nonhuman, declaring their mutual incompletion and need for support, and producing a politics in which the question "how are we to live?" might, at the very least, be posed, and without which it cannot be answered.

Notes

1 *Global Wealth Report 2013* (Zurich: Credit Suisse), https://publications.credit-suisse.com/tasks/render/file/?fileID=BCDB1364-A105-0560-1332EC9100FF5C83.
2 *Human Development Report 2011: Sustainability and Equity: A Better Future for All* (New York, NY: United Nations Development Programme), 17.
3 Thomas Piketty, *Capital in the Twenty First Century*, trans. Arthur Goldhammer (Cambridge, MA: Belknap Press, 2014).
4 Rosa Luxemburg, *The Accumulation of Capital*, trans. Agnes Schwarzschild (London: Routledge, 1953 [orig. 1913]).
5 See David Harvey, *The New Imperialism* (Oxford: Oxford University Press, 2003).
6 See Elena Esposito, *The Future of Futures: The Time of Money in Financing and Society* (Cheltenham: Edward Elgar, 2011); Maurizio Lazzarato, *The Making of the Indebted Man: An Essay on the Neo-Liberal Condition*, trans. Joshua David Jordan (New York, NY: Semiotext(e), 2012).
7 Michael Hardt and Antonio Negri, *Commonwealth* (Cambridge, MA: Harvard University Press, 2009); André Gorz, *The Immaterial*, trans. Chris Turner (Chicago, IL: University of Chicago Press, 2010).
8 Ernst and Young, *Global Semiconductor Industry Study*. Ernst and Young Global Technology Center, http://www.ey.com/Publication/vwLUAssets/Global_semiconductor_industry_study/$File/Global_semiconductor_industry_study_report.pdf (34).
9 See Harvey, *The New Imperialism*, 2003; Zakir Husain and Dutta Mousoumi, *Women in Kolkata's IT Sector: Satisficing between Work and Household* (New Delhi: Springer, 2014).
10 See Paul Taylor, *Respect for Nature* (Princeton, NJ: Princeton University Press, 1986).
11 "Climate Partnership for the Semiconductor Industry: Basic Information, 2008," United States Environmental Protection Agency, http://www.epa.gov/semiconductor-pfc/basic.html.

12 See Jintai Lin, Da Pan, Steven J. Davis, Qiang Zhang, Kebin He, Can Wang, David G. Streets et al., "China's International Trade and Air Pollution in the United States," *Proceedings of the National Academy of Sciences of the United States of America (PNAS)*, 111, no. 5 (2013): 1736–1741, http://www.pnas.org/content/early/2014/01/16/1312860111. For an account of the interstate trade in pollutants from semiconductor fabs in the United States, see Matt Drange and Susanne Rust with the Center for Investigative Reporting, "Toxic Trail: The Weak Points in the Superfund Waste System," *Guardian*, March 17, 2014, http://www.theguardian.com/environment/2014/mar/17/toxic-trail-weak-points-superfund-waste.

13 Corky Chew, "Environmental and Public Health Issues," 83. Microelectronics and Semiconductors, Williams, Michael E., Editor, *Encyclopedia of Occupational Health and Safety*, Jeanne Mager Stellman, Editor-in-Chief (Geneva, Switzerland: International Labor Organization), http://www.ilo.org/iloenc/part-xiii/microelectronics-and-semiconductors/item/924-environmental-and-public-health-issues.

14 IFC/World Bank, *Environmental, Health, and Safety Guidelines for Semiconductors & Other Electronics Manufacturing* (Washington, DC: International Finance Corporation/World Bank, 2007), http://www.ifc.org/wps/wcm/connect/bc321500488558d4817cd36a6515bb18/Final+-+Semiconductors+and+Other+Electronic+Mnfg.pdf?MOD=AJPERES.

15 John Urry, *Offshoring* (Cambridge: Polity, 2014), 10.

16 Dan Rosso, "Semiconductor Industry Posts Record Sales in 2013," *Semiconductor Industry Association*, February 3, 2014, http://www.semiconductors.org/news/2014/02/03/global_sales_report_2013/semiconductor_industry_posts_record_sales_in_2013/.

17 Jim Turley, *The Essential Guide to Semiconductors* (New York, NY: Prentice Hall, 2002).

18 Gord Cope, "Pure Water, Semiconductors and the Recession," *Global Water Intelligence*, 10, no. 10 (October 2009), http://www.globalwaterintel.com/archive/10/10/market-insight/pure-water-semiconductors-and-the-recession.html.

19 KPMG, *2013 Technology Industry Outlook Survey*, KPMG Technology Innovation Center, https://techinnovation.kpmg.chaordix.com/survey/2013.

20 Jalen Chung Frances Huang, "ASE Expects Production at Full Capacity in Q3," *Focus Taiwan*, June 26, 2014, http://focustaiwan.tw/news/aeco/201406260023.aspx.

21 Yu-Tzu Chiu, "ASE Faces Possible Halt Due to Water Pollution," *ZDnet*, December 11, 2013, http://www.zdnet.com/ase-faces-possible-halt-due-to-water-pollution-7000024138/.

22 China Water Risk, *8 Things You Should Know about Water & Semiconductors*, July 11, 2013, http://chinawaterrisk.org/resources/analysis-reviews/8-things-you-should-know-about-water-and-semiconductors/.

23 H.W. Chen, "Gallium, Indium, and Arsenic Pollution of Groundwater from a Semiconductor Manufacturing Area of Taiwan," *Bulletin of Environmental Contamination and Toxicology*, 77, no. 2 (August 2006): 289–296.

24 Mengchang He, Xiangqin Wang, Fengchang Wu, and Zhiyou Fu, "Antimony Pollution in China," *Science of the Total Environment*, 421–422 (April 2012): 41–50.

25 Agency for Toxic Substances and Disease Registry, *Medical Management Guidelines for Hydrogen Peroxide* (Atlanta, GA: Agency for Toxic Substances and Disease Registry, 2014), http://www.atsdr.cdc.gov/MMG/MMG.asp?id=304&tid=55.

26 Elizabeth Grossman, "Toxics in the 'Clean Rooms': Are Samsung Workers at Risk?," *Yale Environment 360* (June 9, 2011), http://e360.yale.edu/content/print.msp?id=2414.

27 Associated Press, "Samsung Promises to Compensate Factory Workers Who Suffered Cancer," *Guardian*, May 14, 2014, http://www.theguardian.com/technology/2014/may/14/samsung-compensate-factory-workers-cancer.

28 See, for example, China Labor Watch, *Tragedies of Globalization: The Truth Behind Electronics Sweatshops* (New York, NY: China Labor Watch, 2011), http://www.chinalaborwatch.org/news/new-350.html.

29 Axel Bruns, *Blogs, Wikipedia, Second Life, and Beyond: From Production to Produsage* (New York, NY: Peter Lang, 2008).
30 See Yochai Benkler's "Commons-Based Peer Production" in *The Wealth of Networks: How Social Production Transforms Markets and Freedom* (New Haven, CT: Yale University Press, 2006), 60.
31 See Michel Bauwens, "The Political Economy of Peer Production," *C-Theory* (2005), http://www.ctheory.net/articles.aspx?id=499.
32 Ernst and Young, *Global Semiconductor Industry Study*, 6.
33 Jeffrey T. Macher, David C. Mowrey, and Alberto Di Minin, "The 'Non-Globalization' of Innovation in the Semiconductor Industry," *California Management Review*, 50, no. 1 (Fall 2005): 217–242.
34 Ali Keshavarzi and Chris Nicol, "Perspectives on the Future of the Semiconductor Industry and the Future of Disruptive Innovation," *IEEE Solid-State Circuits Magazine* (Spring 2014): 77–81, doi:10.1109/MSSC.2014.2317431.
35 See Giovanni Arrighi and Beverly J. Silver (with Iftikhar Ahmad, Kenneth Barr, Shuji Hisaeda, Po-keung Hui, Krishmendu Ray, Thomas Ehrlich Reifer, Miin-wen Shih and Eric Slater), *Chaos and Governance in the Modern World System* (Minneapolis: University of Minnesota Press, 1999).
36 See Claudia Schatan and Liliana Castilleja, *The Maquiladora Electronics Industry and the Environment along Mexico's Northern Border* (Montreal: Commission for Environmental Cooperation, 2005).
37 See EPA, "Border Air Quality Data—Ciudad Juarez/El Paso Area," 2007, http://www.epa.gov/ttncatc1/cica/sites_cj_e.html; see also Andrea Kelly, "Border-Crossing Dust Earns Nogales Air Quality Exemption," *Arizona Public Media*, September 6, 2012, https://www.azpm.org/p/top-news/2012/9/6/15338-nogales-air-quality-plan-takes-into-accounty-dust-from-mexico/
38 Bruno Latour, *Reassembling the Social: An Introduction to Actor-Network Theory* (Oxford: Oxford University Press, 2005).
39 Karl Marx, *Capital: A Critique of Political Economy*, vol. 1, trans. Rodney Livingstone (London: NLB/Penguin, 1976), 993.
40 Franco "Bifo" Berardi, *The Soul at Work: From Alienation to Autonomy*, trans. Francesca Cadel and Giueppina Mecchia (New York, NY: Semiotext(e), 2009), 75.
41 Ibid., 73.
42 IFC/World Bank, *Environmental, Health, and Safety Guidelines for Semiconductors & Other Electronics Manufacturing*, 2007.
43 Angela Moscaritolo, "3 More Foxconn Employees Commit Suicide," *PC News*, May 20, 2013, http://uk.pcmag.com/news/15360/reports-3-more-foxconn-employees-commit-suicide.
44 Marx, *Capital*, 164.

10

RE-THINGIFYING THE INTERNET OF THINGS

Jennifer Gabrys

An increasing number of gadgets are now electronic, or fitted with microchips, and typically networked. The many claims made for the Internet of Things announce that computation is set to become ever more ubiquitous, where "things" equipped with radio frequency identification (RFID) tags and sensors will collect and circulate data for actuating responses through interoperable networks and so become "smart." The Internet of Things is at once heralded as a revolution in organizing our physical worlds as well as an unparalleled source of economic development. With the promised explosion of networked objects, electronics industries have turned to things, after so long fixing their attention on screens, software, and "cyberspace." No more does the distinction of virtual and physical retain its neat bifurcation, since what would have counted as virtual is coursing through and remaking the contours and functionalities of the physical. Things, within the Internet of Things, are the curious creatures to which I turn my attention in this chapter. What are these things in the Internet of Things and what are the characteristics of their emerging materialities? How, as newly electronicized objects, do they manifest distinct material and environmental effects? And how might an attention to these material and environmental effects provide an opportunity for generating new areas of environmental intervention in relation to sustainable media?

Media theory is now increasingly calling for an attention to the materiality of the digital. Where previous studies may have focused on the meaning or signification of media, or transfixed on screens as sites of cultural representation, materiality-based media studies are increasingly in development as key contributions to the field. Friedrich Kittler and Katherine Hayles have taken materiality as a topic of interest, noting that materialities are structuring conditions that inform the very possibilities of communication.[1] However, these earlier studies typically have attended less to materiality as a social, environmental, or political concern, and more as a logistical, structuring, or informing condition. Alongside theorists like Lisa Parks and Matthew Fuller, I have argued for a consideration of what lies beyond the screen, of how hardware unfolds into wider ecologies of media devices, and of how electronic waste may evidence the complex ways in which media are material and environmental, despite our tendency to overlook these interconnected infrastructures, supports, and resources.[2]

Materiality as a topic and research focus now pervades media studies, as much as an obligation and directive not to forget all that media rely upon. At the same time, but along different lines, a more material turn could be found within the industries of digital media as well. "Thingification" is an overtly material approach to the previously "virtual" concerns of digital media, and is an industry strategy that is meant to expand the reach, capacities, and economic growth of the Internet. Where at one time industry claims were made for the resource-free living that might be achieved through the growth of economies spurred through virtual technologies, this deliberate thingification instead makes the case for the ways in which computational logics may make any number of activities and practices within our everyday lives more efficient, sustainable, and safe. The material relations that are the proposed site of intervention by digital media industries are now less about the erasure or elision of material resources, and more about making materialities and environments smarter and more effective, while stretching resources further in the face of increasing scarcity and planetary pressures.[3]

This chapter then asks: What are these things within the Internet of Things and how do they influence, challenge, disrupt, or reroute discussions of materiality within media studies? What consequences do these things have for thinking about the environmental effects and relations generated through the Internet of Things? After first discussing the things and thingification that the Internet of Things generates, I consider the ways in which the Internet of Things is oriented toward enhancing everyday lives, by focusing specifically on the environmental improvements meant to be achieved through these devices. On the one hand, ubiquitous computing has become central to performing new environmental practices such as monitoring environments for pollution, as is the case with citizen sensing.[4] Within these emerging practices, sensor technologies are also entangled with proposals for new efficiencies to be gained, as well as new opportunities to achieve sustainability through ongoing monitoring of resource use. Yet on the other hand and as will be my focus here, the projected rise in computational objects and applications is sure to generate new modalities and distributions of electronic waste.

How do these specific applications and imaginaries of the Internet of Things inform the materialities—and things—that are generated? And what implications do these materialities and things have for media theory and practice? In order to take up these questions, I then discuss how the different approaches to materiality now circulating within media theory and beyond might tune our attention to the thingness of the Internet of Things. I end this chapter by attending to the "The Crystal World" creative-practice project, which materializes a markedly different encounter with digital things. I finally ask how we might further analyze and address the emerging materialities and environmental practices to which ubiquitous computing developments and imaginaries are committing us.

Thingification = Digital Proliferation

In *Digital Rubbish: A Natural History of Electronics*, a study on the materiality of digital media focused on electronic waste, I accounted for the many gadgets that have become electronic by providing lists of objects that required special disposal and waste handling upon end-of-life as electronic waste.[5] The Waste from Electrical and Electronic Equipment (WEEE) directive of the European Commission documents a bewildering array of items—from laptops to toasters—for treatment as a special category of this hazardous waste.[6] These lists of electronic and electrical waste clearly demonstrate that computation has rapidly spread to numerous gadgets in order to shift and change their functionalities. In this light, it could be argued that toasters and desktop computers share the same space of technomateriality.

Yet, in an early assessment of *Digital Rubbish*, one reviewer commented that such technologies, even if fitted with electronic capacities, should not be assessed or discussed as computational devices. Why? Because, unlike a computer, a toaster quite obviously does not have storage capacity. However, I would argue here, as I have also done in *Digital Rubbish*, that digital functionalities are not exclusively located within an object-based architecture of computation, and that computational modalities and distributions may even shift through electronic appliances as banal as toasters and energy monitors. In fact, the WEEE list seems to act as an invitation for further electronicization, containing as it does both electrical and electronic gadgets, thereby suggesting that an iron could as easily be electronic as electrical, that a toaster would surely benefit from having a built-in computer and app-ability, and that a refrigerator, too, would have plenty to talk about if it could be wired up with smart communication capabilities.

The proliferation of computational things within the Internet of Things reads as an itemized list of electronic waste in the making. Focusing only on the home within the expanded Internet of Things "ecosystem" (since urban, manufacturing, and logistics applications are also considerable sites for Internet of Things development), we find that the mute and inanimate objects that surround us are steadily learning how to talk through electronic means. From smart toothbrushes to wired dog collars, interconnected coffee machines to alarm clocks, smart energy meters to thermostats, wired-up crockpots to toasters, app-able garage doors to door locks, smart bathroom scales to toilets, networked smoke alarms to security cameras, smart pill boxes to heart rate monitors, data-generating recycling bins to houseplants, networked light bulbs to weather sensors, smart picture frames to glasses, Wi-Fi shopping wands to wearable fertility thermometers, smart bicycle helmets to smart guns, baby sensors to food scanners, smart air monitors to intelligent faucets, talking shoes to plumbing sensors, Bluetooth gloves to tagged key finders, smart luggage to networked egg trays, smart utensils to connected lamps, as well as wireless sleep sensors, smart fire extinguishers, smart irrigation controllers, wearable cameras, and smart bike tires, the things within the Internet of Things consist of a growing list of intelligent devices that would augment, optimize, and interconnect every aspect of our daily lives.

To what extent might this expanding array of digital things generate different modalities, materialities, and environments of computation? While the essential characteristics and operations of computation are often referred back to John von Neumann's computational architecture that encompasses five aspects of input, logic, memory, control, and output (and so the problem of toaster-storage emerges in this context), arguably this proliferation of things is giving rise to different computational diagrams. In this context, sites of storage may shift to USB data loggers and the cloud. Toasters, refrigerators, and energy monitors may have fewer requirements for localized storage and processing, and instead may acquire greater functionality through links to grocery stores, food expiration registries, power plants, and smart grids. While inputs and outputs might still be present in some form, the operations and actors of inputting and outputting may also shift to distinctively nonhuman registers. Computation, then, may occur not only in toasters, but also across multiple appliances, networks, and sites, such that the distribution and materiality is configured along much different lines than a discrete PC-type object.

The point here is not merely to attend to the increasing computerization and sensorization of gadgets, but also to focus on the ways in which the Internet of Things and the proliferation of sensors shift the registers, materialities, and environments in and through which we access and experience computation. The Internet of Things works toward the networking and interoperability of objects and infrastructures and, in the process, generates distinct environments and materialities of computation. These are not simply environments as spatial zones or assemblages; even more so, they are environments as emerging conditions by which the Internet of Things is able to control and inform the capacities of and relations to these things, as a distinct technological development.[7]

Thingification by the Billions

As the above list demonstrates, the vast array of household objects that are currently being transformed into electronic technologies is not only lengthening, but also beginning to constitute a categorically different media "ecosystem." These objects become potentially operative things within the Internet of Things because of RFID tags, sensors, and devices such as smartphones, so that objects may be interacted with, controlled, and even automated to sense and gather data and to carry out programmed and learned functions. The Internet of Things is just as often referred to as the Internet of *Everything*, since networked and programmed capabilities are meant to inform products, bodies, environments, and systems, where the world is connected through sensors, networks, and a steady flow of data.

The number of devices connected to the Internet is currently estimated to be approximately one and a half to two billion. By 2020, however, this number is forecast to grow to up to fifty billion devices, with many more set to follow.[8] Billions of things are to be networked, interoperating, and forming new interactions across machines, environments, and people. This promised explosion of interconnected things indicates a shift emerging in the ways in which the Internet operates and things talk and interactions occur, with humans comprising a diminishing portion of Internet traffic. Indeed, while the Internet of Things as a concept is often dated to Mark Weiser's work on ubiquitous computing at Xerox Parc in the 1980s and 1990s,[9] and as an actual term is dated to 1999,[10] another pivotal moment in the concept's elaboration is 2008, the year when Internet-based machine-to-machine connectivity surpassed that of human-to-human connectivity.[11]

Inevitably, the explosion of things within the Internet of Things is also promised to bring a considerable opportunity for economic development, with the market estimated to be worth between two to fourteen trillion dollars.[12] Much of this economic potential is meant to be realized through new efficiencies in services, operations, manufacturing, and more. However, economic expansion could also be achieved through the remaking and proliferation of new types of things connected up and made available for new types of interactions—typically at urban and industrial scales, but also through consumer applications. The Consumer Electronics Show in 2014 presented just a small range of some of the new things available for consumption, including an Internet-of-Things-enabled crockpot, controlled by a smartphone, which "puts you in control of your food, which is really exciting."[13] The idea that sensors and actuators provide a locus of "control" is a logic that pervades most Internet of Things applications; sensors, networks, and the data they generate are configurations that enable automated exchanges—across people and machines—and may even provide opportunities for control and "new insights," through the ongoing generation and analysis of thing-based data.

In a 2012 presentation on the Internet of Things to the USENIX Association, Google's "chief Internet evangelist," Vint Cerf, discussed the importance of, and growth in, sensor networks and smart things. He noted that it has been "amusing to see the kinds of things that have been connected to the Internet," where gadgets such as an Internet-enabled picture frame at first "sound about as useful as an electric fork." But Cerf admits that these things can actually be "quite handy," as they become controllable through remote or web-based applications so that images may be delivered to picture frames from friends and family, for instance.

Recalling the proverbial toaster reference within computation, Cerf remembered the days when technologists speculated about what it would be like to communicate with your toaster to say "how burned you wanted your toast to be," and that now sensor networks are providing "the ability to remotely manage and observe" any number of thing-based interactions (including burned toast). This alone is important in relation to monitoring and security, Cerf opines, but he draws a further connection to how such a "feedback loop is going to be important from an environmental point of view, because I would say that we don't always understand the consequences of our actions." He concludes, "this kind of feedback loop may actually help us do a better job of managing our response to environmental problems including global warming."[14]

As with many Internet-of-Things applications, the implicit assumption is that sensor-generated data is needed in order to answer questions and solve problems, including how to be more efficient and how to change behavior through real-time feedback. Smart and sensorized things become sites for realizing new environmental engagements and for encouraging sustainable behavior. "Sustainable media" in this sense involves implementing new electronic infrastructures for controlling environmental systems and problems, and for making the most efficient responses an automatic feature within these networked infrastructures. Ubiquitous computing is thus not only environmental in its *spatiality*, but also environmental in the way in which it would make these systems *sustainable*. The sustainability of media here focuses on the proliferation of computing, however, where environments are remade as computational infrastructures and processes.

Thingification as Enabling and Ennobling Technology

A frequent presentation if not promotion of Internet-of-Things gadgets consists of celebrating the masses of data that will be collected, the new insights that will be gained, and the improvements, often by way of sensor-actuated exchanges, which may be realized through influencing behavior, resource use, and patterns of efficiency. In some cases, this is as basic as "streamlining" experiences so that an alarm clock coordinates timings and talks to a coffee pot in order to ensure coffee is made immediately upon waking. Heating can be set to adjust according to whether one is at home or on the way. Energy meters are one pervasive example of how recurring access to data about energy consumption is meant to influence behavior and bring about a reduction in energy use. Any number of daily practices and relationships become sites where automation and sensor-actuator triggers might "optimize" engagements while influencing behavior. Elsewhere, I refer to this phenomenon as "electronic environmentalism," in order to attend to the ways in which digital technologies have become central to how we identify and act on environmental problems, and offer potential solutions.[15] The Internet of Things is presented as an enabling and ennobling set of technologies that allow for the seamless identification of opportunities to be more parsimonious with resources, for instance, through the sensor-actuator exchanges that intelligent things provide.[16]

Such a logic has pervaded Internet-of-Things prototype and proof-of-concept projects for some time now. A project developed in 2009 through the Senseable City Lab at MIT, "Trash Track," uses this approach by tracking items of trash with electronic tags to provide trash location based on proximity to cellular phone towers.[17] The far-flung journeys of trash are then mapped in order to understand just how far garbage travels across the United States. Drawing attention to the expanded circuits of how waste travels, the project description notes, "TrashTrack focuses on how pervasive technologies can expose the challenges of waste management and sustainability. Can these same pervasive technologies make 100% recycling a reality?"[18]

Trash Track presents an interesting, if potentially contradictory, example of the drive to use electronics to monitor and act upon environmental problems. Computational exchanges are here the basis for observing, documenting, and so apparently overcoming environmental problems. The project description suggests that by using electronics to focus on the "'removal chain'" of waste, there might be realized "a bottom-up approach to managing resources and promoting behavioral change through pervasive technologies."[19] In many ways, Trash Track plays out the complex contradictions of how electronics become enabling and ennobling devices, while also demonstrating the material and energetic inputs that these technologies require. In other words, what the "Trash Track" project reveals is that the process of mapping trash in order to identify these journeys may be one way of revealing the environmental problem of waste, and yet the documentation of these journeys requires another intensive layer of electronics, communication infrastructures, and computational interfaces to bear environmental witness to the movements of waste. Moreover, the link to how this mapping will enable greater levels of recyclability remains rather unclear, and that which counts as "sustainability" remains within a computational problem space. In an attempt to use electronics as sustainable media tools, a remaking and rematerializing of things occurs that carries additional and specific environmental effects and consequences, often without immediately or obviously addressing the environmental problems that would be solved.

What do things become, in this case, when they are electronically animated to perform resource- and time-saving functions while also enabling environmental practices? In what ways do practices of electronic environmentalism, as articulated through the Internet of Things, entangle us within material-political arrangements and practices that require electronics in order to be activated? Thingification presents a considerable (if as-of-yet unaddressed) dilemma in accounting for how sustainability might be articulated through the materialities and exchanges facilitated by the Internet of Things, while also generating new questions about *what sorts of things are these*?

"Re-thingifying" Media Theory

As mentioned in my introduction, media theory, as well as a host of other disciplines, has adopted an interest in all things material—or, in other words, has taken a "material turn." At the same time, this attention to materiality often coincides with an interest in "things," in their ability to influence material-political engagements, and even to have a force of their own. From Bruno Latour and his *dingpolitik* to Jane Bennett and her "vital materialism," as well as the wider developments in new materialism that have emerged as both correctives and supplements to historical materialism, there are now multiple inroads for thinking about and attending to materiality.[20]

This is by no means to conflate these multiple and even diverging approaches to materiality and things, but instead to flag the renewed and ongoing interest in this area. Indeed, given this proliferation of approaches, there arises the very pressing question of how these multiple thing-theories and material philosophies influence our engagements with things when they are as flickering and in process as the Internet of Things. Thingification, to thingify, is a term of possibility and development within Internet of Things sectors: money can be made by connecting things to the Internet. However, within some theoretical arenas to "thingify" has a distinctly different and even pejorative tone, where it suggests an approach that objectifies and reifies (in an historical materialist vein); or an approach that favors things over relations (in a feminist materialism critique). Karen Barad has remarked that "thingification" involves "the turning of relations into 'things,' 'entities,' 'relata,'" and that this "infects much of the way we understand the world and our relationship to it."[21] Relations necessarily give rise to things, in Barad's relational ontology, such that to predetermine things is to compromise an attention to, and investment in, how relations and things emerge together.

From a different perspective, speculative realism has reacted to the privileging of relations as forcing an always-human engagement with things, where things become determined by (typically human) relations. Writings in this area have made the case that things should be allowed to stand alone, even untouched by relations, exuding a thingful integrity.[22] Within these speculative realist registers, one frequently encounters shimmering poetic lists of things that are seemingly discrete, autonomous, and complete. From fireflies and stones to lightning storms and violets, the lists that speculative realists might assemble provide rather different evocations of things than the smart toothbrushes, vibrators, crockpots, bidets, and baby monitors that litter the gadget world of the Internet of Things.[23] Here, however, is a no-less quivering but perhaps slightly more sordid world of things that not only are able to talk to and for "us" as consumer-users, but also are able to undertake their own autonomous operations without human interference. In other words, human interaction within the Internet of Things is not a prerequisite for relationality; but relationality does unfold among things, nevertheless.

One could also think of efforts to "follow the thing," where the "social life" of things might be drawn out through the ways in which things circulate and generate distinct social interactions.[24] Or one could take account of counterproposals that suggest that following the thing is a difficult undertaking when things are in process, falling apart, and generating a complex set of unintended material effects, as things do in the case of electronic waste.[25] Whichever way you encounter them, electronicized things are not without material-political implication. Rather than attempting to settle a case for or against things, I contend that the way in which particular things are mobilized and animated within the Internet of Things has consequences for the sorts of materiality that are addressed, the processes of materialization that are attended to, and the material relations that are animated or obscured. In this sense, I make the case that the Internet of Things requires us to attend not to one version of materiality, but to many; and to consider how things are never fully formed and fixed objects, but always on the go, generating effects that are never without consequence.

One way of opening up materiality to a proliferation and processual set of encounters is then to ask what "counts" as matter, and often this also means asking what "counts" as the empirical. In what ways are certain registers of thingness apparently given or self-evident? How does the apparent brute facticity of certain things direct our attention to consider materiality in certain ways and not others? If products become synonymous with things in the Internet of Things, for instance, how does this influence the very ways in which the givenness of things also assumes an environment where product-things assemble as discrete entities that are always already addressed in their formation as bundled technological and branded objects?

In *Digital Rubbish*, I make the case for addressing electronic waste through a "more-than-empirical" register, and I would suggest a similar proposal for encountering the Internet of Things.[26] As a mode of materiality, electronic waste is always more than empirical *evidence* with which to itemize the resources that support and the discards that result from digital technologies. Instead, electronic waste forces us to encounter empiricity and the self-evidence of materiality differently—not just as processes of materialization that might become realized in distinct and disparate places and objects, but also as material politics with specific effects and affective relations.[27]

"More-than," as a strategy, is a way of accounting for things that might also be characterized as radical empiricism, since things can be understood as plural, processual, relational, incomplete, and even as provocations that open into practice.[28] Things are always more-than-things as immediately encountered. Even the ways in which apparently self-evident things take hold are in themselves tales of material politics, technological arrangements, and environments where thingness remains relatively unquestioned. In other words, a discussion about things should not throw us back into substantialist debates about mind and matter (or derivatives thereof), but rather open up attention to how things come to be, what sustains things, and the effects that things have in the world. This is not an idle philosophical project, but one that has consequences for how relations and things emerge, are mobilized, and transformed. Within the scope of media theory and practice that are increasingly tuned toward material engagements, such an approach also suggests that re-thingification does not simply involve mapping out the static stuff that constitutes any particular media technology, but rather requires attending to the ways in which things attract, infect, and propagate mediatized relations, practices, imaginaries, and environments.

"Wondering about Materialism"

In this more-than-empirical approach to thingly conjugations, new media theories and practices might emerge that involve not ideas applied to matter, not a discursive animation of brute facts, not imaginaries and beliefs as epiphenomena to a more solid matter, and not a bifurcation of nature that might even be put back together through various hybridities. And yet, it often seems that that which "counts" as empirical research continues to plow the same furrow of self-evident facts or matter that forms objects of study. Instead, and as Isabelle Stengers suggests, we have a need for other kinds of narratives and imaginations, to "make present, vivid and mattering, the imbroglio, perplexity and messiness of a worldly world, a world where we, our ideas and power relations, are not alone, were never alone, will never be alone."[29]

Self-evident approaches to materiality can often be strategies of elimination and reduction, Stengers suggests, and it is by encountering materiality in its messiness that we might have cause to "wonder," or in other words, to think about the capacities of things and how they come into formation, how they affect other things, and how they may not simply be doorstops for reason. As Stengers writes,

> if there must be a materialist understanding of how, with matter, we get sensitivity, life, memory, consciousness, passions and thought, such an understanding demands an interpretive adventure that must be defended against the authority of whoever claims to stop it in the name of reason.[30]

Stengers suggests that materialism should not be eliminative by merely focusing on the self-evident, but rather that it should connect to "struggle."[31] By considering the apparently self-evident thingness of the Internet of Things, we might then reconsider what sort of thingness is this, what materialities and practices these things commit us to, and what struggles might emerge or be elided in these contexts.

Re-thingifying the Internet of Things

While the Internet of Things promises to help us achieve greater efficiency and sustainability in many areas of our everyday lives, it then also gives cause to wonder what material entanglements these things generate. While the Internet of Things is meant to continually monitor any number of environmental variables to bring them into a space of data-based management and optimization, these things give cause to wonder about the sorts of environmental awareness and practices that would purportedly be enabled. If we were to move beyond an unproblematic acceptance of the things in the Internet of Things, and begin to ask after how they become things in the first place; or if we were to consider how a forensic tracing of everyday life may document some areas for environmental intervention while eliding or overlooking others, then we might be prompted to consider how to re-thingify the Internet of Things: not as an unproblematic proliferation of enabling and ennobling gadgets, but rather as an emerging set of material problems with which we will inevitably have to struggle.

With the Internet of Things, we are involved in the ongoing remaking of materialities that will sediment into new futures. The thingification that occurs through these "systems of systems" articulate distinct material-political processes and relations that could be attended to in any number of ways, from the effects and practices these newly digital things generate, to the resources they require, as well as the deformation and environmental effects they generate at end of life. Here I turn to discuss a creative-practice project, "The Crystal World," which works with computational objects in ways that are at a slant to the Internet of Things, and which engages with the materialities of digital technologies as a way of intervening within other "systems of systems" that electronics generate. Through an analysis of this project, I suggest that the re-thingification of the material trajectories of electronics might be addressed in ways that account for the distributed effects and relations that these technologies create, and also in ways that attend to the possibilities of things to incite new forms of media theory and practice.

The Crystal World

A somewhat more chemical engagement with computational materiality and minerality, "The Crystal World" was an exhibition and open laboratory developed by Martin Howse, Ryan Jordan, and Jonathan Kemp at the Space White Building in London during the summer of 2012.[32] This project staged an experimental encounter with the materialities and mineralities of digital technologies, not necessarily as they circulate through markets and homes as functioning electronics, but as they return to the earth at various stages of wasting and residue, whether at end of life or in the process of manufacture.

"The Crystal World" project is a sort of electronic chemistry set in reverse, a cookbook for future fossils, an inquiry into what the life of a chemical-material matrix of electronics is outside of the lab, where the array of substances used for making electronics is apparently without environmental, political, or social effect. The project creators deliberately stage an overflow and menacing bake-off with these materials that, in a more sanitized laboratory setting, would appear to be rendered harmless. But this relatively uncontrolled experiment leads one to ask: How do these materials travel in the world? What are their effects? How do they come undone? And how might an attention to these concerns inform the (re)-making of electronics in the first place?

Working with the core materials of electronics, the exhibition and laboratory stripped open, broke down, and reworked the gold, silver, plastic, copper, and assorted other minerals that make up electronics. Dipping circuit boards into acid baths and baking off plastic housing from copper electrical cables, the project might on one level seem to have attempted to excavate the most fundamental material substances of electronics. And yet, in this lab/workshop encounter with electronic materialities, the attempt to salvage these minerals opened up into the wider networks and relations that support the material composition of these devices, whether through mining and manufacture; and that repurpose them at end of life, whether through recycling, repair, salvage, or disposal. The ways in which electronics break down play out not simply as a material performance of new fossils in the making, but also as the instantiation of particular material and environmental practices and politics: someone, somewhere, is working through electronics in these ways, and the opening up of these machines is also a way to open up the environmental and material politics that undergird them.[33]

The artists assembled a 540-page book as part of the project, which includes a wide range of texts across scientific, philosophical, and artistic fields. As one text collected in *The Crystal World Reader*, and drawn from the US National Mining Association, remarks, there are at least sixty-six individual minerals that contribute to a typical computer, and "it should be evident that without many minerals, there would be no computers, or televisions for that matter."[34] This minerality and materiality in the making is not an experience typically made present in our encounters with gadgets. "The Crystal World" front-stages this minerality, where materialities are made, chemical arrangements are strangely crafted, and thingly geologies are transformed through the leakage, sedimentation, and crystallization of computational technologies.

Here, the residue of electronics is transformed into startling forms that are at once fascinating and yet frightening, to the degree to which these strange substances show up as pollutants with lives of their own. The pathways these material technologies and chemicals take are not typically sites of human intervention, since they leach through landfills and within recycling sites in so many random and often unseen forms. In "The Crystal World" project and text, these processes of materialization are not only made evident but strangely aestheticized, and yet, this process happens in a way that draws us closer to thinking about the material politics and in/sustainability of electronics, rather than distancing us from them. Given that these artists draw the title for their installation from a J.G. Ballard novel, I am inevitably also drawn to consider yet another artist influenced by Ballard, who was a quintessential thinker and maker of material geologies, Robert Smithson—someone whose writing is also included in their reader. As Smithson has suggested, sediment reveals the often-overlooked aspects of technology. He prefers to think of technology less as "'extensions' of man," and more as "aggregates of elements," or "raw matter of the earth"; he also considers how rust—the apparent decline of technology—is a "fundamental property" of that technology.[35] As

Smithson writes, "rust evokes a fear of disuse, inactivity, entropy, and ruin." These more pervasive conditions of rust, sediment, and grit are the dynamics that run through technology, since "solids are particles built up around flux, they are objective illusions supporting grit, a collection of surfaces ready to be cracked." If, for Bruno Latour, technology is society made durable, then, for Smithson, durability (somewhat perversely) extends to the sediment and recurring remainder that accretes and informs the very life and death of those technologies.[36]

Conclusion

With the growing electronicization of objects, as well as the emergence of new computational practices and processes, the Internet of Things might soon become integral to new economies and ways of life, as well as to understandings of how environmental practices might be facilitated through ubiquitous computing. An attention to the specific materialities of the Internet of Things brings into focus the entanglements and complexities of our material-media lives and allows for a consideration of how material relations might also influence the emergence of new media-related practices, as well as new media environments.

If we were to extend the logic of "The Crystal World" to the Internet of Things, then we might account for the sediments and remainder of these electronic toasters and intelligent refrigerators in order to gain greater insight into their material arrangements and environmental effects. The trajectories and journeys of electronics in their chemical and environmental forms suggest that these materials and things are not neatly contained, but go on to have environmental and health effects that may linger for indefinite time spans. The things that are re-thingifying in this creative-practice provocation open up not just to new things in the making, but also to the new environments that accumulate in and through these things.

The processes of opening up, breaking down, and reworking electronics—whether in overtly material form or otherwise—generate a wider landscape of material relations that cannot be contained within any single device. When computation is thingified, it is also drawing on and establishing distinct material infrastructures and connections. The materialities and things that emerge with, and through, electronic gadgets suggest that from desktop PCs to distributed ubiquitous computing, computation takes place through extended milieus and settles into distinct forms that may very well outlast us. What do these distributed arrangements and materialities of computation enable, what processes and relations do they set in play and require, and what new environmental effects do they generate?

The actual and anticipated debris of electronics might provide one way that we could tune into these material processes to develop practices that speculate about material politics and relations in order to be less extractive and harmful. But this approach would require a re-thingification of things, particularly the Internet of Things. As I have argued here, such a re-thingification would involve attending to the versions of materiality and thingness that are mobilized as political, environmental, and even inventively practical operations. Re-thingification, in this way, would also involve encountering materiality as a process of things and environments becoming together, and of forming particular conjugations and experiences, as well as giving shape to particular material problems and struggles.

A critical and material media studies might then begin to develop methods and modes of practice that adopt an experimental set of approaches to re-thingification. These approaches would not necessarily consist of pointing to the brute materiality of electronics—or of simply using electronics to map and describe conditions for behavior change. Instead, they might require the development of practices for engaging with electronic media as environmental and material agents. Re-thingified media practices would then attend to these wider environmental and material effects.

Re-thingification would further require encountering materiality in multiples, since the thingness of an Internet-of-Things coffee pot and alarm clock interaction might generate significantly different registers of thingness in a factory in China or a mine in Africa. But such re-thingification is not simply about following things, either. This would be to commit our material investigations to forensic tracings primarily, with less attention to possibilities for practice, intervention, and creative realignment. Re-thingifying the Internet of Things is then as much an invitation to reroute these modes of thingly-ness, particularly as they are now forming instructions for environmental practice, as it is a suggestion for questioning *what sorts of things are these*?

Acknowledgments

Versions of this chapter have been presented at the "Geologies of Value and Vestige" symposium, Kingston University (2013); the "Media Archaeology and Technological Debris," conference at Goldsmiths, University of London (2013); the "Four by Four" lecture series at the California College of the Arts (2013); and the "When All That Is High-Tech Turns Into Waste" seminar at Lancaster University (2014). The research leading to these results has received funding from the European Research Council under the European Union's Seventh Framework Programme (FP/2007–2013)/ERC Grant Agreement n. 313347, "Citizen Sensing and Environmental Practice: Assessing Participatory Engagements with Environments through Sensor Technologies."

Notes

1 See Friedrich Kittler, *Gramophone, Film, Typewriter*, trans. Geoffrey Winthrop-Young and Michael Wutz (Stanford, CA: Stanford University Press, 1999); and N. Katherine Hayles, *How We Became Posthuman: Virtual Bodies in Cybernetics, Literature, and Informatics* (Chicago, IL: Chicago University Press, 1999).

2 See Lisa Parks, "Kinetic Screens: Epistemologies of Movement at the Interface," in *MediaSpace: Place, Scale and Culture in a Media Age*, edited by Nick Couldry and Anna McCarthy (London: Routledge, 2004), 37–57; Matthew Fuller, *Materialist Energies in Art and Technoculture* (Cambridge, MA: MIT Press, 2005); and Jennifer Gabrys, "Media

in the Dump," in *Trash*, edited by John Knechtel (Cambridge. MA: MIT Press, 2006), 156–165.

3 Jennifer Gabrys, "Programming Environments: Environmentality and Citizen Sensing in the Smart City," *Environment and Planning D*, 32, no. 1 (2014): 30–48.
4 For one example of how sensors are used to test and realize new environmental practices, see the Citizen Sense project at http://www.citizensense.net (accessed November 18, 2015).
5 Jennifer Gabrys, *Digital Rubbish: A Natural History of Electronics* (Ann Arbor: University of Michigan Press, 2011).
6 The first 2002 WEEE Directive was put into place in February 2003, and has since been updated to the 2012 Directive. See European Commission, "Directive 2012/19/EU of the European Parliament and of the Council of 4 July 2012 on Waste Electrical and Electronic Equipment (WEEE) (recast)," *Official Journal of the European Union*, July 24, 2012, L 197/38. The WEEE directive is accompanied by the RoHS Directive, which restricts certain hazardous substances in electrical and electronic equipment. The first 2002 RoHS Directive was similarly put into place February 2003, and the new RoHS Directive was adopted in January 2013. See European Commission, "Directive 2011/65/EU of the European Parliament and of the Council of 8 June 2011 on the Restriction of the Use of Certain Hazardous Substances in Electrical and Electronic Equipment (recast)," *Official Journal of the European Union*, July 1, 2011, L 174/88.
7 This argument is taken up more extensively in Jennifer Gabrys, *Program Earth: Environmental Sensing Technology and the Making of a Computational Planet* (Minneapolis: University of Minnesota Press, 2016).
8 Reporting on a European Commission study in the Internet of Things, the BBC writes,

> The commission says that the average person living within the 27-nation bloc has at least two devices connected to the net at present—typically a computer and smartphone. It expects the figure to rise to seven by 2015, with a total of 25 billion wirelessly connected to the net worldwide. By the end of the decade it says that could climb to 50 billion.

See "EU Investigates Internet's Spread to More Devices," *BBC News*, April 12, 2012, http://www.bbc.co.uk/news/technology-17687373 (accessed November 18, 2015).

9 Mark Weiser, "The Computer for the 21st Century," *Scientific American*, 265, no. 3 (1991): 94–104.
10 The term is attributed to Kevin Ashton. See "A Brief History of the Internet of Things," *Postscapes*, http://postscapes.com/internet-of-things-history (accessed November 18, 2015).
11 Dave Evans, "The Internet of Things: How the Next Evolution of the Internet Is Changing Everything," Cisco Whitepaper, April 2011, http://www.cisco.com/web/about/ac79/docs/innov/IoT_IBSG_0411FINAL.pdf (accessed November 18, 2015).
12 Joseph Bradley, Jeff Loucks, James Macaulay, and Andy Noronha, "Internet of Everything (IoE) Value Index," Cisco Whitepaper, 2013, http://internetofeverything.cisco.com/sites/default/files/docs/en/ioe-value-index_Whitepaper.pdf (accessed February 2015).
13 "Internet of Things Highlights: 2014 CES," International Consumer Electronics Show (CES), January 9, 2014, http://www.youtube.com/watch?v=tTHqb5RZTVg (accessed November 18, 2015).
14 Vint Cerf, "The Internet of Things and Sensors and Actuators," USENIX Association, March 19, 2013, http://www.youtube.com/watch?v=hIISiYs7lDo (accessed February 2015). USENIX is the Advanced Computing Systems Association. See https://www.usenix.org (accessed November 18, 2015).
15 Jennifer Gabrys, "Powering the Digital: From Energy Ecologies to Electronic Environmentalism," in *Media and the Ecological Crisis*, edited by Richard Maxwell, Jon Raundalen, and Nina Lager Vestberg (New York, NY: Routledge, 2014), 3–18.

16 For an early, if speculative, example of this, see Bruce Sterling, *Shaping Things* (Cambridge, MA: Mediawork Pamphlet Series, MIT Press, 2005).
17 Senseable City Lab, "Trash Track" (2009), http://senseable.mit.edu/trashtrack/index.php (accessed November 18, 2015).
18 Ibid.
19 Ibid.
20 For instance, see Bruno Latour, "From Realpolitik to *Dingpolitik* or How to Make Things Public," in *Making Things Public: Atmospheres of Democracy*, edited by Bruno Latour and Peter Weibel (Cambridge, MA and Karlsruhe: MIT Press and ZKM, 2005), 14–41; Jane Bennett, *Vibrant Matter: A Political Ecology of Things* (Durham, NC: Duke University Press, 2010); and Rick Dolphijn and Iris van der Tuin, *New Materialism: Interviews & Cartographies* (Ann Arbor: University of Michigan Library, Open Humanities Press, 2012).
21 Karen Barad, "Posthumanist Performativity: Toward an Understanding of How Matter Comes to Matter," *Signs*, 28, no. 3 (2003): 812.
22 For an overview of these speculative realist approaches, see Levi Bryant, Nick Srnicek, and Graham Harman (eds.), *The Speculative Turn: Continental Materialism and Realism* (Melbourne: re.press, 2011).
23 An exception to this is the "Object Lessons" series edited by Ian Bogost and Chris Schaberg. See http://objectsobjectsobjects.com (accessed November 18, 2015).
24 Arjun Appadurai, "Introduction: Commodities and the Politics of Value," in *The Social Life of Things: Commodities in Cultural Perspective* (Cambridge: Cambridge University Press, 1986), 3–63.
25 Gabrys, *Digital Rubbish*, 74–98, 127–143.
26 Ibid., 3–4.
27 For an expanded discussion on material politics, see Jennifer Gabrys, Gay Hawkins, and Mike Michael, eds., *Accumulation: The Material Politics of Plastic* (London: Routledge, 2013).
28 In this condensed list, I am drawing on Adrian Mackenzie's discussion of radical empiricism and things, and he is further informed by William James and Elizabeth Grosz. See Mackenzie, *Wirelessness: Radical Empiricism in Network Cultures* (Cambridge, MA: MIT Press, 2010); James, *Essays in Radical Empiricism* (Lincoln: University of Nebraska Press, 1996 [1912]); and Grosz, *Time Travels: Feminism, Nature, Power* (Durham, NC: Duke University Press, 2005).
29 Isabelle Stengers, "Wondering about Materialism," in *The Speculative Turn*, 371.
30 Ibid., 372.
31 Ibid. For an expanded discussion on this point of encountering materiality through "diverging" approaches, see Jennifer Gabrys, "A Cosmopolitics of Energy: Diverging Materialities and Hesitating Practices," *Environment and Planning A*, 46, no. 9 (2014): 2095–2109.
32 Martin Howse, Ryan Jordan, and Jonathan Kemp, "The Crystal World" (2012), http://crystal.xxn.org.uk/wiki/doku.php?id=the_crystal_world:space:opening (accessed November 18, 2015).
33 Martin Howse and Jonathan, Kemp, "The Crystal World," *Mute*, September 26, 2012, http://www.metamute.org/editorial/articles/crystal-world (accessed November 18, 2015). For a discussion of electronics as fossils, see Gabrys, *Digital Rubbish*.
34 National Mining Association, "Minerals in Typical Computers," in *The Crystal World Reader and Manual of Speculative Apparatus*, edited by Anonymous (January 2012), 137.
35 Robert Smithson, *The Collected Writings of Robert Smithson*, edited by Jack Flam (Berkeley: University of California Press, 1996), 17.
36 Smithson, *The Collected Writings*, 100–107; and Bruno Latour, "Technology is Society Made Durable," in *A Sociology of Monsters: Essays on Power, Technology and Domination*, edited by John Law (London: Routledge, 1991), 103–131.

11

SO-CALLED NATURE

Friedrich Kittler and Ecological Media Materialism

Jussi Parikka

To talk of sustainable media begs the question what do we mean by "sustainable" but also what do we mean by "media"? Do we refer to media as communication, discourses, patterns of meaning created, circulated, and controlled in political economic contexts? Do we think of media as the work of mediation between people? What count as media? Only the list of media we encounter in media studies courses such as cinema, radio, television, and, well "new media"—that eclectic mix that, broadly speaking, addresses digital culture? Or do we see it as a question of the material settings—media not just as a cultural reality of communication but at the same time, a material reality of technologies that, to put it bluntly, are made of *something* and demand *energy*. What if media are approached as devices, infrastructures, and material channels of communication? This is not meant to dismiss the meaning-creating reality of communication or to become naïvely technodeterminist, but to underline that the reproduction of data that underpins human communication happens on levels that are material and inherently connected to issues of ecology and the environment. This begs for media theoretical positions that take this sort of materiality into account.

Besides the empirically important mapping of the material weight that media technologies produce, we need to talk of the theoretical weight that media ecology is able to offer to discourses of sustainability. The past years of media theory have seen a significant new interest in media ecological thought in ways that also refer to Félix Guattari's "three ecologies" and other related concepts.[1] However, one could say that some of the ecological writing has been surprisingly hesitant to address environmental issues. In other words, what are the questions, issues, and fields where media ecology and environmental concerns overlap?

This chapter is an attempt to situate issues of materiality and ecology in ways that address one of the key media theorists of past decades. Friedrich Kittler's media theory has been one of the most significant inspirations for conceptualizations of media materiality, but we need also to critically ask how the "green issues" might have been overshadowed, in the midst of "gray media matter" (if we want to refer to the environment with this slightly worn-out color-coded rhetoric). Let's use the word green as shorthand, but acknowledge nonetheless the multiplicity of colors that define the milieu and recent cultural theoretical takes on it.[2]

As elaborated in this chapter, Kittler's thinking focuses on the materiality of machines, their scientific networks of emergence, and the link between a close reading of technology from software to hardware with a nonhermeneutic methodology that is indebted to Michel Foucault. He introduces several famous figurations that continued in Foucault's wake, including the term "so-called Man" to designate how our understanding of what the human being both stands for and is capable of, is conditioned by media technological networks in which the entity achieves consistency.[3] Technical media subsumes the Man in its own networks, undermining any fantasy of self-awareness and intention in a move where Kittler acts like a media analyst of the unconscious of the age of advanced technology.[4] And yet, one might well ask whether such accounts of materiality and technological conditions of existence are sufficiently aware of the environmental conditions of media. It is also in this context that we need to ask the question about the materiality of hardware. In other words, should we also pay closer to attention to the "so-called Nature"—the environment as conditioned by the technological condition, as well as itself conditioning the existence of technology by becoming a resource for advanced technological culture? This also means its (real) subsumption as part of the destructive drive of resource depletion and other related material patterns that are restructuring technopolitical relations of power. Any talk of sustainability has to critically ask also whether the political economy of contemporary technological culture is what exactly should *not* be sustained.

Beyond Kittler, there are several different notions of materiality that have been circulating in international media and cultural studies discussions during the past few decades. In a cultural and media studies context, materiality is meant to focus on something different than the same word within a German *Medienwissenschaft* context. In Britain, on the axis of universities from Lancaster to London with the center at Birmingham in the Midlands, "materiality" quite often calls to mind a lineage starting from Raymond Williams's cultural studies emphasis and its modern variations. Lawrence Grossberg's spatial materialism was also a significant take on the theory conversations.[5] University departments still carry a strong legacy of cultural studies and, even in media studies, the technical specifics of media culture are often left to colleagues in the sciences and engineering on the other side of the faculty lines. In such a heritage, materiality is often related to (human) experience, and has brought about an important political agenda that was also visible in Stuart Hall's massive role in connecting media and cultural studies to the everyday life.

In such traditions, the construction of meaning is a central feature through which to understand the living materiality of media practices, and media materialism reverts to concrete practices of cultural production in specific spatiotemporal settings.[6] Therefore, it is no wonder that the teachings of the likes of Kittler did not always find fertile ground in the United Kingdom, despite his admiration for certain sites of British culture: the Cambridge of Alan Turing and Pink Floyd, Bletchley Park, and the London of rock engineering culture and studio technologies.[7] But Kittler's appreciation did not extend to either British cultural studies and political economic research or the rising hegemony of English-language international universities.[8] Indeed, as a theoretician, he was invested in writing an alternative history of media and materiality, even if the latter refocused in his later work on the love life of the Greeks, read now as part of media history and the birth of the alphabet as the original digital binary (vowels and consonants) system. As an internationally known media theorist, Kittler is both lauded and reviled, especially in the Anglo-American world, for his emphasis on the primacy of technology. This Kittler is interested in the triangle of war, science, and engineering as indexes for what humanities should focus on in the Turing age.

For people trained in contextual thinking, some of Kittler's demands feel anachronistic. Kittler's methodology is partly recognizable for those knowledgeable about Foucault, but also puzzling when it comes to ideas such as those expressed in *Optical Media*, which takes Claude Shannon's mathematical theory of communication of 1948, and transposes it as a method through which to map media history before the term "media" was even properly used in the way in which we understand the term. This became the model of seeing technocultural reality and history through the set diagram of Source-Coding-Channel-Decoding-Reception. The engineered model is detoured around the semiotic understanding it received, and returned to its mathematical roots—and, therefore, also those roots planted as the initial stages of an investigation into media reality.

As John Durham Peters notes in his introduction to the English translation of *Optical Media*, there is a specific understanding of a materialistic media methodology in Kittler's relation to time. The difference from British cultural studies, for example, is not always evident, even on the level of topics: it is as if institutions, people, and political agency were constantly haunting Kittler's agenda. Indeed, instead of too easily succumbing to misinformed clichés about Kittler as a determinist who thinks that the only thing that trumps technology is the military:

> *Optical media* is clearly interested in the development of institutions such as Edison's Lab, the ways that marketing imposes compromises between consumers and engineers, the unique historical conditions that enabled the emergence of photography in the 1830s, and the ways that war financing affected the development of television. His celebration elsewhere of the technological bricoleur who, with soldering gun in hand and DOS screen in view, reconfigures the user-friendly interfaces of dumbed down technologies, certainly suggests an activist role for at least those with an engineering bent. Indeed, in his 1990s campaign against Microsoft Windows, Kittler

> could sound a bit like a cultural studies type praising the agency of ordinary folks who rewrite the dominant code for their own purposes.[9]

For sure, he wanted to ensure that media studies did not become a footnote to everyday journalism and popular culture, noting how "the problem is that by now one can hardly distinguish media studies from the self-evidence of everyday life."[10] His ontological stance was different—and decisively focused to understand the scientific and technological contexts of knowledge production. Kittler was also aware that knowledge production was also about excluding certain groups from that knowledge; for him, that group were the end-users and consumers. Technology is set as a central part of the modulation of such powers, with individual inventions as well as networks of the material world as preferred case studies to wider sets of issues of cultural history.

As a media theorist, Kittler articulates technological materiality as part of the humanities' agenda before Digital Humanities. "For the humanities, there is nothing nontechnical to teach and research."[11] In this way, he also posits new notions of media ontology and temporality. Hence we need to ask if there are ways to expand Kittler's media historical interests to something I will call "nonmediatic media materiality"—which leads to recognizing a different way to mobilize the media theoretical conceptual arsenal as a means of investigating issues relevant to sustainability. This approach is not interested in offering a commentary on Kittler's theory, but rather to use that theory to investigate the relevance of materiality for environmentally aimed media theory. This means critically asking how to extend Kittler's inspiring accounts of media and materiality, and how, on an international level, his work is giving rise to further variations on the theme of media ecology. Indeed, Kittler has influenced much of postcultural studies media theory and media ecology over the past several years in Britain.[12] Beyond this link, however, I want to open up the possibility of thinking about ecology more literally.

The briefly mentioned methodological stance of reading media reality and history through technologies as well as notions of information processing might have to be topped up with something that is harder than hardware, more persistent than functional machines: namely, issues of waste, resources, and the mineral materiality of media technological culture that persist in terms of their role in global media industries of production and discarding of technologies. Indeed, the broken, toxic, and obsolete do not feature strongly in Kittler's approach, but are increasingly coming to define contemporary media culture.[13]

The Place of Media Matter

To put this in broad terms, Kittler was devoted to analysis that paid close attention to the materiality of contemporary media technologies. His investigations extended the poststructuralist agenda to technological conditions, and articulated the theoretical work of Foucault and even Lacan in relation to media and science.

The conceptual arsenal familiar from French poststructuralism became infused and entangled with media history of technological systems, which take the place of power systems as well as structurations of the self. The so-called Man became Kittler's paradigmatic version of the poststructuralist subject, articulated as an expression of a variety of forces that, in this case, were explicitly technological too.

What separates Kittler from otherwise appropriate references to established traditions of media ecology in Canada—primarily McLuhan and Innis—is his penchant for a particular style of writing and love of technical specificity. Kittler's approach provides a systematic way of dealing with media historical topics in the present. For later commentators, this explains why his theory's relation to the anthropotechnics tradition from Leroi-Gourhan to Simondon, and even to Stiegler, continues to attract interest. This could also provide a link to a lot of contemporary media ecological writing. Naturally, for Kittler, the absence of media from media theory was one of the worst possible starting points: "It's dreadful when media scholars pontificate about computers without ever having looked underneath the lid."[14] A narrative of his early career might best be told through the technologies he used to write with, from typewriters to electric typewriters, to computers that expressed the internalized Nietzschean mantra, and complemented his 1970s building of synthesizers.

In this context, one can already speak of the two forms of Foucauldian influence of the 1970s mobilizing into what was later called media archaeology: (1) a descent in the historical sense, through the archival, as genealogical mapping of the contemporary moment, and (2) and in the technological sense, an "under-the-hood" methodology of media theoretical excavation that relates to what Nick Montfort later called a move beyond "screen-essentialism," toward software and hardware realities.[15] There is much more to visual culture than the mere screen and its images, which connects a range of material agendas in media studies across the Atlantic. The empirically visible is not the whole story of materially real, which conditions what we can see and/or hear.[16]

This double movement of history-technology characterizes much of Kittler's influence both in terms of media studies and practice-based research. The latter especially is often undertaken outside of universities; much of the Kittlerian ethos can still be seen in hacktivist practices. In any case, the creative tension between history and technology constantly focuses on the materiality of the object and its networks in Kittler's media history. So what is the place of materiality in his media historical account? What are the suggestions offered that characterize the idea of media materialism as it pertains to the imagined brand of German media studies?

Put simply, the analysis of technical media cultures is an epistemological operation that also *produces* lists of media: "What are media?" is a question answered through the various operations in which it is being exercised. Methodological choices are always ways of producing the objects that are analyzed and, as such, they become an archive of what counts as media. In a similar way to what McLuhan achieved, German-based media studies such as Kittler's and many more recent ones (e.g., in the discussion of *Kulturtechniken*, cultural techniques) have successfully expanded the definition of media beyond mass media to practices of technology that both condition formations of knowledge and serve as a historical ontology for current investigations.[17] For McLuhan in *Understanding Media* over fifty years ago, this meant extending the definition of media to allow for analyses of roads and housing, clothes and clocks.[18] The approaches became methodologically welcoming in the sense that media studies as a discipline appropriated not only the methods of historiography (Foucault and new historicism led to media archaeology) but also those of anthropology and other humanities fields.[19] Pushing the boundaries of media studies has even created a discourse of media studies in the past tense, as well as claims that the notion itself has become redundant: "What were the media?" becomes as pertinent a question as the claim that "there *is* no media."[20]

Rather than expressing a nihilistic sentiment intended to close down media studies departments—something that one might suspect in the UK, for instance, given the long-term political attacks by conservative stakeholders against such new disciplines—the idea that there is no media actually aims to capture the pragmatic aspects of how to approach the materiality of the object. Indeed, instead of closing the discipline down by definition—an ontological project—Eva Horn posits the lack of media in media studies as a radical opening:

> Perhaps such an anti-ontological approach to media, a radical opening of the analytical domain to any kind of medial process, has been more productive and theoretically challenging than any attempt, however convincing, at answering the question of what media are.[21]

Siegfried Zielinski's most recent book, *[. . . nach den Medien]*, seems to develop a similar approach underlying a crisis of media studies with regard to its conceptual basis:

> Now the media exist in superabundance, there is certainly no lack. For the thoroughly media-conditioned individuals they cannot possibly be the stuff that obsessions are made of any longer. What has now turned into a given that is at one's disposal, is utilized and defended as property but it is no longer a coveted object of desire. In this specific sense the media have become *superfluous/redundant*. Through the monumental exertions of the twentieth century, they have also become *timeworn/shopworn/used up/spent*.[22]

Zielinski laments the disappearance of the critical moment from the promise of media, and calls attention to the abundance of media-related discourse that is emptied of any particular critical force. Indeed, his observation targets a contemporary analysis of the globalization of digital economy as a floating signifier through which "media" can assume almost any meaning. However, what if one takes the last part of his critique literally and begins to examine all the used up and spent media technological devices, infrastructure, and materials? In other words, there is a correspondence in terms of overwhelming inflation and redundancy of media-related discourse, and the abundance of media as material objects, systems, and things; the material rhetorical link between the timeworn/shopworn both in discourse and its nondiscursive side-effects that is demonstrated with the accelerations of used-up electronics.

Concepts of and research on media obsolescence, dead media, and more recently, "zombie media," take up this speculation, as do an increasing number of accounts in media analysis, which acknowledge the environmental hazards that technologies carry with them.[23] At first, the issue of environmental dangers of chemicals and metals feels removed from discussions of media materiality, but it actually helps expand our thinking about media materiality. It also offers a crucial ethical, or ethico-aesthetic, aspect to the theoretical discussions of media materialism. This is also an alternative to the traditional sustainability discourse by emphasizing the multiple ecologies in which environmental issues are to be considered.

In short, media materiality—and this is a necessary simplification for my purposes here—has been successfully expanded and intensified by media theoretical accounts such as Kittler's and, therefore, a new arena of media research has opened up that eschews worn-out statements about the primacy of mass media. However, media studies has also concentrated on specific ways of understanding materiality and yet ignoring some other conditions.

In Kittler's work, especially his writings from the 1980s and 1990s, this approach implied a focus on the engineering and scientific contexts of technical media, with his occasionally heavy emphasis on the military. I will not recapitulate Kittler's main points in detail here. It will suffice to recall his early notes on the need to update Foucault's archaeology to methodologically suit the technical media age (instead of focusing solely on books and archives); his suggestion to update Lacan by incorporating the symbolic logic of information technology; and his meticulous close readings of analog and digital technical media. In the words of Markus Krajewski, who provided a short memorial reminder on Kittler:

> One of his most prominent demands was that nowadays people should not only be obliged to learn to read and write. It is now also one of the required fundamental cultural techniques to understand how computers can be directed. After writing whole books about the subtle effects and inner-political aspects of alphabetization [. . .], Kittler demanded an equal level of knowledge in writing computer sources among students, scholars, and intellectuals.[24]

The nucleus of this knowledge comprises the painstaking cultural techniques of coding, addressing C and Assembler, algorithmic thinking, and the wider computer infrastructure that can be argued to be governing procedures of knowledge production. Similar to the way in which writing in the age of word processing is subsumed under the logic of executable commands, and the whole corporate logic in which technicity is hidden, the historical method and the archival moment are to be subsumed only to the logic of technical media.[25] As suggested earlier, it is not only the choice of hardware objects based on the primacy of the materiality of the technological—and hence the pejorative accounts that condemn Kittler as a hardware fetishist masquerading as a media historian—but also the methodology itself that is based on this primacy: the historical methodology that begins with Shannon and Weaver's mathematical theory of communication.[26]

In different ways, related ideas have guided post-Kittler developments in media archaeology, too, including how historical time has been transposed as technological time; the *Eigenzeit* of the machine as a specific instance of microtemporality, as Wolfgang Ernst suggests, leads from a media-theoretical intervention to media-historical methods.[27] However, there is another sort of media materiality attached to different temporalities that escapes most of these accounts. Indeed, what if there is a way of tapping into media temporality other than that of the machine time, and what if there is another materiality of the media that, paradoxically, is not only related to media devices and infrastructures of operational media? Should we also attend to the broken and discarded machines, to obsolescence, instead of Kittler's own speculative stance, which remains an anecdote, not a fully fledged argument? "The fact that Foucault and I had such an interest in functioning machines—as opposed to broken ones, like those in [Gilles Deleuze and Felix Guattari's] *Anti-Oedipus*—may have been what united us."[28] Should we pay attention to material things that are not yet media in the traditional sense? This might imply a focus on rocks, minerals, and metals as a metallurgical prelude to media before its time. And it might imply also a needed interest in things broken, too.

Ecology as Nonmediatic Media Materialism

So far I have argued that Kittler's media theoretical framework has paved the way for a reconsideration of media materiality. However, it has its own set of preferences where media materiality homes in on the technological, whereas there are certain contexts that are increasingly important to considerations about which materialities and temporalities of media we choose to focus on. Indeed, media studies as a discipline is founded on objects that it, itself, has posited as media.[29] This sense of materiality exists in a systems-theoretical loop. But it remains clear, as a matter of environmental importance as well, that we must consider an expanding list of objects, including media events, as relevant to our analyses. These new objects also necessitate broadening media cultural inquiry in terms of a range of media objects that do not relate to media as used, practiced, or as meaningful—but in terms of the material continuum between the environment and technology.[30] In short, this new perspective corresponds to the environmental urgency of electronic waste as a global problem, as well as other environmental factors that condition digital technologies as global effects entangling with the soil, the air, the atmosphere: e.g., cloud computing's energy use and CO_2 emissions, that all comprise one set of material conditions that extend media materiality to media ecology.

In this context, media ecology does not refer to ideas stemming from the Canadian school of media studies from McLuhan to Neil Postman, but rather to the concrete connections that media as technology has to resources (as in the Heideggerian concept of "standing in reserve") and nature. As noted earlier, the media ecological context has been increasingly raised in recent media theory, screen studies, and media history. Aside from an interest in the current ecological problems caused, for instance, by toxic screen technologies, media ecology also implies rethinking the definitions that constitute the borders of media studies: what objects and processes can media studies incorporate as part of its theoretical and empirical work? For example, when it comes to issues of the environment and sustainability, is there a point of closure beyond which media studies loses its focus?

At least these meditations push media studies to reflect on when it stops being about "media" in the clearest sense. Providing an overarching definition of media becomes difficult in this sort of a methodological context of material media studies, such as Kittler's. Media studies becomes more of an analysis of what *operations* media consist. This point, clarified by Winthrop-Young, opens media studies to a new list of objects that count as media, and provides a way to understand the constitution of key concepts in humanities—such as body, sense, meaning, truth, communication—underpinned by media techniques.[31]

In a similar way to Kittler, we might use the theoretical seeds as an impetus to broaden this agenda further toward the nonmediatic basis of technical media: from minerals to the scientific development of synthetic materials, and to the afterlife of media technological waste, we encounter both an alternative materiality and an alternative temporality of media. Through a selection of what sort of objects media studies deals with, one opens up a different sort of media cultural and historical horizon. For instance, one could consider media studies to include objects such as plastics, wood, plywood, copper, aluminum, silver, gold, palladium, lead, mercury, arsenic, cadmium, selenium, hexavalent chromium, and flame retardants.[32] A list such as this one, which can be found in electronic waste management manuals as well as in a document distributed to health officials enumerating the toxic materials inside information technology, contains materials that leak not only into nature but also into media theoretical accounts.

At the other end of the spectrum, one could include rare earth minerals necessary for the specific effects of audiovisuality. Coltan is the conflict mineral that appears in accounts by several recent theorists, but there are many others that are of importance for geopolitical and material reasons: lanthanum, cerium, neodymium, yttrium, and praseodymium are examples of what helps make our media digital. Aside from the official mines (in China, Africa, etc.), there is also the widespread illegal business of rare earth mineral mining. Along with new mines, one can also reuse older mines: an old copper mine might turn out to be rich in indium, which is also part of visual culture in the form of LCD and LED displays. This is a different sort of remediation culture than some earlier uses of the term have indicated.[33] The geopolitical stakes of the material and energy economics are also revealed to be increasingly dependent on extending neocolonial ties of power.[34]

An alternative collection of media objects—a kind of media culture of the materialities of media from minerals to discarded and waste objects—includes a change of horizon toward media materiality in the context of sustainability that cannot merely reproduce the sustainability of the current economic system of capitalism that produces this state of things, but a more radical sustainability that looks at it from the perspective of the Earth, the planetary condition. This perspective does not focus on the use-time of media anymore, or even on the time of inventions. Its epistemological shift is one of an episteme that expands to the grounding scientific cultures in which a wider set of materialities is to be discovered. There are depths that underpin even the technoscientific discoveries, which of course themselves become frameworks to see and hear the earth in new ways. This means, for instance, the importance of copper and copper mines to networking technologies; minerals to processing; different materials for encasing; and a whole range of structurations that on a material level offer the existence of technical communications.

Kittler does not entirely neglect the question of how technological materiality has a relation to natural resources, although in his early work Nature is present in other ways. *Discourse Networks 1800/1900* can be seen as a massive work of cultural historical study that picks up on the romantic notion of alienation from Nature by Culture (i.e., language), which Kittler reveals to be a complex set of discourse tropes tied into institutional reforms in education and gender roles (the educating "Mutter's mouth" is one such figure in his analysis).[35] Kittler's work focuses on the transpositions and remediations of Nature in different discourse networks; Nature becomes writing and figuration of the Woman; then, with technical media, Nature becomes registered, inscribed as the Real through the possibilities of technical media; with computers, Nature becomes simulatable in new ways, according to the Turing principles concerning computable numbers.[36] Furthermore, Kittler wanted to move from the discourse of the body to the realm of materials and physics. This tendency is evident, for example, in the *Optical Media* lectures of the late 1990s: "The only thing that remains is to take the concept of media from there—in a step also beyond McLuhan—to where it is most at home: the field of physics in general and telecommunications in particular."[37]

However, when we enter into this field of physics, we need to ask: why stop here? If we take a further step forward, and combine a Kittlerian media history of physics with Manuel Delanda's call for a thousand years of nonlinear history, we enter a new temporal horizon. In short, Delanda's project mapped nonhuman agencies, including rocks, which played a role in the world of human affairs through architecture and other formations. His physics and complexity theory-related accounts even flirt with the idea of a social history of mineralization.[38] The point becomes even clearer if we relate it to media waste. Scholars such as Jennifer Gabrys have argued that media waste pushes dead media and obsolescence discourse into new interesting areas. We should expand the media history discourses and practices of framing time inside museums and collections of objects to failures, breakdowns, and the broken. Gabrys formulates it as follows:

> Any museum or archive in which electronics are held is a collection of repeated obsolescence and breakdown. But failure is only one part of this story. Whether in a state of decay or preservation, obsolete devices begin to express tales that are about something other than technical evolution.[39]

I have worked on something similar to Gabrys's argument in terms of "the general accident of digital culture."[40]

Gabrys suggests including speculative questions such as: Can we gain a new perspective on media history from the point of view of dust? What is the temporality of dust—and then, more significantly, what are the specific temporalities that media technology produces on the level of materials and toxins? This is one specific attempt to tackle the issue of speed and, instead of increasing speed, we might examine the slowness of material decomposition. Gabrys underscores another aspect of media temporality, in which the surface of ephemerality, affiliated with software cultures and information understood mathematically, is embedded in specific materialities of modernity: such media materials include plastic, copper, mercury, and lead, which are "substances thicker and more enduring than any transcription of ones and zeros," as well as beryllium, cadmium, and brominated flame retardants.[41] This approach involves a mix of temporalities, which do not proceed merely from the alternatives of acceleration versus obsolescence, but arise out of a more complex mix of materialities and time. As Gabrys puts it,

> [m]aterials are caught in a tension between the quick and the slow. Ephemerality can only hold at one level; it instead reveals new spaces of permanence. Throw away plastic to discover it lasts for an ice age. The balance of time shifts. The instant plastic package creates new geologies. We now have mountains of congealed carbon polymers. Entirely new landscapes are built up around the fallout from the momentary and the disposable. So this is not just a story about the vaporization of "all that is solid"; rather, it suggests that new forms of solidity—new types of "hardware"—emerge with the program of disposability.[42]

In other words, in response to accusations about the hardware fetishism of media historical methods, one might say that *we have never been hard enough.* This reply refers to the fact that aside from the hardware of media technologies, there is the hardness of which the hardware is composed: namely, the aforementioned lists of materials, toxins, and others that contribute to an ecological and environmental awareness of technical media. Richard Maxwell and Toby Miller elaborate on this point in detail in *Greening the Media*, which offers a researched perspective on the chemical and material basis of media, ranging from mass-manufactured paper to networks and technological screens.[43] This question is also about an ontological perspective: what is the ontology of media historical materiality, what is its focus, and on which "metal" does materiality focus? In short, we need concepts to account for the multiple materialities at play when we consider media history—an ontology of multiple levels of materiality that allows us to speak with ease about the different aspects of media across code, interfaces, hardware and, I would add, environmental concerns, and even labor.[44] Green(er) media technologies is a topic that has moved from the policy and awareness agendas of nongovernmental organizations (NGOs) to the academic work of media scholars. However, aside from concrete empirical work and a challenge to contemporary media policy, the topic of green media technologies may be able to offer theoretical innovation.

As an example of where artistic projects elaborate the potential for new material variations, consider the long lineage of media technologies from the perspective of the recent project by the artist duo Cohen Balen. The H:AlCuTaAu (2014) is an odd design-type of an "unobject" that underlines the material basis of media technologies in minerals. This artificial hybrid mineral consists of aluminum (Al), copper (Cu), gold (Au), tantalum (Ta) (along with some added whetstone) and is completed from already existing technological objects to form an odd "ecological" circle of materiality. It underlines the technological and geopolitical significance of different sorts of minerals as well as the implicit disappearing labor that has gone into the communication media. This sort of materiality is usually embedded in supply lines and logistics, alternative geographies of media from mining to smelting to factories, and as such constitutes the other side of the communicational. Furthermore, besides the material constitution, the politics of energy around contemporary culture relate to a wider media studies agenda, as demonstrated recently by Sean Cubitt.[45] Materials of media and their energy are part of a longer chain of production and dispossession, increasingly related to the geopolitical stakes of mining sites, energy production, and the planetary claim of territories of this raw side of materiality. Complementing this from a media and sound arts perspective, Douglas Kahn has furthermore pitched the idea of a natural history of media.[46]

Discussing media, ecology, and waste constitutes one pole of the discussions related to the Anthropocene—one of the concepts that itself is a curious example of epistemological concepts/objects shifting across disciplinary boundaries in twenty-first-century academic debates.[47] But perhaps Kittler's media historical impetus toward materiality can provide a way to address the Anthropocene and its media cultural contexts too. Similar to the way in which historians like Dipesh

Chakrabarty have called attention to the need to reconsider the field of history in relation to its Other—that is, natural history—media history should also be able to reconvene around shared topics and concerns that attach it to ecology. This does not imply the media ecology we inherited from McLuhan and Postman, but rather one that has been influenced by the more material-sensitive Kittler and that can been pushed toward an environmental research field. More "green"—or dirty—media materialism might present new ways to think about media culture and continue the agenda of the likes of Kittler in ways that produce media analytical objects that do not home in on media only. Media ecology would imply an agenda wherein media materialism shifts from a perspective on devices, technological inventions, and the agency attributed to the scientific basis of technical media, to the materials and material sciences on the outskirts of these interests. Instead of speaking of networks, we could speak about the copper inside cables as the residue of industrial cities or the gutta-percha insulating the networks in the nineteenth century; instead of screens and computer graphics, vectors and bit mapping, we could talk about the chemical basis of cathode ray tubes (CRT), liquid-crystal displays (LCD), and plasma screens. Indeed, knowledge of screen technologies here becomes a question of energy consumption: for example, the knowledge that LCD technology's reliance on mercury-vapor fluorescent backlights consumes less energy than electroluminescent phosphors, a detail pointed out by Cubitt.[48] LCD technologies make a difference to CRTs, a screen media technology that might have accompanied much of our post-World War II lives as part of what theorists then referred to as the simulation, spectacle, and emergence of visual culture—but which, also, are intensive risks in terms of their components (lead, toxic phosphors, barium) and energy consumption.[49] Cubitt implicitly raises the media historical point we are after: should we try to radicalize Kittler's media historical methodology and push media studies out from its sole focus on media, to the world of chemicals and materials, and even energy consumption? This would mean that a more environmental media ecological materialism would not imply a theory that is any less hardware oriented—perhaps this field would be even harder, with its focus on the rocks, minerals, and materials of which media is constituted, and whose histories are those of minerals, the soil, and geological formations. Conceptual figurations such as so-called Man are to be complemented with so-called Nature, which unfortunately does not refer only to the technological conditions of understanding, mapping, and exploiting the earth but also to the grim scenario of mass extinction in which we are living.

Acknowledgments

Many thanks to the editors for their perceptive comments and to everyone else who read and helpfully commented on ideas relating to the text. An earlier version of the text was published in *Archiv für Mediengeschichte* 13 (2013).

Notes

1 Félix Guattari, *The Three Ecologies*, trans. Ian Pindar and Paul Sutton (London: Athlone Press, 2000).
2 Jeffrey Jerome Cohen (ed.), *Prismatic Ecology. Ecotheory beyond Green* (Minneapolis: University of Minnesota Press, 2013).
3 Friedrich Kittler, *Gramophone, Film, Typewriter*, trans. Geoffrey Winthrop-Young and Michael Wutz (Stanford, CA: Stanford University Press, 1999).
4 Geoffrey Winthrop-Young and Michael Wutz, "Translators' Introduction," in *Gramophone, Film, Typewriter*, xxxiii–xxxiv.
5 See Stephen B. Crofts Wiley, "Spatial Materialism. Grossberg's Deleuzean Cultural Studies," *Cultural Studies*, 19, no. 1 (2005): 63–99.
6 Nick Couldry, *Inside Culture: Re-Imagining the Method of Cultural Studies* (London: Sage, 2000), 11–12.
7 On related topics, see Geoffrey Winthrop-Young, "Cultural Studies and German Media Theory," in *New Cultural Studies. Adventures in Theory*, edited by Gary Hall and Clare Birchall (Edinburgh: Edinburgh University Press, 2006), 88–104.
8 See Friedrich Kittler, "Farewell to Sophienstraße," trans. Bernard Geoghegan and Christian Kassung, forthcoming in *Critical Inquiry*.
9 John Durham Peters, "Introduction: Friedrich Kittler's Light Shows," in *Optical Media*, edited by Friedrich Kittler (Cambridge: Polity Press, 2010), 7.
10 Friedrich Kittler, "The Cold War Model of Structure. Friedrich Kittler Interviewed by Christoph Weinberger," *Cultural Politics*, 8, no. 3 (2012): 379.
11 Friedrich Kittler, "Universities: Wet, Hard, Soft, Harder," *Critical Inquiry*, 31, no. 1 (Autumn 2004), 251.
12 See Matthew Fuller, *Media Ecologies. Materialist Energies in Art and Technoculture* (Cambridge, MA: MIT Press, 2005).
13 For developments of such ideas see, for example, Charles Acland (ed.), *Residual Media* (Minneapolis: University of Minnesota Press, 2007).
14 Kittler, "The Cold Model of Structure," 379.
15 Nick Montfort, "Continuous Paper: The Early Materiality and Workings of Electronic Literature," http://nickm.com/writing/essays/continuous_paper_mla.html (accessed July 2015).
16 This was also Deleuze's emphasis in his reading of Foucault, the new archivist. Gilles Deleuze, *Foucault*, trans. Seán Hand (Minneapolis: University of Minnesota Press, 1988).
17 Bernhard Siegert, "The Map Is the Territory," *Radical Philosophy*, 169 (September/October 2011): 15.
18 Marshall McLuhan, *Understanding Media: The Extensions of Man* (London: Routledge, 2001).
19 Bernhard Siegert, "Cacography or Communication? Cultural Techniques in German Media Studies," trans. Geoffrey Winthrop-Young, *Grey Room*, 29 (2008): 27–29.
20 See Claus Pias (ed.), *Was waren Medien* (Zürich: Diaphanes, 2011); Eva Horn, "There Is No Media," *Grey Room*, 29 (2008): 6–13.
21 Horn, "There Is No Media," 8.
22 Siegfried Zielinski, *[. . . nach den Medien]. Nachrichten vom ausgehenden zwanzigsten Jahrhundert* (Berlin: Merve, 2011), 16. Translation from: Siegfried Zielinski, "The Media Have Become Superfluous," *Continent* Issue 3, no. 1 (2013), http://www.continentcontinent.com/index.php/continent/article/view/136 (accessed September 2013).
23 On zombie media, see Garnet Hertz and Jussi Parikka, "Zombie Media. Circuit Bending Media Archaeology into an Art Method," *Leonardo*, 45, no. 5 (2012): 424–430. For references to some recent literature on media, the environment, and waste: Jennifer Gabrys, *Digital Rubbish: A Natural History of Electronics* (Ann Arbor: University of Michigan Press, 2011); Jussi Parikka (ed.), *Medianatures: The Materiality of*

Information Technology and Waste (London: Open Humanities Press, 2011), http://www.livingbooksaboutlife.org/books/Medianatures (accessed September 2013). Richard Maxwell and Toby Miller, *Greening the Media* (Oxford: Oxford University Press, 2012). Sean Cubitt's work has been pioneering in this field, where issues of media waste and the environmental load of advanced technologies must be considered in relation to global media governance. See, e.g., Cubitt, "Current Screens," 21–26.

24 Markus Krajewski, "On Kittler Applied: A Technical Memoir of a Specific Configuration in the 1990s," *Thesis Eleven*, 107, no. 1 (2011): 35.

25 See Friedrich Kittler, "There Is No Software," in *Literature, Media, Information Systems*, edited by John Johnson (Amsterdam: G+A, 1997), 149.

26 Kittler's reception in the Anglosphere has often been characterized by pejorative accounts. These are critically mapped and discussed by Geoffrey Winthrop-Young:

> And since there is a lot to fear in Kittler he has acquired an impressive litany of labels: techno-determinist, anti-humanist, reactionary modernist, military fetishist, disgruntled literary scholar suffering from a strong case of physics envy, Eurocentric Heideggerian clone, and so on.

Geoffrey Winthrop-Young, "Krautrock, Heidegger, Bogeyman: Kittler in the Anglosphere," *Thesis Eleven*, 107, no. 1 (2011): 8.

27 Wolfgang Ernst, *Gleichursprünglichkeit: Zeitwesen und Zeitgegebenheite technischer Medien* (Berlin: Kadmos, 2012); Wolfgang Ernst, "From Media History to *Zeitkritik*," trans. Guide Schenkel, *Theory, Culture & Society*, 30, no. 6 (November 2013): 132–146.

28 Kittler, "The Cold Model of Structure," 380.

29 Wolfgang Hagen, "Wie is eine 'eigentlich so zu nennende' Medienwissenschaft möglich?," in *Was waren Medien*, edited by Claus Pias (Zürich: Diaphanes, 2011), 94.

30 On "medianatures," see Jussi Parikka, "Media Zoology and Waste Management: Animal Energies and Medianatures," *Necsus—European Journal of Media Studies*, no. 4 (Autumn 2013), http://www.necsus-ejms.org/media-zoology-and-waste-management-animal-energies-and-medianatures/ (accessed April 2014).

31 Geoffrey Winthrop-Young, "Cultural Techniques: Preliminary Remarks," *Theory, Culture & Society*, 30, no. 6 (2014): 13.

32 V.N. Pinto, "E-waste Hazard: The Impending Challenge," *Indian Journal of Occupational and Environmental Medicine*, 2 (August 12, 2008): 65–70.

33 Jay David Bolter and Richard Grusin, *Remediation: Understanding New Media*. (Cambridge, MA: MIT Press, 2000).

34 Sean Cubitt, "Integral Waste," *Theory, Culture & Society*. Online First-version, June 27, 2014. The importance of logistics as a media infrastructural arrangement of contemporary global production is underlined by scholars such as Ned Rossiter. See Rossiter, "Coded Vanilla. Logistical Media and the Determination of Action," *South Atlantic Quarterly*, 114, no. 1 (2015): 132–152.

35 Friedrich Kittler, *Discourse Networks 1800/1900*, trans. Chris Cullens with Michael Metteer (Stanford, CA: Stanford University Press, 1990).

36 Ibid.

37 Kittler, *Optical Media*, 32.

38 Manuel Delanda, *A Thousand Years of Nonlinear History* (Cambridge, MA: MIT Press, 2000).

39 Gabrys, *Digital Rubbish*, 104.

40 Jussi Parikka, *Digital Contagions: Media Archaeology of Computer Viruses* (New York: Peter Lang, 2007).

41 Ibid., 87.

42 Ibid., 88.

43 Maxwell and Miller, *Greening the Media*.

44 David M. Berry, *Critical Theory and the Digital* (London: Bloomsbury, 2014), 60–63.

45 Cubitt, "Integral Waste."

46 Douglas Kahn, *Earth Sound Earth Signal: Energies and Earth Magnitude in the Arts*. (Berkeley: University of California Press, 2013). This theme is elaborated in Kahn's forthcoming book project too.
47 See also a variation on the theme of the Anthropocene: Jussi Parikka, *The Anthrobscene* (Minneapolis: University of Minnesota Press, 2014). See also Jussi Parikka, *A Geology of Media* (Minneapolis: University of Minnesota Press, 2015).
48 See Cubitt, "Current Screens."
49 Ibid., 22–28.

PART IV

Scaling, Modeling, Coupling

12

THINK GALACTICALLY, ACT MICROSCOPICALLY?

The Science of Scale in Video Games

Alenda Y. Chang

> I meant no harm. I most truly did not.
> But I had to grow bigger. So bigger I got.
>
> —The Once-ler, in Dr. Seuss's *The Lorax*

> The greatest shortcoming of the human race is our inability to understand the exponential function.
>
> —Dr. Albert A. Bartlett[1]

Fixtures of the domestic interior, computer and video game systems can seem woefully remote from the proliferating concerns of the outer world. But while early Christian doctrine and Greek natural philosophy alike designated the world as a Book of Nature, today the physical world is intensely mediated and remediated not just by textual forms, but by photos, movies, soundtracks, apps, tchotchke, and even digital games. In this vastly expanded media universe, the Book of Nature warrants reimagining—but as what? As our material and virtual lives continue to converge, perhaps nature too has become the stuff of pixels and polygons.

To date, the dual categorization of games as technology and as recreation has shielded them from vital questions about how they model natural environments. While growing numbers of academics and game designers have worked to establish games' artistic, educational, and social value, few have yet asked pointed questions about how games exhibit and influence human environmental understanding. For a moment, let us entertain the provocation that the video game is the book's appropriate successor as metaphor for the world. To speak of the Game of Nature, facetiously or no, would be to acknowledge not only the growing influence of virtual realities on our everyday lives, but also the modern tendency to treat nature as the stage for everything from leisure to business to near-constant anxiety about future outcomes. Gamers and nongamers alike have been deeply impacted by computer-driven practices that arguably achieve their clearest manifestation in games, among them graphic immersion, telecommunication, social networking, and the exchange of virtual currency. What's more, those of us who enact our daily routines deeply conscious of a desire to avert the worst of climate change predictions travel through a time and space that are undeniably game-like, in that they are profoundly shaped by cause and effect and sometimes stultifying calculations of risk and reward. What a ludic world metaphor makes visible is this growing overlap between real and virtual worlds, whether through painstakingly rendered three-dimensional digital environments or our propensity to map conventions derived from one to the other, and vice versa.

Here at the nexus between media and ecology, questions of dimension invariably arise. Environmental thinking calls for suprapersonal awareness of the impact and extent of one's actions: What's the big deal if I rely on bottled water rather than tap water? How many thousands of kilometers of fiber-optic cable must be laid to satisfy our desire for broadband connectivity? How much is too much, and when is enough enough? Environmental representation confronts an additional edifice of ideals around realism and transparency: does verisimilitude demand one-to-one mapping, or does simplification allow patterns to emerge more readily? In my mind, games are tailor-made to develop scalar environmental consciousness, for instance by operationalizing relations between the local and the global. As the following brief review of the scientific literature on scale in ecological journals will suggest, video games share certain fundamental qualities with field and laboratory experiments, and both players and scientists exercise their creative and rational faculties to make sense of the worlds around them.

First, several caveats are in order. Like sustainability, the watchword of this volume, scale means many things to many people. Groups as diverse as mapmakers, human geographers, field ecologists, and computer scientists use the term but seldom agree on its definition, in part because scale is inherently relational. As Willard McCarty has said of models, scale implies at minimum a ternary structure between measure-taker, that which is being measured, and the measure itself.[2] In practice, this has meant that scale serves equally well as an instrument of revelation and distortion. Scalar arguments have accordingly been both the clarion call and the bête noire of environmental movements—picture the Leonardo DiCaprio–narrated *The 11th Hour* (dir. Leila Connors-Peterson and Nadia Connors, 2007) and the bumper-sticker mantra "think globally, act locally" alongside Malthusian declamations of population growth and the gleeful globalization rhetoric encapsulated by Dr. Seuss's Once-ler and his "figgering on biggering."[3] For my purposes, scale is less significant as a graded system of measurement than as an acknowledgement of interspecies and Latourian interobjective relativity. That is to say, scale connotes dependence as well as magnitude. The word may conjure the undue tidiness of architectural miniatures and the conformation of environments to users, but it also makes possible a thinking through of excess, a thinking *beyond* in the realm of hyperobjects.[4]

In what follows, I hope to reveal some of ecology's potential affinities with digital interactive media.[5] At the same time, this chapter is also intended as an initial inquiry into the variations in scalar representation across media. While I want to resist characterizing the transition from print and photography to the moving image (*Powers of Ten*) to the sandbox simulation game (*Spore*) as progressive teleology, I will argue that contemporary games offer quantitatively and qualitatively distinctive opportunities for the representation of pressing ecological quandaries. Where Ralph Waldo Emerson once described poetry as possessing a rare ability to "magnify the small" and "micrify the great," thereby transforming our perception of the material world, I present games as both an aesthetic and an ethical means to engage in world design and management, one especially well suited to exploring questions of sustainable action and scope.[6] Understanding how scale is defined and instantiated in our media culture should be of paramount importance in an age vexed by parochialism, transnational corporate ambitions, and borderless phenomena of minute and massive proportions, from toxic contamination of living tissue to ocean eddies and extreme weather.

Leveling Up (and Down) in the Lab and in the Living Room

> Graphical excellence is nearly always multivariate.
>
> —Edward R. Tufte[7]

Experts across the disciplinary spectrum express mixed feelings about scalar models, seeing them as necessary and illuminating frameworks on the one hand, and falsely subordinating constructs on the other.[8] We might expect scientists to have fewer qualms, but lately many have been at pains to historicize and disambiguate the notion of scale. A case in point: though today, references to ecology routinely evoke local communities, regions, and an entire planet in crisis, the concept of scale did not become popular in scientific circles until the 1970s.[9] Biologist John Wiens could write in the late 1980s that "'Scale' is rapidly becoming a new ecological buzzword," even while chiding fellow ecologists for being slow to adopt scalar thinking (relative to colleagues in physics, math, geography, and atmospheric and earth science).[10] Over two decades later, a recent literature review conducted by Brody Sandel and Adam Smith demonstrates that scale remains both poorly understood and applied as a variable in experimental design. As early as 2002, NASA researcher Jennifer Dungan and her coauthors recommended avoiding the word scale completely, in favor of more specific and universally agreed-upon terms.[11]

These scientific misgivings about scale are summarized in the following five cautionary principles, which in turn imply why games may embody a desirable alternative to actual experiments:

1 Scale operates at both spatial and temporal levels.

As Wiens notes in his early article in *Functional Ecology*, the inclination is to favor the spatial over the temporal or to examine the two aspects in isolation, rather than in tandem. Of course, such reduction flies in the face of the equally spatial and temporal complexity of human-wrought environmental change—from disrupted patterns of animal migration and biotic homogenization through globalization to ecologist Eugene Stoermer's and chemist Paul Crutzen's dubbing of the current geological epoch as the Anthropocene.[12] Human agency is dramatically altering the very scale of events even as human perception has proven itself blind to catastrophic change that occurs in geologic time.[13]

2 Multiscalar analysis is better than monoscalar analysis.

Scientists also tend to design experiments confined to only one spatial scale, even though this limits the validity and generality of their findings. According to Sandel and Smith, scale is a "lurking factor" that frequently is not acknowledged or studied due to logistical constraints and the field ecologist's traditional reliance on the quadrat.[14] More resources must be expended to collect data at multiple scales, while restricting the "grain" (or unit size) and "extent" (or range) of one's experiment is always tempting because it reduces the number and impact of confounding variables.

3 We design experiments on anthropocentric rather than "effective" scales.

It probably comes as no great surprise that scientists usually design ecological studies on scales appropriate to human experience and perception, rather than the species or subjects in question. Yet ecologists Göran Englund and Scott Cooper, among others, explicitly warn against arbitrary and anthropocentric selection of scale. Instead, ideally, experimental design should involve "matching the scale attributes of organisms, processes, and the abiotic environment. Often this amounts to preserving the 'effective scale,' which describes the scale of the system as experienced by the organisms."[15]

4 We favor biotic, rather than abiotic, explanations.

Wiens suggests another problem directly related to anthropocentric design, namely ecologists' penchant for focusing on biological rather than physical processes. At smaller scales, we might reasonably expect biological interaction to dominate results, but at larger scales, factors such as atmospheric and geological effects could loom larger.

5 We assume continuity when discontinuity, or nonlinearity, is actually the norm.

> Perhaps the most important lesson from this survey of the ecological literature on scale is scientists' acknowledgement of nonlinearity or discontinuity as the governing principle of many natural states and processes. Essentially, one can never assume that what holds true at one scale will hold at another. Sometimes faulty statistical extrapolation and aggregation error are to blame, but, in general, Wiens reminds us, "the continuous linear scales we use to measure space and time may not be appropriate for organisms or processes whose dynamics or rates vary discontinuously."[16] He gives the example of insect diapause, one kind of animal dormancy in which certain species go into a period of growthless inactivity in response to unfavorable external conditions.

Why do these five closely linked observations regarding complex scale-dependence matter? In short, our tendency to select and measure at terrestrial, biological, and human scales and to describe phenomena as orderly and continuous series severely limits our understanding of nonhuman agency and experience. Depending on the scale of observation, too, the same factors may have differing "explanatory power."[17] The challenge then becomes moving past recognition of these issues to more responsible kinds of evaluation and engagement. As some ecologists have put it, how do we match large-scale questions and small-scale data, or "scale up" from "experimentally tractable scales" to the realities of expansive natural environments?[18]

Among the options available to us, those kinds of overzealous cartography intent on creating a one-to-one map of the world are the least appealing.[19] Far more intriguing are the varied media forms that dramatize scalar dependencies without sacrificing the capacity for wonder. Consider the whimsical Japanese game *Katamari Damacy* (2004), in which you play a tiny cosmic prince charged with rebuilding the universe by rolling up matter (Figure 12.1). Using a sticky "katamari" ball to pick up mass, you start with relatively small items but soon progress to larger and larger ones as the ball swells in size. Each shift in effective scale is signaled by a visual blurring, as the playable world stretches to new dimensions, and obstacles at one scale—hedges, pylons, and parking structures, to name just a few—become katamari fodder at another. Along these lines, *Katamari Damacy* cheekily invites us to raise our scalar consciousness to absurd heights, while literalizing the interconnectedness of all things. The game's title even translates to English as "clump soul."

Yet whether or not games like *Katamari Damacy* make available novel ways of conceptualizing the world and our place within it remains to be seen. Historically, other works have attempted to leverage new technologies of visualization to take

FIGURE 12.1 Rolling right along in *Katamari Forever* (2009) for the PlayStation 3, one of many games in the *Katamari* series.

Source: screen capture by author.

the universe's measure with the same trademark curiosity, albeit a good deal more sobriety.

Exponential Vision and the *Powers of Ten*

In 1977, Charles and Ray Eames released the short film *Powers of Ten*, an approximately nine and a half-minute educational journey through space and the human body that uses the distinguishing framing device of an expanding and contracting white square, whose sides are determined by a power of ten, to demonstrate the differences in scale between astronomic and atomic levels of inquiry.[20] In its iconic opening scene, the film begins from a vantage point just a few feet above a man and woman picnicking by Lake Michigan in Chicago.[21] Looking down at the couple, as if pinned to the airy nothingness above them, the film gradually expands out to the then known boundaries of the universe (1,024 meters), then returns at accelerated speed to the blanket, only to plunge deep into the cells of the man's resting hand. Eventually, the film reaches the inverse magnitude of 10^{16} meters, or the scale of an individual proton.

As enticing as it may be to dismiss the film as propaganda for the triumphal march of science, with its obsessively tidy vision and authoritative male narrator (the voice of physicist Philip Morrison), the bulk of the film notably takes place beyond the limits of unassisted sight, venturing deep into both conceptual and pictorial speculation. As such, the film testifies not only to a centuries-old scientific desire for all-encompassing observation but also the fundamentally imaginative character of scientific epistemology. Alex Funke, a key contributor to the

1977 film, described the production staff's creative protocol for dealing with the twin limits to knowledge and imaging as follows:

> In preparing for the film, we first sought out at every power the very best pictures available, then asked workers in that particular realm what we might see if the imaging were a hundred, a thousand times better. We had the raw material [. . .]. Then in each case *we made the imaging more than real* through adding, by hand, the details of what might (or should) be there.[22]

Like the atomic landscapes or topographical maps now produced by nonoptical technologies like the scanning tunneling microscope, many of the images used in *Powers of Ten* are less direct imprints of actuality than mediated constructions, or enhanced renderings of the real. At the macrocosmic scales, the film dissolves between artful composites of satellite and observatory photos and visualizations of data garnered outside the visible spectrum via radio, ultraviolet, and infrared astronomy; at the microcosmic scales, the film relies heavily on scanning electron and transmission electron microscopy, but also takes representational liberties. "When there were only mental models, we made physical ones," explains Funke.[23] Thus the film's pointillist depiction of a proton at the interior of a nucleus "is no photo but an abstract symbol of the physics we just begin to comprehend."[24]

Based on an illustrated, young-adult book by Kees Boeke called *Cosmic View: The Universe in Forty Jumps* (1957), and itself the model for numerous later adaptations, among them games and online applications,[25] *Powers of Ten* and its rapid history of permutation in many respects recapitulates the longer account of scientific vision outlined by historians of science Lorraine Daston and Peter Galison. The film not only attests to the idiosyncrasy and inventiveness of scientific vision, but within the relatively brief span of fifty years or so, the film, its immediate precursor, and its many successors also make evident the decisive effects of medium and time period on scientific visualization. Any such film, made today, would not only have to cope with the increased scale of astronomical and biological observation, but would also have to contend with images of city, planet, and cell that have since proliferated and grown more fraught. While the original *Powers of Ten* is an unapologetic paean to the scientific imagination, depicting a world where couples lounge contentedly near "bustling" freeways, seemingly sandwiched between two wondrous worlds of undiscovered matter, the intervening decades have borne witness to growing environmental concerns and fears. A modern audience shown aerial views of Lake Michigan, Chicago highways, and the troposphere might be more likely to associate them with the invasive zebra mussel, automotive congestion, and greenhouse gases than with idyllic summer relaxation. Similarly, peering into the recesses of the cell and the atom today is likely to conjure debates over genetic modification, cloning, and nuclear energy—the common litany of post–World War II anxieties over the nature and extent of scientific progress.

Furthermore, given the pace of scientific and technological innovation since the film's original release, *Powers of Ten* belongs to a now bygone era of exceptional visualization. In the late 1960s and early 1970s, when astronauts on various NASA Apollo missions took some of the first full-view pictures of our planet from outer space, several of the resulting images, most famously the "Blue Marble," became icons of the nascent environmental movement.[26] In that historical period, one noticeably less saturated by satellite imagery, the sight of Earth suspended in the void of space highlighted the planet's singular fragility. Since 2005, however, satellite imagery of the Google Earth variety has become a staple of daily media use, moving beyond government, particularly military usage, to become the quotidian basis for everything from maps and driving routes to weather and traffic monitoring. In the contemporary moment, *Powers of Ten* loses much of its initial novelty, for now anyone with a smartphone or broadband-enabled computer can replicate the film's visual maneuvering from the terrestrial to the atmospheric.

Yet *Powers of Ten* captures a pivotal moment in the history of scientific visualization, in its bypassing of traditional print media in favor of cinematic animation. In the volume that followed the film, Eames supporters Philip and Phylis Morrison emphasized the superiority of the moving over the still image:

> No visual model can convey unaided the full content of our scientific understanding, the less if it is restricted to the static. [. . .] The limitation of the static image is not simply that it lacks the flow that marks our visual perception of motion: Real change in the universe is often too slow or too fast for any responses of the visual system. The deeper lack is one of content. A single take belies the manifold event.[27]

For the Morrisons, the advantages of film and its fledgling companion video derive from their capacity to present not only movement, but also change over time, leading them to conclude in this same passage that "Film and the video processes together constitute the most characteristic form of art in this changeful period of human history."

From our vantage point in the new millennium, we might well wonder what the Morrisons (not to mention the Eameses, who were well known for their interest in toys) would have made of video games, media that were still in their infancy at the time of the film's 1977 release. Could a game capture "the manifold event" even more readily than the conventional moving image? If so, would games then constitute the most characteristic form of art in this changeful period of human history?

"Your Personal Universe in a Box"

> Tired of your planet? Build a new one as you embark on the most amazing journey ever.
>
> —Spore packaging

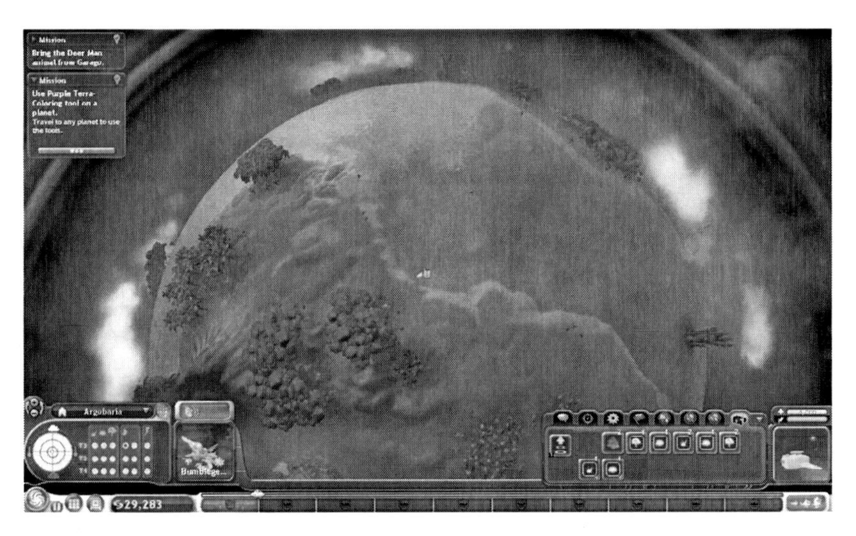

FIGURE 12.2 A spaceship (center) hovers over a planet's surface in *Spore*'s final stage.

Source: screen capture by author.

Developed by game luminary Will Wright and the studio Maxis, and published by Electronic Arts (EA) in 2008, *Spore* was ambitiously touted as an evolution-simulation game featuring five stages of species development: cell, creature, tribe, civilization, and space (Figure 12.2). Historically speaking, *Spore* represents the culmination of nearly two decades of Wright's work in the game industry, most of it on *Sim* games like *SimCity* and *The Sims*. The collective *Sim* games conveniently suggest something of Wright's broad-ranging fascination with environmental modeling, in their unresolved tension between an emphasis on environmental or biocentric concerns—how to manage a planet, an ant colony, an urban landscape—and an equal anthropocentric fascination with how the agents within those landscapes carry out their lives. In visual and experiential terms, the various *Sim* games also represent different points on a scale of magnification, from the global perspective of *SimEarth: The Living Planet* (1990) and the metropolitan perspective of *SimCity* (1989) to the neighborhoods and single-family dwellings of *The Sims* (2000–present) and the backyard dirt colonies of *SimAnt* (1991). However, *Spore*'s clearest predecessor in the *Sim* franchise is *SimEarth*, which invites players to supervise evolutionary scenarios running the gamut between open-ended experiments and theoretically pre-scripted paradigms like the "daisyworld" proposed by Andrew Watson and James Lovelock, in which the world and its inorganic and organic actors are posited as a holistic, self-regulating entity greater than the sum of its parts.[28]

Graced with all the computational and graphical advantages of the intervening nearly twenty years, *Spore* appears to offer an unprecedented level of virtual ecological detail. One could easily spend hours within the game's Creature

Creator, shaping one's image of the ideal species, and the game unobtrusively enfolds aspects of sciences ranging from geology, zoology, and ecology to climatology and astrobiology. Gameplay in the space phase, in particular, touches on long-cherished environmental principles like ecosystem stability and habitat renewal. For instance, one of the primary tasks of the stage is to render environmentally challenged planetoids hospitable enough for colonization, which requires substantial attention to both climate and species diversity. Players also carry out missions, some of which deal overtly with environmental crisis, such as the alarming directive, "Save planet *X* from ecological disaster!"

On the one hand, these tasks helpfully entreat the player to take on the mantle of environmental steward for colonized worlds; on the other, the espoused version of ecological care drastically oversimplifies life's complexity and threatens to perpetuate the myth that humans can exercise surgical precision in diagnosing and addressing environmental ills. Saving a planet, it turns out, often means hunting down and violently exterminating "infected" organisms using your spaceship's onboard laser. Restoring balance to an ecosystem translates into filling vacant animal or plant niches in a planet's food chain via the indiscriminate abduction of species from other planets. Even planetary climate correction becomes almost trivial given the ready availability of futuristic tools. Not enough atmosphere? Not a problem—toss an atmosphere generator at the planet surface and watch clouds of reassuring-looking gases drift into the troposphere. Climate too chilly? Rain a toasty meteor shower down on the world or apply a heat ray to begin a warming trend. Using weapons-grade lasers, high-tech rays, and elaborate mechanical gizmos to bludgeon a planet's climate into shape makes a mockery of the delicate "butterfly effects" espoused by chaos theoreticians to describe the extremely sensitive dependence of final states on even seemingly unrelated or minor initial conditions. In sum, *Spore* does pose ecological lessons, but those lessons verge on environmental slapstick.

Not surprisingly, after the game's much-heralded release scientists quickly realized that *Spore* had fallen short of its advertised marks. *Science* magazine's John Bohannon called the game a massive disappointment in terms of its potential for science education, even after granting that its primary aim was to please rather than inform. After playing *Spore* with a team of scientists to evaluate its scientific merits, Bohannon ultimately flunked the game, lamenting that it got "most of biology badly, needlessly, and often bizarrely wrong," particularly in its treatment of evolution.[29] Two of the scientists who helped to assess the game, evolutionary biologists Ryan Gregory and Niles Eldredge, similarly concluded that "*Spore* is essentially a very impressive, entertaining, and elaborate Mr. Potato Head that uses the language of evolution but none of the major principles."[30] Foremost among *Spore*'s many evolutionary inaccuracies is the complete lack of consequence for player death! *Spore*'s much ballyhooed version of evolution is, in fact, closer to the long discredited theory of Lamarckian evolution (in which an individual organism can develop and pass on adaptations during its lifetime) or evolution's creationist-tending nemesis, intelligent design (where players are the universe's unseen architects). In the eyes of scientists, *Spore* deploys evolution primarily as a marketing gimmick; the theories of genetic succession are less the guiding force for actual game mechanics than rhetorical trimming around the digital dollhouse play for which Wright has become famous.

In comparing *Spore* to a glorified Mr. Potato Head, Gregory and Eldredge acknowledge that the protracted, random nature of real evolution is noticeably at odds with the logic of immediate customization that makes *Spore* so attractive. The game's versatile Creature Creator software has arguably proven more popular than the game itself; it is a feature, or a subgame, that threatens to render the rest of the game a mere showcase for the well-crafted avatar. In a similar fashion, the bulk of the game's command interface attends to matters of aesthetic preference. Should you find yourself displeased by the lumpy contours of your planet or its dull sandy color, you can use special tools to level terrain, form "cute" canyons and crystalline mountains, or turn the sea purple, the atmosphere red, and the land cyan. Incredibly, none of these changes seems to affect life on the planet, implying at some fundamental level that cosmetic alteration need not affect environmental health. Like the Creature Creator, which essentially equates evolution with deliberate customization in a digitally enhanced production mentality at odds with the vagaries of actual evolution, *Spore*'s building-, vehicle-, and planet-design menus emphasize the malleability of matter, not so much its ontological essentialism as its receptivity to the expression of individual preference.

Criticisms like these reveal an unavoidable tension in using games to model environmental processes—the necessary give-and-take between player freedom, or agency, and ecological constraint. The worlds that Wright creates tend to be sandboxes more than slides, that is, open-ended systems inviting experimentation rather than goal-oriented spaces centered on measurable achievement. Reviewers and players disappointed by *Spore*'s lackluster gameplay have therefore sometimes generously allowed that *Spore* is less a game than a software toy, and Wright himself has called *Spore* a "philosophy toy," designed to lead younger generations to insights via self-directed investigation.[31] At the same time, *Spore*, like most of Wright's games, is recognizably a "God game," meaning that players act as omnipotent beings whose every action influences the universe in which they operate. Tellingly, Wright has said that he wanted players of *Spore* to feel like George Lucas, not Luke Skywalker—that is, the architect of fantastic worlds rather than an individual within them. *Spore* accordingly reflects Wright's valorization of human agency and intentionality. From an environmental standpoint, *Spore* models the strain between envisioning nature as either a design space or a problem space, or a place of invention and expression versus an arena fitted with recognizable troubles and solutions.

While it might be tempting to read *Spore* as an exercise in frivolous and, in the end, noncommittal play, Wright has publicly vaunted the game's potential to underscore environmental objectives. During his demo of *Spore* to TED

Conference participants in March 2007, Wright used his in-game spaceship to pump huge amounts of carbon dioxide gases into a planet's atmosphere, thereby raising its ocean levels, swamping his own cities, and increasing the temperature of the planet to a point where the oceans evaporated and the surface burst into flame—clearly not a winning strategy so much as a curiosity-driven experiment. Having done this, Wright casually remarked:

> What's interesting to me about games in some sense is that I think we can take a lot of long-term dynamics and compress them into very short-term kind of experiences, because it's so hard for people to think fifty or a hundred years out, but when you can give them a toy and they can experience these long-term dynamics in just a few minutes, I think it's an entirely different kind of point of view, where we're actually mapping, using the game to remap our intuition. It's almost like in the same way that a telescope or microscope recalibrates your eyesight. I think computer simulations can recalibrate your instinct across vast scales of both space and time.[32]

Wright thus asserts that games can act as intellectual and spatiotemporal prostheses, in language heavily reminiscent of Marshall McLuhan's contention that media act as extensions to humankind. *Spore* evidently has the power to reveal to us the dramatic consequences of our current follies, here the overproduction of greenhouse gases that trap the heat of the sun's rays and lead to global warming. But what may be more important is the residual impression that environmental catastrophe of this sort is neither unforeseeable nor inevitable.

Unlike the iterations of a purely scientific model or the preset narratives of film or science fiction, a game like *Spore* offers both repetition and difference, as directed by the personal choices of the player—what gamers would call replay value. *Spore*, which by design has been freed from the constraint of a single "win state," opens up an ethically unencumbered space in which players can spool out countless environmental futures, from pastoral empires to admittedly morbid fantasies of ecological disaster. Put more broadly, what *Spore* perhaps does best is it gives players the ability to experience and affect procedural change at scales ranging from the microscopic to the galactic and from the short- to the long-term, effectively heeding the scalar warnings of ecologists. The game's deliberate open-endedness forces players to ponder the benefits and drawbacks of interaction at each level. Like scientists who study and model real-world environments, players in virtual worlds may find themselves struggling with a similar set of questions: what is the value of remaining at one scale, and when is it necessary to move beyond that scale to examine the relations or transgressions that occur across the artificially imposed boundaries of hierarchical thinking?

The effective visualization of ecological states at a range of scales has become crucial, given contemporary environmentalism's trouble with representing largely intangible problems and its internal frictions between what we could call the microenvironmental and macroenvironmental approaches to pressing ecological problems. Although games inevitably participate in flows of material and capital and so-called attention economies that place them squarely within the ongoing debate over local, as opposed to global, modes of thinking and living,[33] digital games can obviate the perceived choice through multiscalar play. Both Ursula Heise and Timothy Morton have suggested that environmentalists have lost sight of the large-scale nature of environmental challenges in their well-intentioned espousal of place. Heise wonders about the aesthetic possibilities of a deterritorialized environmental vision, while Morton exhorts us toward expansive, even cosmological thinking. The local and the global are, after all, only imagined extremes, beyond which lie countless microorganisms, elementary particles, and most of the known universe. If we could, like a Sporovian spacecraft, escape the anthropomorphic drag of the local-global dualism, perhaps our earth-friendly bumper stickers would urge us to Think Galactically, Act Microscopically.

Playing the Game of Nature

As unorthodox as the thought may be, video games may be even better suited to scientific visualization than the conventional moving but noninteractive image,[34] and support for such a proposition may be found well outside the bailiwicks of game scholars and educational software developers. Daston and Galison, and Colin Milburn, for example, have independently identified the same trend in scientific imaging—away from depiction toward fabrication, at a point where the formerly distinct boundaries between recording and producing have been breached. For my purposes, the value of what Milburn, Daston, and Galison believe to be a recent paradigm shift from ocular to tactile science, or perhaps the unexpected convergence of visual and haptic epistemologies, lies in its evident recapitulation in less rarefied media contexts. Well outside the elite research laboratory, in millions of ordinary living rooms and home offices, computer and video games have popularized the same qualitative shift beyond vision toward interactivity, in roughly the same period (as Milburn chronicles, nanotechnology flourished from the 1980s onward). The player of a game like *Spore* is thus akin to the archetypal scientist of the latest representational epoch described by these authors—one who melds creativity and intuition with the efforts of instrumental science.[35]

Games are not easy solutions to vexing scientific problems or palliative alternatives to real-world action, but there are felicitous similarities between gameplay and ecological work. In fact, the same ecological literature that earlier outlined the difficulties with scalar modeling also reveals some of these foundational correspondences. First, all games are at some level exercises in controlled experimentation, or what ecologists would call "perturbation experiments," meaning that some factor is manipulated in certain units while other units are left as unmanipulated controls. By necessity, too, both ecologists and game designers must distill the richness of real-world systems into manageable experimental structures through modeling and dimensional analysis, though perhaps only ecologists feel they must account for the effects of the reduction in complexity. A third method that scientists within and beyond ecology use to reconcile scales, which has several counterparts in games, is ground-truthing. Ground-truthing refers to observations made, usually on land, which are then applied to calibrate remote-sensing devices like satellites through confirmation or denial of their measurements. Players of games regularly engage in a kind of ground-truthing when they consult overhead mini-maps or other navigational aids, and a growing number of alternate-reality games and GPS/GIS-based activities like geocaching also explicitly meld game fiction with investigations at actual coordinates.

In one area, however, games are far better suited to ecological modeling and experimentation than actual scientific studies—namely, system breakdown. Working ecologists cannot destroy or arbitrarily change real-world environments without very good reason, and often they must opportunistically wait for events like fires and oil spills to study things like ecosystem response to extreme conditions. In games, quite the opposite holds. The ludic impulse encourages sometimes methodical, sometimes rambunctious trial and error, and in some games, disasters, both natural and manmade, are only a click away. In some ways, this extends the insights in Jesper Juul's recent volume on the art of failure in video games to suggest that failure can be seen not just in the sense of personal fault that Juul dwells on, but also the kind of systemic failure we invoke when speaking of unhealthy bodies or ecological units.[36] For now, we might tentatively call these four modes of ecology-game crossover experimentation, modeling, verification, and failure, all of which demand thoughtful interplay between scales, players and designers, and environmental experience and speculation. *Spore*'s modest successes in this area owe much to the influence of *Powers of Ten*, not least because *Spore* embeds its player in neatly nested experimental domains, treating developing life at successive orders of magnitude, from the microscopic to the macrocosmic. Wright also pays unmistakable homage to the film in the game's culminating space stage, by allowing players to control in-game perspective through the use of their mouse wheels—scroll the wheel forward and your spaceship descends from orbit through layers of atmosphere to the chosen planet's surface, where you can skim the ground to search for native flora and fauna or engage city populations. Scroll the wheel backward and your spaceship lifts off and returns to the microgravity of outer space. Keep scrolling, and the game perspective widens from planet to solar system and finally to the entire galaxy, where in much accelerated time, you can watch spinning celestial arms crammed with the twinkling lights of dying stars.

Unlike viewers of *Powers of Ten*, however, *Spore* players may navigate between these different scales at will, guided by such sundry motivations as curiosity, goal-oriented achievement, aesthetic preference, or perhaps even morbid or whimsical brands of environmental schadenfreude, as when Will Wright nonchalantly destroyed a game planet for his TED conference audience. In ecological jargon, we might also designate game environments or virtual worlds as mesocosms, that is, arenas of a size usefully intermediary between field experiments and laboratory conditions. Like portions of a field sectioned off for study, or partially enclosed waters, game ecologies toy with select variables within environments that remain close to but apart from life. The best games, like the most successful ecological experiments, tread a fine line between bounded tidiness and inclusive reality, heightening our awareness of mechanism while providing ample outlets for our energy and curiosity. Games are inherently multiscalar—melding the quantitative and the qualitative, the experiential and the analytic, the computational and the graphical—and a universe of questions awaits.

Notes

1 "Professor Emeritus Al Bartlett—Physics at University of Colorado at Boulder—Articles on Exponential Growth, Peak Oil and Population Growth, Sustainability, Renewable Resources and the Environment," http://www.albartlett.org/ (accessed June 30, 2015).
2 Willard McCarty, "Modeling: A Study in Words and Meanings," in *A Companion to Digital Humanities*, edited by Susan Schreibman, Ray Siemens, and John Unsworth (Oxford: Blackwell, 2004), para. 3, http://www.digitalhumanities.org/companion/ (accessed September 29, 2014).
3 From Dr. Seuss, *The Lorax* (New York, NY: Random House, 1971). Over the years, environmentalists have had decidedly mixed attitudes to the so-called population problem, including the many uses and abuses of Thomas Malthus's *Essay on the Principle of Population* (1798), Garrett Hardin's "The Tragedy of the Commons" (1968), Paul Ehrlich's *The Population Bomb* (1968), and Alan Weisman's *Countdown* (2013). Often, "scale" is a tempting euphemism for optimum population, or sustainable human numbers.
4 See Timothy Morton, *The Ecological Thought* (Cambridge, MA: Harvard University Press, 2010), 1. For Morton, hyperobjects are "things that are massively distributed in time and space relative to humans" and therefore impossible to grasp in their entirety, such as radiation, black holes, and global warming.
5 Unfortunately, ecology's overuse by ordinary people, well-meaning environmentalists, and diverse academics has evacuated it of its original, largely scientific meaning. Ecology textbooks today bear little imprint of previous and ongoing contests over its scope and meaning, while everyday use of the term has transformed it into a flaccid descriptor connoting everything from interdependence to the study of all things natural. For more on ecology's history and its ties to cybernetics, see the chapter by Erica Robles-Anderson and Max Liboiron in this volume.
6 Ralph Waldo Emerson, "Nature," in *The Collected Works of Ralph Waldo Emerson, Volume I: Nature, Addresses, and Lectures*, edited by Robert E. Spiller and Alfred R. Ferguson (Cambridge, MA: Harvard University Press, 1971), 32.
7 Edward R. Tufte, *The Visual Display of Quantitative Information* (Cheshire, CT: Graphics Press, 2001).
8 For the latter, see Chris Tong, "Ecology without Scale: Unthinking the World Zoom," *Animation*, 9, no. 2 (2014): 196–211. Tong recommends jettisoning scalar models because they reify unfortunate spatial and classificatory hierarchies.
9 According to Brody Sandel and Adam B. Smith, although they first find the more specific search phrase "spatial scal*" in an article from 1987. See Sandel and Smith, "Scale as a Lurking Factor: Incorporating Scale-Dependence

in Experimental Ecology," *Oikos*, 118, no. 9 (2009): 1284–1291, doi:10.1111/j.1600-0706.2009.17421.x.

10 John A. Wiens, "Spatial Scaling in Ecology," *Functional Ecology*, 3, no. 4 (1989): 385.

11 J.L. Dungan, J.N. Perry, M.R.T. Dale, P. Legendre, S. Citron-Pousty, M.-J. Fortin, A. Jakomulska et al., "A Balanced View of Scale in Spatial Statistical Analysis," *Ecography*, 25, no. 5 (2002): 626–640.

12 Technically speaking, we still live in the Holocene, although the Anthropocene has attracted numerous adherents. In fact, Erle C. Ellis and Navin Ramankutty argue that humans have had such a tremendous impact on the Earth's surface that we should now think in terms of "anthromes" rather than climate- and vegetation-defined biomes. See Ellis and Ramankutty, "Putting People in the Map: Anthropogenic Biomes of the World," *Frontiers in Ecology and the Environment*, 6, no. 8 (2008): 439–447.

13 Consider Paul Martin's prehistoric overkill (of megafauna) hypothesis, described in Elizabeth Kolbert, *The Sixth Extinction: An Unnatural History* (New York, NY: Henry Holt, 2014), 231–232.

14 A quadrat is a sampling plot typically measuring around 1 square meter.

15 Göran Englund and Scott D. Cooper, "Scale Effects and Extrapolation in Ecological Experiments," *Advances in Ecological Research*, 33 (2003): 170. This idea of effective scale recalls early naturalist philosopher Jakob von Uexküll's theories about organismal time and space, most notably captured in his description of the tick. See Jakob von Uexküll, *A Foray into the Worlds of Animals and Humans*, trans. Joseph D. O'Neil (Minneapolis: University of Minnesota Press, 2010), 44–52.

16 Wiens, "Spatial Scaling in Ecology," 389.

17 Englund and Cooper, "Scale Effects and Extrapolation in Ecological Experiments," 170. For instance, some influences that figure largely at small scales, like experimenter disturbance or predation, decrease and grow almost negligible at very large arena sizes, in what Englund and Cooper call an asymptotic or logistic relationship.

18 N. Underwood, P. Hambäck, and B.D. Inouye, "Large-Scale Questions and Small-Scale Data: Empirical and Theoretical Methods for Scaling up in Ecology," *Oecologia*, 145, no. 2 (2005): 177–178.

19 I cannot help but think of Jorge Luis Borges's "On Exactitude in Science," in *Collected Fictions*, trans. Andrew Hurley (New York, NY: Penguin, 1998). Borges's notoriously pithy story could be fruitfully applied to many contemporary projects, from genome sequencing to database-driven big-data initiatives.

20 A shorter, black-and-white "rough sketch" of the film was released in 1968, made for the Commission on College Physics. The 1977 color version was funded by IBM.

21 In actuality, the live portion of this memorable picnic scene was filmed in Los Angeles, for greater production control. The picnic scene in the 1968 sketch took place in Florida.

22 Philip and Phylis Morrison and the Office of Charles and Ray Eames, *Powers of Ten: A Book about the Relative Size of Things in the Universe and the Effect of Adding Another Zero* (New York, NY: Scientific American, 1982), 145 (my emphasis).

23 Morrisons, *Powers of Ten*, 145.

24 Ibid., 101.

25 The film has inspired many reincarnations, for instance the artists' collective Futurefarmers' *A Variation on the Powers of Ten* and Cary and Michael Huang's web-based applications *The Scale of the Universe* and *The Scale of the Universe 2*.

26 Most notably, Stewart Brand lobbied for the release of these planetary images and used them on a number of the covers of his counterculture magazine, *The Whole Earth Catalog*, published between 1968 and 1974.

27 Morrisons, *Powers of Ten*, 8.

28 The daisyworld theoretical models attempted to provide support for Lovelock's Gaia hypothesis by demonstrating that the population behavior of living species (in this case, black or white daisies with different albedos, or reflective properties) could explain the

planet's apparent ability to regulate its own atmospheric and surface temperatures in response to varying solar luminosity.

29 John Bohannon, "Flunking *Spore*," *Science*, 322, no. 5901 (2008): 531.

30 Ibid.

31 Will Wright, "*Spore*, Birth of a Game" (presentation, TED2007, Monterey, CA, 9 March 2007).

32 Ibid.

33 Many environmental specialists, among them Lowell Monke, David Sobel, and Mitchell Thomashow, have expressed serious misgivings about digital media, aligning them with a distracted inattention to lived space.

34 A worthwhile point of comparison would be the viral "molecular movies" described by Bishnupriya Ghosh in this volume.

35 See Lorraine Daston and Peter Galison, *Objectivity* (New York, NY: Zone Books, 2007), 46; Colin Milburn, *Nanovision: Engineering the Future* (Durham, NC: Duke University Press, 2008). In Daston and Galison's words, "practitioners of trained judgment professed themselves unable to distinguish between work and play—or, for that matter, between art and science. [. . .] surrendering themselves to the quasi-ludic promptings of well-honed intuitions." Milburn's nanotechnology professionals are no exception; some of the most iconic images to emerge from nanoscience have had less to do with function and "serious" research than an artistic sense of play, for example, Donald Eigler and Erhard Schweizer's now-famous creation of the IBM logo using xenon on nickel.

36 Jesper Juul, *The Art of Failure: An Essay on the Pain of Playing Video Games* (Cambridge, MA: MIT Press, 2013), 45.

13

TOWARD SYMBIOSIS

Human-viral Futures in the "Molecular Movies"

Bishnupriya Ghosh

> Our general notion of the structure of the universe leads us therefore to expect that we might well meet with things which are not so alive as the sunflower and not so dead as a brick, and the phenomena that we study under the heading "filterable viruses" suggest that we now have sight of some of this intermediate group.
>
> —A.E. Boycott[1]

Somewhere between a brick and a sunflower rose a strange thing whose status as a living organism was unclear until the cracking of its genetic code. Writing in *Nature* in 1929, one researcher, a professor of pathology at the University of London, defined the virus as an "intermediate" organism, inorganic and organic, dead and alive. The rumination was hardly unusual for the time. After joining the ranks of the life sciences around 1800, the discipline of biology reverberated with vigorous debates about the essence of living things until 1944, when the physicist, Erwin Schrödinger, provided a simple but lasting distinction of living things. Schrödinger maintained that living things were defined by their capacity for self-regeneration (to grow, repair, and reproduce), their fight against entropy, and their tendency toward a sustainable equilibrium.[2] His distinction recharged evolutionary inquiries into the virus: Was the virus a living organism, if it could not live without its host? Could it be one of the first organisms (a pre-DNA cellular form) in a four-billion-year-old primordial soup, a relic with primitive RNA? Or was it just an organism that had lost the evolutionary war, a fugitive from the host genes that had degenerated into a parasitic lifestyle?

Half a century later, the question of what constitutes "life" has returned with full vigor as scientists turn to designing and building "life itself" or the biological substrates of living organisms, molecule by molecule. In the cool confines of the lab, the virus "lives" in wetware (sections, cultures, samples) and in software (digital files), still a stranger seeking refuge in living hosts. But now there are sophisticated means of mediatizing the virus in order to understand and potentially to transform its behaviors. This chapter concerns one of these means: scientific visualizations, or SciVis, that "make the unseen visible," as the mantra goes from the first SciVis congregation of 1987 (the Visualization in Scientific Computing Conference).[3] Within the larger field, I focus on 3-D "scientific animations" born of collaborations among academic researchers, biotech corporations, and digital animators that are affectionately dubbed "the molecular movies" (because they feature molecular-level events). Ostensibly representations of primary data, at their most expressive, they conjecturally fill in the gaps in research, and speculatively bring the nonobservable into existence. Hence they function as collaborative platforms for scientists and animators to produce the experimental data necessary for designing source materials (digital files for synthetic compounds) to be then actualized as marketable "biologicals" (biotechnological products). Backed by industry, the molecular movies are tied to biotech and biomed markets and, therefore, to research seeking to intervene in vital processes; in the case of human-virus relations, the stress falls on blocking, altering, modifying, or interrupting the processes of pathogenic virulence. If the broader goals of research on viruses are to alter the terms of debilitating or deadly biological partnerships, then the molecular movies are critical to such an evolutionary agenda.

The perspective from evolutionary biology situates the molecular movies as sustainable media. I track the role that scientific visualizations of molecular events play in the repair of "unsustainable" human-virus relations—that is, pathogenic relations in which the virus, as obligate parasite, depletes host resources so completely that they become nonrenewable. When the depletion is not immediately deadly (in contrast to the case of Ebola, for example), the hope is that molecular-level repair, together with the regeneration of the host's vital capacities, can build sustainable relations between parasite and host. Scientific visualizations are constitutive of that repair and regeneration, of biomedical and biotechnological interventions necessary to sustain complex living systems. Certainly technological repair and regeneration has immediate or short-term consequences, enhancing a patient's capacity to live longer. But successful changes in human-virus interactions also have long-term implications for organismic relations. We have learned this most acutely from drug-resistant bacteria that adapted and modified their behavior in response to biomedical interventions. In the case of viruses, since many cannot be eradicated with drugs (there is no equivalent to penicillin to intervene in human-virus relations), the biotechnological ambition is to alter—cut, snip, reconfigure—segments of viral DNA, so as to change viral behavior. Molecular-level change is the baseline for changing human-virus organismic relations, enabling the host to "live with" a once-hostile pathogenic parasite. In the closing section of this chapter, I suggest such a move *toward symbiosis* is equally a call for ecological sustainability. The actualization of that call in scientific visualization folds the molecular-cellular event into the deep timescales of the planetary. It is in this context that I pose the question: to what extent are molecular movies featuring human-virus interactions "sustainable media"?

The speculations of the early twentieth century return with renewed vigor: What is life? Is the virus alive? Can its host survive the virus touch? And with those questions the organismic scale of evolutionary biology returns to haunt the single molecular event that is the classic tale in a molecular movie. Take the example of a 1.42-minute 3-D animation of the flavivirus (the pathogenic agent of dengue fever) entering a human cell, in which the viewer encounters the slow transport of the teal-colored sphere into the cell, set to a minimal extra-diegetic score. There are "close-ups" of proteins and recessed images of membrane. Red proteins unfold like coral blooms (Figure 13.1a), then fold and twist like insects (Figure 13.1b) in their work of fusing the cellular and viral-caspid membranes.

The denatured artifice of the digital image is made explicit in the strongly expressive production: the extra-diegetic score, the elegant and controlled slow motion, and the "subjective" visual grammar.[4] "My proteins are always red," Janet Iwasa, one of the collaborators on the film told me, when I interviewed her in January 2013. Distributed through digital platforms, *Dengue Viral Entry* was a collaboration among Gaël McGill, the director of the molecular visualization outfit at the Harvard Medical School (HMS) Center for Molecular and Cellular Dynamics; Stephen Harrison, faculty in Structural Biology at HMS; and Janet Iwasa, at that time a postdoctoral student at the HMS Department of Cell Biology and now in the Department of Biochemistry at the University of Utah. In the interview, Iwasa spoke enthusiastically about the dengue film made for Boston's WGBH television station, and her personal favorite, *Clathrin-Mediated*

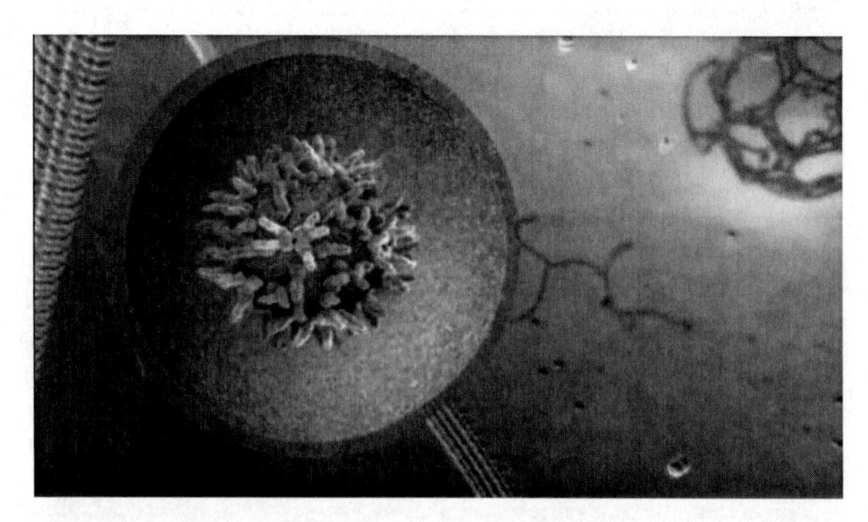

FIGURE 13.1A Dengue virus proteins undergoing change.

Source: Dengue Viral Entry, credit: Janet Iwasa.

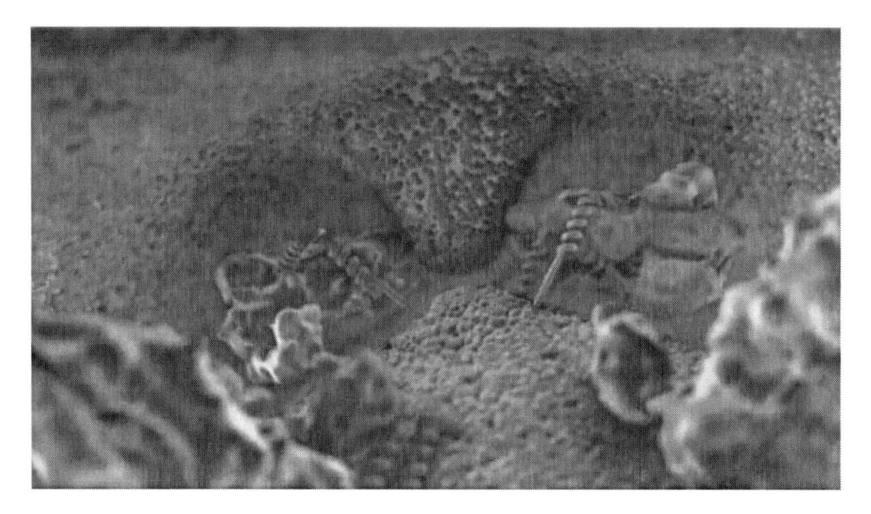

FIGURE 13.1B Folding glycoproteins fusing membranes.

Source: Dengue Viral Entry, credit: Janet Iwasa.

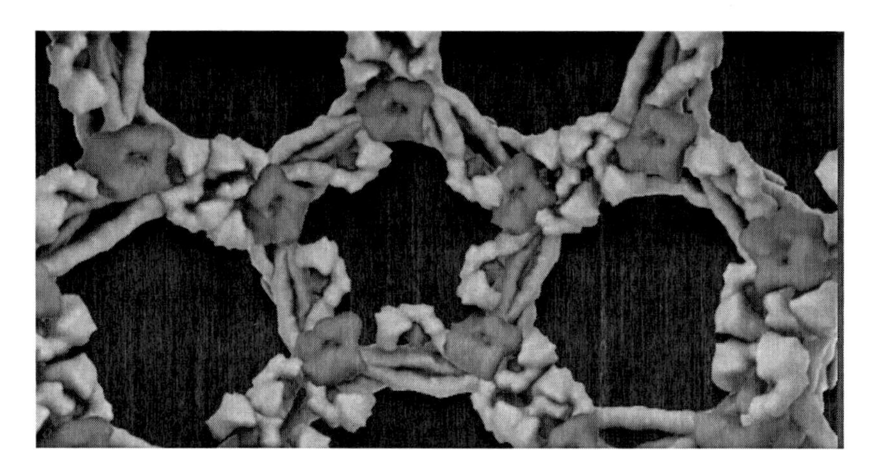

FIGURE 13.2 Clathrin Lattice.

Source: Clathrin-Mediated Endocytosis, credit: Janet Iwasa.

Endocytosis, made for research purposes. In both, a personalized visual grammar underscores the strongly interpretive dimension of the works. In *Dengue Viral Entry*, that dimension is self-consciously presented through the restrained voice-over that sequences the digital images into a cellular story. While *Clathrin-Mediated Endocytosis* does not editorialize through a voice-over, it has the more resolutely expressive extra-diegetic score: it is set to Rimsky-Korsakov's "Flight of the Bumblebee," suggesting, Iwasa maintained, the intricate cultural labor of making the lattice from sixty separate strands (Figure 13.2).

Such shared features establish continuities between films made for slightly different target spectators/users: the first made for the public (students, researchers, citizen-scientists), and the second for the lab-based study of physiological signals. Both explicitly generate wonder, be that at the technological bravado of wandering to the body's interior or more heavy-handed generic cues that encourage spectators to bask in a fantasy of intergalactic travel. The latter is an emergent aesthetic in viral entry film: as viral orbs circle and implode inside the human cells, some critics are beginning to theorize an "asteroid aesthetics."[5] The cellular membrane remains a *partially* visible curved line on one edge of the frame, a recurring visual design folding the extracellular environment into the intergalactic one. Against these recesses, the curve serves as anthropogenic autoreference for the distinction of the human cell/earth in this encounter with a foreign, possibly hostile, macromolecule. With such scalar switches, the human-virus relation becomes a molecular-cellular-organismic event. To be sure, the perspectival switches between scales are clearly a sign of the times, given advances in optical technologies (telescopes to microscopes) that can efficaciously manage distances, depths, and sizes.[6] But beyond technological capacity, the scalar switches *perform* "emergence" as an event unfolding nonlinearly at multiple levels. In molecular movies on human-virus interactions, we see the emergence of the molecular event that is at once an evolutionary one.

Perhaps in marvelous odysseys of this sort we are witnessing something like a "cellular fantastic," the successor to what Akira Lippit once theorized as the early twentieth-century "optical fantastic."[7] If the conjunction of X-ray, nuclear fission, and psychoanalysis sent a shiver down the spine when the body's interior was exposed to light, we are at another such crossroads today in the age of bioengineering/biotechnology/biomedicine. Perhaps the palpable sense of the *wonder* in these scientific animations has everything to do with the new plasticity of the body, whose interior can be redesigned, rebuilt, and manipulated. Wondrous technological feats are possible, as research and citizen-scientists embrace the speculative science—in this instance, the science of designing and building new human-virus relations.[8]

Making Molecular Movies

Scientific visualization as a historical process has always been crucial to studying viruses, since these microorganisms were originally identified as "filterable agents" passing through Louis Pasteur's porcelain Chamberland filters. Virological research moved beyond classification to morphological study only after the first micrographs of the virus appeared in 1938. The optical image dominated its study until the mid-twentieth century, until the digital transcription of viral informatic codes turned the virus into a bioinformatic actor. Enter the molecular movies. By the last decades of the twentieth century, the boom in the business of designing and building "life itself" interlocked technological and biological processes ever more tightly. As biological substrates became subject to the mechanisms of "reproduction, flattening, and patenting" as were other media forms, Sarah Kember and Joanna Zylinska argue that media scholars turned to study "life itself" as a medium.[9] Eugene Thacker, one of the first scholars to analyze the processes of mediation that altered and modified biological substrates, locates "turning information into flesh"—DNA into digital file and then back into synthetic compounds—at the historical conjunctions of molecular biology and the information sciences in the 1970s. Thacker's groundbreaking *Biomedia* (2004) mainly elaborates informatic mediatizations, rewriting the code of "life," and not representational media, even as he readily admits that there are branches of biological research in which expressive modulations of primary data are indispensible to reproducing life itself. Scientific animations exemplify such media.

The buzz is that Hollywood has come to the research lab. There are "molecular movie stars" at annual SciVis meetings, and virtuoso special-effects animators who divide their time between filmmaking and 3-D lab instruction.[10] One of Iwasa's collaborators, Gaël McGill, straddles what was once the "media arts" (image compositing, special effects, storyboarding) and once lab-based scientific research. McGill founded Digizyme, a company for designing molecular and cellular visualizations, and hosts an online directory of molecular and cell animations.[11] Most recently, he leads a project funded by the National Science Foundation (NSF) to develop a free software toolkit for scientific researchers, "Molecular Maya" (a streamlined version of Maya, the software used by animators at Pixar). McGill's entrepreneurial outfit services the new interdisciplinary field of molecular visualization, which demands that some scientists learn techniques of animation hitherto taught as media production. Trained as both cell biologist and animator, a researcher like Janet Iwasa is in high demand. As she recounted in the January 2013 interview, biologists with different specializations frequently approach Iwasa to collaborate on visualizations of competing or incipient hypotheses. The idea is to present these hypotheses on online interactive platforms to the larger scientific community. What excites Iwasa is the creative dimension to the collaborations, the expressive freedom she has to animate speculations. For *Clathrin-Mediated Endocytosis*, Iwasa generated storyboards from discussions with an expert in the field (Tomas Kirchhausen), researching databases for already available datasets, and reading relevant scientific papers. For the *Dengue Virus Entry* animation, however, the storyboards were provided for Iwasa, who then rendered the frames and created the digital assets. Since the latter was pitched not at researchers but for a public platform,[12] Iwasa's film was postprocessed (adding sound, motion through video capture, and additional effects) at the XVIVO design company known for the award-winning *Inner Life of the Cell* (2006). The films give us a glimpse into an emerging media industry in the business of molecular movies, some for scientists only, and others for a wider public. *Dengue Virus Entry*—featuring a flavivirus, a group that includes yellow fever and West Nile—exemplifies the latter: scientific animations as edutainment for schoolchildren, university students, science buffs, museum audiences, and other citizen-scientists. As such, it commences with visual text on dengue as incurable, imbuing the images we are about to see with historical and therapeutic value. The former, *Clathrin-Mediated Endocytosis*, is of the other kind: a molecular movie for the "working through" of an incipient scientific hypothesis,[13] sequencing the logic of existing but partial data into what remains still unverified.

For scientists, molecular movies open to new frontiers in laboratory research. As Stephen Harrison (Iwasa's collaborator) notes, scientific animations are transforming the study of structural biology, once the leading science for morphological studies of the virus.[14] Reflecting on the current state of his field in 2004, Harrison pushes for coordinating new technologies such as "super resolution microscopes," "new image-analysis software," and more "flexible fluorescence tagging" for the better articulation of molecular structures and their functions in the cellular context.[15] The hope is for a more integrated approach to the physiological dynamism of cells; that hope, Harrison maintains, lies in visualizing the 3-D architecture of molecules (structural biology) *and* their dynamic functions in the cellular context (cell biology). Physiological signals, for instance, alerting the human immune system to viral proteins at the cellular membrane captivate structural and cell biologists. The consequent push to recombine protein associations in order to manipulate "recognition modules" mandates the integration of the study of individual molecular machines (the purview of structural biology) into the larger scale of studying cellular interaction domains (the concern of cell biology); in a film like *Dengue Virus Entry*, the result is the design of a molecular-cellular event. Since physiological signals are "transient, contingent, and complex," what is crucial to their study is greater insight into the *morphing* structures of protein molecules and those conformational biochemical changes in extracellular environments that allow attachments between membranes.[16] Such a task seems ever more feasible with live cell imaging, high-resolution optics, and image-processing technologies.

Collaborating on a scientific visualization, then, makes it possible not only to share research, but also to test new hypotheses from any one specialization in the context of another—in short, to gain a systemic snapshot of the molecular event. The new media not only generate new media industries, but also intellectual and economic investments in altering vital processes of reproduction. It becomes possible for scientists, artists, and industrialists to collaborate on scuttling pathogenesis, on interfering in human-viral interactions. The robust marketplace for biotechnological and biomedical products looms large, orienting the lab toward its industrial context. And so, in the *most* lab-based studies devoted to the virus—the field of virology[17]—the virus is disarticulated into biological substrates to be altered, cut, snipped, and manipulated; it is no longer a foreign submicroscopic particle to be destroyed and eradicated.

The Virological Enterprise

Lab-based virology is generally concerned with how regular cellular processes are interrupted or misdirected in encounters with viruses. The key is to know how to intervene in cellular disturbances that create mass cell death in living organisms. In the case of pathogenic viruses, after the first medical interventions that sought to eradicate them, clinical virology became increasingly preoccupied with experimental data as booming biotech industries upped the ante on harnessing and optimizing viral biopotentials. Lentiviruses' ability to cut and paste their RNA into human DNA, for instance, presented opportunities for the (still controversial) use of such retroviruses as "viral vectors" for gene therapy.[18] If such adventures put an onus on optimization as a social and economic good, they fuel virological research behind both biomedicine's attempt to defang the pathogenic enemy (inhibiting its entry into cells, neutralizing its recoding of human DNA) and biotechnologies' search to employ the virus as strategic ally (as in the case of gene therapy with viral vectors). As we shall see, it is the new mediatizations of the virus that make possible virtual and actual manipulations of existing human-virus interactions.

Just how they achieve this requires some explanation. To be sure, scientific animations are still in the business of producing technologically enhanced images from "direct observation," the staple procedure for scientific research in the biological sciences. They construct a digital image composited from a range of datasets, including optical images that constituted the first mediality of the virus. That story is well known: Ernst and Helmut Ruska's transformative introduction of the electron microscope to lab-based virology in 1938, an optical instrument with shorter wavelength than the light microscope and, therefore, capture-capable of (thousands of times) higher resolution images. Under the electron microscope, for the first time, the submicroscopic virion, a nucleic acid strip with an outer coat of protein (the capsid), was clearly visible in the needle-like biophysical structure of the tobacco mosaic virus (TMV), the model organism[19] of virological study for many decades. Researchers reveled in the new technological capacity to *see* and to *touch* the crystallized form, to probe and move viral particles, with an electron beam that excited positive and negative charges of histological dyes. That excitement gave way to the celebration of the gene as the gauge of biological specificity in the mid-twentieth century, and optical mediatizations of the virus took a backseat to digital ones.[20]

For a microbial species hounded by the definitions of what constituted "life," such a turn first meant a shift from morphological considerations to the biochemistry of their gene expression; subsequently, in the further confluence of the information sciences and molecular biology (already an interdisciplinary field),[21] the organism's DNA code became the measure of its biological specificity. The new focus on the informatics of biological substrates, reducing the thing to its core elements, opened the doors to the extraction and manipulation of those substrates, with wide commercial implications. Crucially, the new computational processes that were now a part of virological study enabled the prognostic modeling at the level of molecular change. Now the will to design and build living organisms became the perfect complement to the will to see and touch them.

How do we understand the role of digital images in this scenario? To make digital images, academic researchers, lab technicians, and computer experts first transcribed biophysical and biochemical data from direct observation into a digital file, extending existing understandings of the processes involved in human-viral behaviors into numerical abstraction; the resultant data streams were reprocessed to produce digital "assets," images made to technical specifications that would make observable the unobservable. Manipulating the digital image to visualize alterations in specific processes of pathogenesis began with numerical modifications to primary data, and therein lay the possibility of prognostic modeling. What if *this* set of gene instructions or *that* biochemical threshold could be changed to alter the course of human-viral biological events such as viral entry into the human cell? What would the process look like as a multileveled systemic event? Prognostic models sought to organize these digital assets into logical sequences so that the process, and not necessarily the nature of the object, became legible. In the molecular movies, narratives emerge from this logical sequencing, simulations of change for grasping the systemic impacts of digital manipulation. The digital files were stored for future actualization in biological products. A "missing" protein or depleted chemical compound could be introduced into the thick of the human-viral relations, for instance, in order to alter the course of destructive pathogenesis. This is precisely the new frontier in HIV research focused on a group of "elite HIV controllers," one in three hundred, who produce six proteins that allow them to live with HIV without long-term progression into AIDS *and* without drug therapy. Scientists on all fronts—geneticists, structural biologists, biochemists, systems biologists—are engaged in modeling their systems so as to make a biotechnological intervention. Were it possible to *optimize* such living systems, HIV could well become a domestic partner.[22]

Driven by epidemic exigencies, such virological imperatives for virulence amelioration are part and parcel of the move toward evolutionary cooperation with microbial life. Even as the molecular movies appear to focus on a molecular-cellular event, evolutionary cooperation between human and (pathogenic) virus haunts the scene. The asteroid aesthetic is the formal trace of this nonlinear event horizon, and indicates an *integrative* approach to altering, modifying, and manipulating life at the molecular level. Hence molecular movies are expressive accounts of the general lean toward nonlinear complex systems theory in the biological sciences. The disciplinary armature is systems biology, a relatively new discipline that is the late twentieth-century successor to genomics. It would be well nigh impossible to offer even the briefest of genealogies for the field.[23] But for my purposes here, the computational integration of high throughput data from multiple sources (from structural biology, genomics, cell biology, biochemistry); the emphasis on nonlinear causalities (circuits, switches, scaffolds, and feedbacks of cybernetic systems); and the attention to biological events occurring at the same time, but at different levels, within living systems—all privilege the relational complexities necessary for human-microbial planetary cohabitation.

Systems biologists run "perturbations" or simulations of biological events on multiple fronts for a prognosis of systemic adjustments, modifications, compensations, or reorientations; they rarely present a single solution, for there are many adjustments to be made in order to change a single process. Altering a physiological signal, for instance, puts everything from protein synthesis to modifying metabolic pathways on the table. Hence marketable outcomes are expensive propositions, leading to the wry characterization of a "low input, high throughput, no output biology."[24] And yet, the turn toward an integrative approach is precisely what is exciting to virologists, equipped with different methods, procedures, and techniques, who study the same process at different scales. As collaborative simulations, the molecular movies become research platforms for the systemic analysis of viral processes and not just the virus studied in isolation. In them, the virus—the premiere subject of virology—emerges as a complex epistemic object, at once a submicroscopic particle, a foreign genetic substrate, a protein-lipid combo, and a nonhuman organismic agent whose study is irreducible to one domain of knowledge.

Symbiosis as Sustainable Future

As virological research, how are the molecular movies sustainable media? We have seen how they are in the business of altering, modifying, and manipulating the vital processes—and especially those involved in deadly pathogenesis. More often than not, these scientific visualizations lure us away from the messy unpredictability of "nature" to the cool confines of the virological lab, where scientists and animators speculate sustainable human-virus futures. But if we look beyond the molecular event to think about interlocking complex systems, the molecular movies open into the deep timescales of evolutionary biology. Working toward biotechnological repair and regeneration beneficial to both virus and host, they become crucial for ecological sustainability. Through the cracking of informatics codes, in the molecular movies, we encounter the virus as ecological actor.

The ecological conception of the virus as a resilient form of life is hardly new, but it has emerged as a dominant one in the last decades of the twentieth century. An obligate parasite,[25] a bit of nucleic acid with a protein coat and no cell walls, the virus "comes alive" upon attaching to a living host. Hence it is an especially *needy* organism, whose restless approach we fear with every flu season and every pandemic. The iconic microbial threat, ever since Joshua Lederberg (the scientist who coined the term "microbiome") christened it "the deadliest threat to mankind" in 1989,[26] two major conceptions of the virus circulate in contemporary postindustrial cultures: a bioinformatic agent whose informatic dispersion, rewriting human DNA code, hitching a ride on human cellular resources, replicating and multiplying, fascinates and intrigues, and a latent ecological threat, erupting as "emergence"[27] or a multileveled event, whose evolutionary resilience terrifies and awes its human other. The latter is a hard-earned historical lesson from the sudden resurgence of deadly viruses in the 1980s (Ebola, Marburg, and HIV—the last, the most). In that story, HIV, responsible for thirty-four million deaths worldwide, holds an unusual place. For even though we now know that HIV was globally resurgent as early as 1971,[28] it took more than twenty-five years for its spread to be discursively constituted as a "pandemic."[29] By that time the truth of a lost "war on germs," once that dream appeared well positioned to win after the mass manufacture of penicillin during World War II, had sunk in.[30] At the apocryphal conference on "emerging infectious disease" held in Washington DC, in 1989, when scientists put the blame for deadly virulence on ecological devastation, causing upheavals of coevolving ecologies, journalists rushed to report "coming plagues."[31] Evolutionary biologists joined environmentalists in promoting biological partnerships between humans and microbes as the best option for sustainable planetary futures.

In a larger project, *The Virus Touch: Living with Epidemics*, I argue that the ecological conception of the virus marks an epistemic shift toward living symbiotically with microbes that has been underway in the last thirty years. Big science projects such as the Human Microbiome Project (HMP) launched in 2008 mark the apex of that epistemological shift. Generating the same degree of excitement within scientific communities as the international Human Genome Project did a decade ago, the HMP plans to sequence approximately nine hundred microbial genomes of bacteria, viruses, and fungi from samples collected from specific sites of the human body.[32] At the 2012 meeting of the HMP's international consortium, researchers presented the positive correlation between high biodiversity of microbes in the human body and the condition of good health. Clearly those invisible microbes, feared as "our" worst enemies, are proving to be eminently beneficial, even necessary, to human survival. As a microorganism that lives in obligation to other organisms, the virus has emerged as something of a biological exemplar, admired for the secrets of intelligent adaptation. This relatively new orientation toward evolutionary cooperation, rather than competition, has significant implications for environmental sustainability: it orients research toward taming deadly virulence through biotechnological intervention, so that humans and pathogenic viruses may develop a biological partnership that does not destroy any one of them.

To learn to live with a parasite christened "poison" (as the Latin term, *virus*, signifies) when discovered in 1892 involves the fundamental realization that the virus is not well served by its deadly antagonism to the human.[33] Contrary to popular mythologies, it is not in the parasite's interest to kill the host organism, for it needs that host in order to regenerate and maintain itself. Its primary motivation, then, is to find a mutually beneficial balance, a struggle toward symbiosis in the dynamic context of constantly changing relations between living organisms, on the one hand, and between living organisms and their environment, on the other. Most evolutionary biologists argue symbiosis places organisms at a selective advantage. Not only do organisms enhance their capacities through evolutionary innovation, but they also modify initially antagonistic relations over time for mutual benefit. Angela Douglas, one of the foremost scholars on the subject,[34] argues that symbiosis is a *derived* state, a gradual evolutionary transition from antagonism (including virulent pathogenesis) to mutually beneficial relations including a stable, managed partitioning of resources. One partner—usually the host—takes control, imposing sanctions and controlling transmissions for both partners, even as both organisms develop novel capacities (a lateral, not hereditary, transfer of properties) in order to maintain an ecological equilibrium. If symbiosis is a dynamic process unfolding in the deep timescales of organismic evolution, then symbiosis must include a broad spectrum of relations, among which "symbiosis-at-risk," as in the case of a parasite that gives little to the host, is one stage—a first step on the evolutionary ladder. But what does this imply for pathogenic viruses like HIV that mobilize host resources to such a degree that they destroy the host?

Departing from de Bray's formulation that parasitism should be included within the definition of symbiosis, Douglas and her contemporaries argue that, to be considered symbiotic, organismic relations should be mutually beneficial to the participants for the major duration of their lifetimes. This does not mean that parasitism is not symbiotic, but that pathogenic parasitism is not. This is especially so with swift and deadly virulence, such as Ebola behavior, for instance, that leaves no time for the amelioration of virulence to commence.[35] Less virulent parasites are at a selective advantage, in this regard, since they can continue to use the host's resources; the host experiences depleted vital states, but it survives. But microbiologists also note that a hitherto symbiotic microbe can *become* pathogenic as a result of multileveled upheavals in coevolving ecosystems. This is evident in the emergence of zoonotic viruses that skip the species barrier, some of the deadliest microorganisms on the planet today. HIV, for one, coexisted with animal populations in Cameroon for a hundred years until the butchering of bush meat enabled its transfer into human populations.[36] In such cases, epidemiologists maintain that *unsustainable* relations—relations in which the resources that maintain living systems are so depleted that they become nonrenewable—habitually result from human interventions into the dynamic and delicate equilibria of existing ecosystems. As obligate parasites, viruses tend toward symbiosis until there are massive upheavals in their habitat; then, new resources, new hosts, and new pathways become available, but the ensuing human-virus relations are not always sustainable. Often sustainability mandates technological intervention to address radical disrepair. Thinking at the close of the Anthropocene, the "age of man" that has radically interrupted and refashioned natural processes,[37] microbiologists and evolutionary biologists are more than ever aware of the need to arrest destructive practices (e.g., the introduction of invasive plants that lead to biotic homogenization) and to repair, even engineer, sustainable ecologies, wherever possible (e.g., the bioengineered restoration of coral reefs). The hope is that such technological intervention will slow down the destruction of planetary systems, from the geologic to microbiologic. Seen in this light, the molecular movies are one stage in the biomedical and biotechnological repair of human-virus relations, processes of mediation constitutive of human intervention into pathogenesis. As sustainable media, they work toward symbiotic futures: at best, mutually beneficial partnerships, and, at worst, uneasy truce with hitherto pathogenic viruses.

Notes

1 A.E. Boycott, "Transition from Live to Dead: The Nature of Filterable Viruses," *Nature*, 3090 (January 19, 1929): 94.
2 Erwin Schrödinger, *What Is Life? Mind and Matter* (Cambridge: Cambridge University Press, 1944).
3 Industry, academics, and government officials gathered to explore potentials for the conjunction of computational methods with 3-D graphics, and to help researchers understand complicated models. See the introduction to Clifford Pickford and Stuart K. Tewksbury, *Frontiers of Scientific Visualization* (New York, NY: Wiley-Interscience, 1994).
4 I am not suggesting that scientific animations that draw from diverse locational and dynamic datasets and "build" the blanks through algorithmic and imaging technologies institute an artifice not found in scientific films with a will to indexical realism. The latter are eminently aesthetic, as Kirsten Ostherr notes of early twentieth-century microcinematographic documentaries on cellular processes. See Ostherr, "Animating Informatics: Scientific Discovery through Documentary Film," in *A Companion to Contemporary Documentary Film*, edited by Alexandra Juhasz and Alisa Lebow (Oxford: Wiley-Blackwell, 2015), 280–297. See also Hannah Landecker and Christopher M. Kelty, "Seeing Things: Microcinematography to Live Cell Imaging," *Nature Imaging*, 6, no. 10 (October 2009): 708, and also Landecker and Kelty, "A Theory of Animations: Cells, Film and L-Systems," *Grey Room*, 17 (Fall 2004): 30–63. These authors point to the denaturing of "life itself" in the reconstruction of cellular processes in early scientific films. In the context of this scholarship, Iwasa's account of what she does as animator is instructive in thinking about scientific visualization as an expressive practice: an *embrace* of artifice in the pursuit of empirical knowledge. The indexical persists in the precise methods of producing primary data: the wetware processes of hydrating/freezing of cells, the specialized tagging of cellular components, and the advanced optical techniques of capturing tilted 2-D images that are later rendered into a 3-D vector field, in order to create the composite digital image. Yet despite indexical constraints and technical specifications, Iwasa underscores the fact that the "error bar" in scientific animation is always high, since indexical images are almost always accompanied by designed elements that materialize a hypothesis, diagnostic or prognostic.
5 See, for instance, Tom Seiber's account of interactive modules for citizen-scientists that depict voyages into the body's interior as an intergalactic adventure. Seiber, "Playable Virus: HIV Molecular Aesthetics in Science and Popular Culture," *Animation*, 9 (July 2014): 261–276.
6 The term "scalarity" typically refers to the ability to move between multiple levels. The increasing proliferations of sophisticated optical technologies, I am arguing, enhance this capacity, making optical scaling more reflexive and frequent than ever before. For an elaboration of the point, see William Bynum, *A Little History of Science* (New Haven, CT: Yale University Press, 2012), and Geoff Anderson, *The Telescope: Its History, Technology, and Future* (Princeton, NJ: Princeton University Press, 2007).
7 Akira Lippit's account of the "optical fantastic" triangulates the development of X-ray technologies, the splitting of the atom, and psychoanalysis, in *Atomic Light (Shadow Optics)* (Minneapolis: University of Minnesota Press, 2005).
8 For citizen-scientists, games such as *Foldit* use competitive play as a research method, enabling a nonlaboratory investigation. In just over a week after the game hit the market, players had twisted and rotated the ribbon—representing the chemical structure

of the Mason-Pfizer monkey virus (M-PMV) as a sequence of amino acids—into a solution to the problem of identifying the protein's molecular structure, potentially enabling new treatments for HIV, including molecular medications and gene therapy. See "Foldit Online Protein Puzzle," *Scientific American*, http://www.scientificamerican.com/citizen-science/foldit-protein-exploration-puzzle/ (accessed July 2015). See also Dean Praetorius, "Gamers Decode AIDS Protein That Stumped Researchers for 15 Years in Just 3 Weeks," *Huffington Post*, September 19, 2011, http://www.huffingtonpost.com/2011/09/19/aids-protein-decoded-gamers_n_970113.html (accessed July 2015).

9 Sarah Kember and Joanna Zylinska, *Life after New Media: Mediation as Vital Process* (Cambridge, MA: MIT Press, 2012).

10 See Morgan Ryan, "Molecular Movie Stars," *American Scientist*, September 2009, http://www.americanscientist.org/issues/pub/molecular-movie-stars (accessed July 2015). Ryan foregrounds the rise of the new expert: "Hollywood-based Eric Keller [is] a well-known 3-D instructor and special-effects animator who must budget his time between molecular projects and (his scientific colleagues pridefully note) next summer's action blockbuster."

11 See http://www.molecularmovies.org (accessed November 22, 2015).

12 Other animations for public platforms include Iwasa's work in Jack Szotak's lab at Harvard University (e.g., animations of single-strand RNA molecules surrounded by a fatty acid membrane) that find their way into online and physical sites such as Boston's Museum of Science; for some examples, see http://molbio.mgh.harvard.edu/szostak web/ (accessed November 22, 2015).

13 An incipient hypothesis includes the scientific hunch, the epistemological cousins of scientific intuition. In "Toward an Understanding of Intuition and its Importance in the Scientific Endeavor," Lois Isenmen distinguishes between the hunch and the intuition, claiming a place for both in experimental research as the motor of scientific advance. See *Perspectives in Biology and Medicine*, 40, no. 3 (Spring 1997): 395–403.

14 Harrison emphasizes structural observation as a staple in biology, the legacy of the Romanian cell biologist, George Emil Palade (who won the Nobel Prize for Physiology or Medicine in 1974) and Austrian-British molecular biologist, Max Perutz (who won the Nobel Prize in 1962 for his work on cellular proteins).

15 See Stephen C. Harrison's "Whither Structural Biology?," in *Nature: Structural and Molecular Biology*, 1, no. 11 (January 2004): 12–15. For a history of technological and laboratory procedures from cell-tagging to the ultra-centrifuge, see Hans Jörg Rheinberger's *An Epistemology of the Concrete* (Durham, NC: Duke University Press, 2010).

16 Harrison, "Whither Structural Biology?," 12.

17 In some respects, virology was late to the table because of its dependence on technological mediations of the virus. The debates over whether or not the virus was a living organism congeal around research in clinical virology, a field of scientific expertise blossoming with Thomas M. Rivers's classic *Filterable Agents* (Bailliere Tindell, 1928) that established the virus as an obligate parasite. Virology would move inexorably toward the technological capture of the virus, transforming it into an intriguing epistemic object. But before the 1930s, its sibling disciplines, immunology and epidemiology, had already taken root.

18 Recombinant DNA technology developed in the 1970s contributed to the stabilization of dependable viral vectors to intervene in infection and to alter the human genome. The drawback to viral vectors is that some retroviruses integrate the altered DNA at a random location in the host genome, disturbing the function of cellular genes and thereby increasing the risk of cancer. See Natasha McDowell, "New Cancer Trial Halts U.S. Gene Therapy," *New Scientist*, January 2003, http://www.newscientist.com/article/dn3271 (accessed July 2015).

19 Living things tailored to experimental purposes that can metonymically generate understandings of a whole class. See Rheinberger, *An Epistemology of the Concrete*, 7.

20 There are many histories to the discovery of the gene, starting from George Mendel, but the watershed point is James D. Watson and Francis Crick's definition of the molecular structure of DNA in 1953.

21 In his "Prologue" to *Epistemologies of the Concrete*, Rheinberger maintains that nucleic acids as macromolecules, as the active agents of life, were identified as early as the 1930s, well before Crick and Watson's 1953 evidence of the double helix form of DNA. Hence there is already a shift toward finding the material substrate of life, even at the height of excitement over morphological identification of living organisms enabled by new technologies.

22 A large-scale genetic study overseen by Bruce Walker (Ragon Institute, Boston) showed how helper T cells could clearly "read" five or six amino acids lining the structural pocket; responding like a bull to a red flag, the T cells of the HIV controllers immediately sent in the killer T cells to destroy the infected cells. The massive gene-sequencing project, involving three hundred investigators worldwide, located the instructions for making this protein in chromosome 6 (usually the location for immune function genes) of the HIV controllers. With the 2010 publication of the "genome-wide association study," it was clear that the "cure" lay in further investigations of cellular signaling and thereby a better understanding of the relation of the genome to its immediate environment. See Bruce Walker, "The Secret of Elite HIV Controllers," *Scientific American*, 306, no. 7 (July 2012): 44–51.

23 Whatever the intellectual genealogy one might produce for complex systems theory in the biological sciences, one of its key preoccupations is with *second-order systems* characterized by circularity, running on feedbacks between systems. One of the key figures in the biological sciences, Heinz von Förster, who directed the Biological Computer Laboratory at University of Illinois, Urbana-Champaign (1958–75), emphasizes the looping nonlinear causality of the living system's interface with its environment as follows: "When we perceive our environment, it is we who invent it." See Heinz von Förster, *Observing System* (Seaside, CA: Intersystems, 1981), 5; see elaboration of this work in Bruce Clarke and Mark Hansen, "Introduction: Neocybernetic Emergence," in *Emergence and Embodiment: New Essays on Second-Order Systems* (Durham, NC: Duke University Press, 2009), 1–25.

24 The pithy ironic description of the new discipline is taken from an article that speculates on "systems vaccinology" as a future industrial application: Bali Pulendran, Shuzhao Li, and Helder I. Nayaka, "Systems Vaccinology," *Immunity*, 33, no. 4 (October 29, 2010): 516–529, http://www.sciencedirect.com/science/article/pii/S1074761310003663 (accessed July 2015).

25 An obligate parasite is an organism that cannot live without a host (that is, it cannot process all the cellular components it needs to regenerate itself) as opposed to a facultative parasite that can live independently but becomes a parasite under certain conditions (e.g., a flea).

26 Lederberg made the pronouncement at the 1989 conference sponsored by the National Institutes of Health and Rockefeller University, 1989, and followed his observations in a book, co-edited with Robert E. Shope and Stanley C. Oakes, *Emerging Infections: Microbial Threat to Public Health in the United States* (Washington, DC: Institute of Medicine, 1992) (the Institute for Medicine's landmark publication).

27 "Emergence" (from the Latin *emergere*, "to appear") is a capacious term for multileveled occurrences across scales of action, human and nonhuman, that resists linear causality and is therefore difficult to predict. The term was first used in the ecological context by French-born American microbiologist, René Dubos, in his 1959 classic, *The Mirage of Health: Utopias, Progress, and Biological Change* (New Brunswick, NJ: Rutgers University Press, 1987). New conjunctions between cybernetics and molecular biology have revived the classical concept of emergence: see Manuel Delanda, "Emergence in History," in *Philosophy of Simulation: The Emergence of Synthetic Reason* (New York, NY: Bloomsbury Academic, 2011).

28 There are many accounts of "first sightings," some moving as far back as 1959 (cases now disproven by David Ho); the first case in the United States was Robert R., who died in 1969. Usually, early cases are the pre-1981 cases (1981 is when AIDS became known to the medical profession).

29 At the level of international organizing, AIDS was seen to be an epidemic erupting in multiple contexts as early as 1986. The 1992 International AIDS Conference in Amsterdam was the first to announce a global agenda (eight thousand participants).

30 See Melinda Cooper, *Life as Surplus: Biotechnology and Capitalism in the Neoliberal Era* (Seattle: University of Washington Press, 2008), 76–77.

31 The most memorable among these was Laurie Garrett's *The Coming Plague: Newly Emerging Diseases in a World Out of Balance* (New York, NY: Farrar, Straus and Giroux, 1994), a book that generated consternation upon its 1995 appearance in paperback.

32 Among other things, we learn that the total weight of microorganisms in the human body is as little as 200 grams, even as their cells outnumber human cells ten to one. See Michael Balter, "International Human Microbe Program Looks Ahead," *Science Magazine*, March 22, 2012, and Carl Zimmer, "How Microbes Defend and Define Us," *New York Times*, July 13, 2010, http://www.nytimes.com/2010/07/13/science/13micro.html?_r=3&pagewanted=all& (accessed July 2015).

33 When Dutch scientist Adolf Mayer showed that a "soluble, enzyme-like" sap mottled tobacco leaves, the first conceptions of the virus as a biochemical agent—seeping, leaking, spreading—had taken hold. Soon after, Russian botanist, Dmitri Ivanovski observed a toxin causing "wildfire" in 1892, and Dutch botanist and microbiologist Martinus Beijernick named the "contagious living fluid" a "virus" in 1898. And so was born the first virus, the tobacco mosaic virus, whose ability to cause economic ruin (the destruction of tobacco plants) motivated further research. Within a year of the TMV, foot-and-mouth disease became the first viral infection observed in animals, quickly followed by yellow fever as the first among humans; and at the turn of the twentieth century, there was a growing ecological perception that these microbes became pathogenic when they skipped the species barrier, for measles (paramyxovirus) in humans had been tracked to distemper in dogs.

34 In her latest work on symbiosis (*The Symbiotic Habit*, Princeton, NJ: Princeton University Press, 2010), Angela Douglas returns to the persistence of this behavior among organisms in light of new thought on the microbiota crucial to immune function and the pragmatic promotion of symbiosis (reintroducing indigenous plant species in an effort to defragment habitats) as bulwark against deleterious anthropogenic effects. Following *Symbiotic Interaction* (1994) and *The Biology of Symbiosis* (1987), the recent book ventures into the role of human-ecological and medical interventions in the processes of symbioses, and, for our purposes, includes a reevaluation of certain organisms originally considered pathogenic as potentially symbiotic in the evolutionary future. See Douglas, *The Symbiotic Habit*, 8.

35 Ibid., 29.

36 See Alan Whiteside, *HIV/AIDS: A Very Short Introduction* (Oxford: Oxford University Press, 2008).

37 The ecologist Eugene Stoermer coined the term, while the Nobel Prize–winning atmospheric chemist famous for his thesis on the hole in the ozone layer, Paul Crutzen, popularized it (see Paul Crutzen and Eugene Stoermer, "The 'Anthropocene,'" *Global Change Newsletter*, 41 (May 2000): 17–18). The five key anthropogenic factors of the Anthropocene relevant to the discussion of symbiosis are large-scale extinction, biotic homogenization through invasive species, habitat fragmentation, climate change, and the modification of biochemical cycles. Since it is not possible to put the clock back, scientists now argue for biotechnological and bioengineered projects that work toward a newly sustainable ecological equilibrium. See Douglas, *The Symbiotic Habit*, 14.

14

COUPLING COMPLEXITY

Ecological Cybernetics as a Resource for Nonrepresentational Moves to Action

Erica Robles-Anderson and Max Liboiron

We live in an era of ecological crisis. Climate change, plastic pollution, radiation drift—our most pressing environmental concerns are on a planetary scale. They are experienced everywhere and yet they are difficult, if not impossible, to see. An extraordinary amount of resources devoted to addressing ecological crises are spent simply trying to depict these crises and trying to teach people how to properly interpret the representations. The hope is that if only the general public and policy makers could *see* what is happening, they would better understand, and this understanding would lead to action. Scientists, activists, and policy makers advocate for *higher* resolution climate models and *more complete* descriptions of the locations and effects of ocean plastics, or radiation drift from Fukushima, or extreme weather. Bigger, better, clearer pictures are the key to informed action. They account for more variables, simulate more mechanisms, and inform the construction of better models that reflect our total understanding.

This pursuit of representational fidelity aligns with the long-standing technocractic legacy of what historian Paul Edwards calls "the closed world," or "global surveillance and control through high-technology military power."[1] With the detonation of atomic bombs at the end of World War II and subsequent threats of planetary nuclear annihilation during the Cold War, it became increasingly possible to imagine that humans could act at planetary scale. Such action required global infrastructures like computing networks for gathering information about everywhere. It also required new kinds of whole world sciences that would provide knowledge paradigms for reasoning about the state of the entire earth.[2] When both the planet and its knowledge systems are imaged as global and enacted as such, ecological crisis also becomes understood through the computational infrastructures undergirding military action.[3]

In this chapter we revisit the legacy of two of the most "whole world" of the whole world sciences: ecology and cybernetics. Both imprinted the popular imagination with ideas about holism, systems, balance, and equilibrium. Their legacies are entangled in the development of global knowledge infrastructures. They are remembered as intellectual disciplines that helped gather, license, represent, and finally act upon global knowledge. What failed to endure was their more radical critique of the technocratic genre of ecological knowledge. By looking back at mid-century cybernetics and ecology, we'll show that in their efforts to describe complex systems these sciences rejected reductive, linearly causal concepts of the world. As they encountered the limits of representation implicit in the knowledge structures of traditional disciplinary sciences,[4] they created ways of describing "systems" and "loops" where many different kinds of entities interacted to maintain stability. Ecology coined the "ecosystem" concept and cybernetics turned to models of self-regulation to find mechanisms like "feedback" by which systems maintained "homeostasis" in spite of environmental change.

We follow a strand of ecological cybernetic thought in which practitioners' ambitions to address massively complex, dynamic, and contingent circumstances led them away from the hallmarks of representational knowledge: translation, prediction, accuracy, resolution, and the accumulation of data. Instead, they articulated a knowledge regime that skipped representation in favor of getting complex systems to talk to one another directly. They offered a vision of complexity managing complexity through co-regulation, or coupling. By attempting to pair ponds with factories or children with algorithms, cyberneticists modeled systems speaking to one another to accomplish action without an interpretive middleman vetting facts and knowledge beforehand. This break with dominant modes of knowing and acting offers resources for critical media scholars and environmental activists interested in addressing ecological crisis without relying on bigger, better, clearer pictures.

How Ecology Got Its System

Ecology developed at the beginning of the twentieth century at the intersection of biology, natural history, and botany. Traditionally, life sciences were concerned with documenting species living in particular areas, and then asking why these species were present. Scientists focused on generating lists of vegetation *or* animals. In the rare cases where both were recorded these lists excluded living and nonliving variables like soil or climate that were key to ecological change.[5] As scientists became interested in changes in species and vegetation occurring over time, methodologies that relied on listing no longer sufficed. In 1935 Arthur Tansley coined the term "ecosystem" as a universal concept to address methodological problems emerging within ecology:[6]

> The whole webs of life adjusted to particular complexes of environmental factors, are real "wholes," often highly integrated wholes, which are the living nuclei of systems in the sense of the physicist [. . .] including not only the organism-complex, but also the whole complex of physical factors

> forming what we call the environment of the biome—the habitat factors in the widest sense. Though the organisms may claim our primary interest, when we are trying to think fundamentally we cannot separate them from their special environment, with which they form one physical system. It is the systems so formed which, from the point of view of the ecologist, are the basic units of nature on the face of the earth.[7]

By the 1940s ecologists treated the environment and its parts as a holistic system.[8] Ecologists generally agreed that living systems changed over time even as they remained relatively stable, but they disagreed about the nature of these seemingly contradictory dynamics. Ecologists like Henry Gleason argued that ecological interactions are uneven, even coincidental, and probabilistic rather than deterministic.[9] Living systems are so complex and indeterminate that accounting for all interactions within them is impossible. Tansley believed "ecosystem" captured this sense of complexity while making the study of interrelations feasible. Ecosystems were "systems we isolate mentally . . . the isolation is partly artificial, but is the only possible way in which we can proceed" with research.[10] Part theory, part thing-in-the-world (constructed as they may be), ecosystems "form one category of the multitudinous physical systems of the universe, which range from the universe as a whole down to the atom."[11] An ecologist could study a small system, or part of a system, and extrapolate discoveries to other scales, but an exhaustive description of the system was never a possibility.[12]

During the late 1930s and early 1940s, Raymond Lindeman became the first ecologist to study an entire ecosystem as an ecosystem.[13] By organizing a lake ecosystem into a food chain (also called trophic levels), he measured energy transfers between species.[14] This method let Lindeman trace the ecosystem's behavior at a minute level, allowing previously invisible and unknown relationships to emerge. Most famously, he discovered "the ten percent law," describing the efficiency of energy transfer between trophic levels. When predators eat herbivores, ten percent of the matter consumed becomes flesh. The rest is lost to respiration and other processes, a shockingly inefficient outcome given assumptions at that time.

Lindeman's models were at the cutting edge of ecology and yet he never purported that they represented the whole system. His calculations allowed him to map ecosystem *relations* and he considered his diagrams approximations.[15] Precision was not required. Imprecise data was fine for building low-resolution models since even low-resolution models proved that energy moved in an ecosystemic fashion. More than forty years later, ecologists still agreed that "an ecosystem model, no matter how sophisticated or difficult to produce, is but a shadow of its prototype."[16] Early ecologists never aspired to completeness. Metaphor, analogy, and approximation sufficed because their goal was to describe the general shapes of systems in terms of their basic mechanics. Lindeman's work demonstrated that even low-resolution models revealed previously unsuspected relationships. Thus, ecological "maps" of complex relationships—node and arrow models that give the "gist" of relations rather than precise descriptions—became the hallmark of the ecological imagination (Figure 14.1).

Historian of science Lorraine Daston distinguishes between precise and accurate scientific representations: precision is "the clarity, distinctness, and intelligibility of concepts; accuracy refers to the fit of numbers to some part of the world, to be ascertained by measurement."[17] Daston argues that early modern scientists often employed precise rather than accurate definitions of novel phenomena. For example, Van Leeuwenhoek's claim "that there 'were no less than 8 or 10,000' living creatures teeming in a drop of water . . . spoke to this penchant for precision without accuracy."[18] Likewise, early ecology was precise, rather than accurate. It pointed to processes discernable within a new perspective and showed the general shape of complexity in an overall system.

Today, ecology and ecosystem are floating signifiers that usually mean whatever interpreters want them to mean, including different and often contradictory concepts in different fields and discourses. In their earliest deployments, however, these terms designated living systems composed of many moving parts that were capable of sustaining themselves in dynamic conditions. Even though early

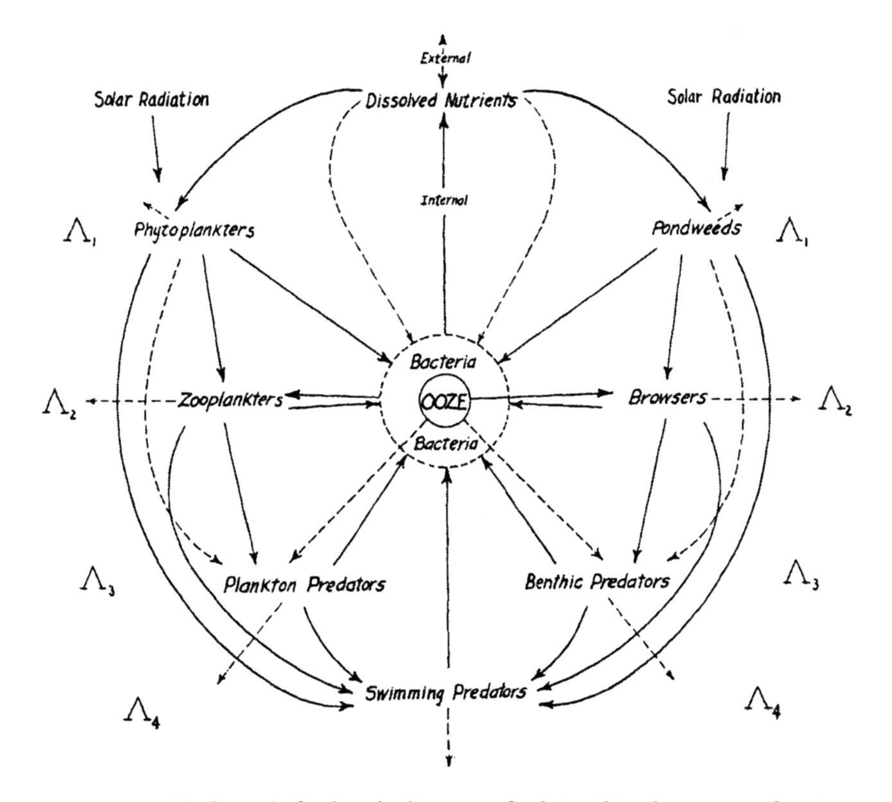

FIGURE 14.1 Lindeman's food web diagram of relationships between soil, microorganisms, animals, plants, and chemicals.

cyberneticists did not collaborate with ecologists for nearly a decade, they were already engaged in ecological thinking because they were working on analogous questions.

How Cybernetics Got Its Loop

Cybernetics, or "the entire field of control and communication theory, whether in the machine or in the animal," coalesced into a research paradigm during World War II.[19] In 1941, Norbert Wiener, one of the field's progenitors, joined the Radiation Laboratory at Massachusetts Institute of Technology (MIT) as part of a project funded by the National Defense Research Committee. There, he collaborated with control engineer Julian Bigelow and physiologist Arturo Rosenblueth to develop an anti-aircraft artillery system that would become emblematic of the cybernetic paradigm. The chief difficulty in building an anti-aircraft system was not the physics. After all, projectiles in flight are mainstay examples of classical Newtonian physics. The problem was the "machine in the middle," or the human being. How could an automatic system accurately predict the behavior of the pilot in the plane? How could a machine forecast the actions of a system that is simultaneously mechanical, electrical, and psychological, whose parts are both human and machine?

In 1943, Rosenblueth, Wiener, and Bigelow published a seminal paper entitled "Behavior, Purpose, and Teleology," arguing that systems could be studied in terms of their relationship to their environment.[20] Environments provided *inputs*, external events that modified the system, and systems generated *outputs*, or changes in the surroundings. Their framework focused on behavioral loops that bound systems to their environments. Thus, cybernetics adopted a *black box ontology*, bypassing any ambitions towards complete descriptions of the system's inner workings to focus instead on how parts interacted to produce sustainable wholes.[21]

Like early ecology, early cybernetics concerned itself with the mechanisms by which systems sustained themselves in changing conditions. The trick was in understanding self-regulation without recourse to the inner workings of the entity in view. Rosenblueth, Wiener, and Bigelow accounted for the self-regulatory process in terms of *feedback* and *homeostasis*, concepts borrowed from control engineering and physiology, respectively. Feedback names the capacity for a system to take past performances as inputs, thus monitoring actual rather than expected performance to adjust future conduct. Homeostasis refers to the processes by which systems self-regulate to maintain stable conditions in shifting external environments. The classic example is temperature regulation in warm-blooded animals.[22]

These concepts let cyberneticists go forward with unknown elements in their systems; the inner workings of the machine in the middle did not matter so long as feedback loops could be constructed. The anti-aircraft system could consider the plane's position (input), then fire (output), and then consider the plane's new position (input) before firing again (output). From recursive loops of inputs

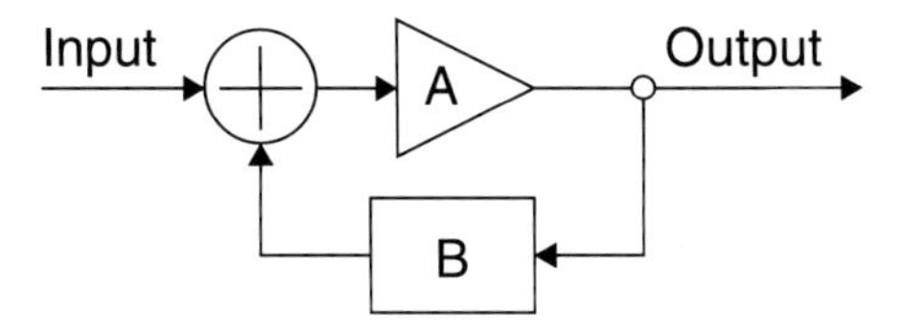

FIGURE 14.2 Cybernetic loop.

Source: Wikimedia Commons. https://commons.wikimedia.org/wiki/File:Ideal_feedback_model.svg, created September 30, 2007.

and outputs, homeostasis emerged. Cybernetic systems were stable despite—and because of—change.

Cybernetics offered a behavioral rather than functional framework for scientific inquiry. What mattered were whole-systems performances rather than complete models of the inner workings of parts. Practitioners could black box parts of systems and focus instead on effects, or outputs, and control by communication, or inputs. Abstracting away inner workings opened the door to a radical inclusion of different entities that could be studied within a common domain. This accounts, at least in part, for the diverse phenomena studied by early cyberneticists: anti-aircraft guns, ataxia, purpose tremors, wave filters, heart flutters, neuronal nets, and rather macabre experiments that involved splicing cats to engineer feline-mechanical circuits.

Second Wave Cybernetics: What Cybernetic Ecology Performs ~~Looks~~ Like

In the Cold War climate of the 1950s, political and environmental threats to the cessation of life on a planetary scale gave ecological and cybernetic projects a sense of urgency. The watchword for both was *entropy*. As social, environmental, and political systems seemed to become more disorganized and precarious, scientists in both disciplines wondered how life—order and organization—was maintained despite the second law of thermodynamics, the tendency towards disorganization.[23] Bernard C. Patten and E.P. Odum, prominent ecologists and staunch cybernetics advocates, wrote that "the balance of nature" is an organizational design that "creates order where there could be chaos."[24] In "An Introduction to the Cybernetics of the Ecosystem: The Trophic-Dynamic Aspect," Patten borrowed concepts from cybernetics to analyze the life-sustaining organization of ecosystems:

> A general theory of ecosystem dynamics is developed from basic considerations of order and disorder as elucidated by modern information theory. Disorder is proportional to entropy; order to information, which is negative entropy. . . . This is in opposition to the second law of thermodynamics which militates that ultimate absolute disarray of all energy matter.[25]

Likewise, Norbert Wiener's *The Human Use of Human Beings* (1950), which introduced the general public to cybernetics, cautioned that "life is an island here and now in a dying world. The process by which we living beings resist the general stream of corruption and decay is known as *homeostasis*."[26] Scientists turned to homeostatic systems to understand how to resist entropy at the level of the organism, machine, civilization, and ecosystem.

These concerns intensified as cybernetics entered its second wave in tandem with popular environmentalism. A growing catalogue of the disastrous effects of pesticides, radiation, and industrial chemicals entered public discourse through works like Rachel Carson's 1962 *Silent Spring*. These accounts made clear that humans had inadequate means for thinking about and controlling their relationship to the environment. Left unchecked, our behaviors risked the survival of humankind. The idea that humans were runaway elements of a larger system that we would eventually crash crystalized in the 1972 book *The Limits to Growth*, which used computer modeling to predict an exponential increase in human population and subsequent depletion of the earth's food supplies.[27]

In 1968 cyberneticist Gregory Bateson convened a conference aimed at addressing the problem of balancing runaway human-ecological systems. The conference entitled "Effect of Conscious Purpose on Human Adaptation" opened with a paper on environmental crisis delivered by Barry Commoner, a founder of the modern environmental movement. Commoner used the example of nitrate pollution to explain how human actions affected larger ecosystems. His call to action dovetailed with Bateson's goal of expanding cybernetics to three kinds of homeostatic systems: the individual human organism, the human society, and the larger ecosystem.[28] At each scale, consciousness was considered "an important component of *coupling* of these systems."[29]

This concept of *coupling*—hooking systems together to constitute each other's environment—brought self-regulating systems into a more complex version of a steady state. Each system remained a black box to the other system while regulating its neighboring system just the same. The trick was getting them to recognize each other's inputs and outputs, or communicate. Bateson described the problem through an analogy to Lewis Carroll's description from *Alice in Wonderland* of Alice playing croquet:

> Alice is coupled with a flamingo, and the "ball" is a hedgehog. The "purposes" (if we may use that term) of these contrasting biological systems are so discrepant that the randomness of play can no longer be [. . .] known to the players. Alice's difficulty arises from the fact that she does not "understand" the flamingo, i.e., she does not have systemic information about the "system" which confronts her. Similarly, the flamingo does not understand Alice. They are at "cross purposes." The problem of coupling [humans] through consciousness with [their] biological environment is comparable. If consciousness lacks information about the nature of [humans] and the

> environment, or if the information is distorted and inappropriately selected, then the coupling is likely to generate meta-random sequences of events [or chaos].[30]

Clarifying the rules of croquet will not make the game more playable. The real stakes of this game, and by analogy this model of science, were communicative, not representational. Alice cannot understand the inner workings of hedgehogs any more than hedgehogs understand the inner workings of little girls. What Alice needed was neither more precision nor greater fidelity, but rather a means of coupling in the hedgehog's terms in a way that allowed her to win the game. The emphasis on regulation through coupling and communication was a hallmark of second wave cybernetics.

Stafford Beer, perhaps the most prolific second-wave cyberneticist, spent much of his career concerned with what he called "exceedingly complex systems." These systems were so indeterminate and unknowable that they could never be accounted for through the traditional paradigms of determinate science and linear causality. Instead, Beer advocated the regulatory paradigm common to both cybernetics and ecology. He drew upon the holistic visions of ecologists like James Lovelock and Jan Smuts to urge a shift towards systems thinking:

> The image of life crawling around in its multiplicity of species, looking for a "biological niche" to which it can somehow adapt, is replaced by the creation of an environment that implies such life. It is an environment in which tiny changes in the salinity of the oceans, the composition of the atmosphere, and indeed all other major variables make life impossible.[31]

In the 1950s Beer began experimenting with what he called "fabrics" for regulating exceedingly complex systems. His hope was that these fabrics might proffer a more robust means of regulating firms, factories, or even nations.[32] Rather than try to blueprint, say, an ideal factory, he treated the factory as a system built of recursive parts that worked in the looping fashion of input and output to respond to a changing world. The trick was to enroll some other already homeostatic system into the position of regulation: to couple Alice to the hedgehog.

Beer surveyed a range of chemical and biological candidates (computing was still too primitive to be considered) such as chemical systems, lipid membranes, human, and other vertebrates. He invented a game where children solved simultaneous linear equations without understanding any of the underlying mathematics. The game demonstrated that systems relations amplified the intelligence of their constituent parts thereby proving animals like mice, rats, and pigeons should also suffice. Beer investigated social insects like bees, ants, termites, and smaller organisms like the larva of yellow fever mosquitoes and *Daphnia*, a freshwater crustacean.

The inner circuitry of these fabrics never needed to be (and never could be) designed in detail. Otherwise, Beer argued, "What virtue is there in self-organizing capacity?"[33] In fact, fixed circuitry was a liability. Unable to adapt or respond to new situations, fixed systems are not stable over the long term. Beer argued that what is needed is "not blueprints, but high-variety undifferentiated fabric which we can *constrain*. We do not want lots of bits and pieces which we have to put together."[34] It was better if fabrics were "constrained by their own experience, an 'epigenetic landscape,'" since no designer could possibly know what the future required.[35] Comfortable with—even celebratory of—radical indeterminacy and not-knowing, Beer thought about how specific systems characteristics might be compatible with humans. Knowledge was gained through trial and error, coupling different things together over and over. Hence the bull in a tea cup with Beer looking on (Figure 14.3); his job was not to know, but to introduce and to prod.

In a series of experiments on *Daphnia*, dead leaves and iron filings were added to their tanks in the hope that if the crustaceans ingested sufficient filings they would respond to magnetic fields. Inputs could then be transmitted to colonies via electromagnets. Outputs would be the colony's adaptive behaviors, as embodied iron filings changed the electrical characteristics of their space. In theory, the system would retain sufficient freedom to continue evolving through processes of self-perpetuation and regulation. Beer's experiments never succeeded because the colony kept collapsing due to a steadily increasing suspension of tiny magnets in the water. His next move was to enroll an even more complex system, an entire pond.

FIGURE 14.3 The difficulty of coupling complex systems.

Source: S. Beer, 1951.

Beer filled tanks with water samples collected from Derbyshire and Surrey. The samples contained the usual suspects: hydra, *Daphnia*, a leech, microorganisms, other biota. Such a system should self-stabilize so long as enough trophic levels were in the sample. Still, the problem of enlisting the system into the position of a controller in some recognizable manner remained.

Historian of science Andrew Pickering writes that the

> sheer oddity of trying to use a pond to manage a factory dramatizes the point that ontology makes a difference. If one imagines the world as populated by a multiplicity of interacting exceedingly complex systems . . . then one just might come up with this idea.[36]

Pickering characterizes cybernetics projects as "non-modern ontological theater," where emblematic projects demonstrate how departing from the traditional techno-scientific vantage produces radically different knowledge structures, practices, and entities.[37] Beer's work inverted the traditional hierarchy between designers and raw materials. Rather than treating materials as "inert lumps of matter that have to be assembled to make a meaningful system," his work treats the world as a place already filled with lively, vibrant matter just waiting to be enrolled.[38] For Pickering (following Heidegger) cybernetic materiality emphasizes relationships of *revealing* rather than *enframing* between humans, machines, nature, and the world. For Heidegger, technology enframed nature as a *standing reserve* of resources for human tasks.[39] Knowledge preceded action. Here, knowledge *is* action. It is revealed in performance rather than mapped in advance.

A concern with coupling moved cybernetics further from the stakes involved in knowledge accumulation models that trade upon representational completeness and towards more performative interests. The science of control grew, at least in part, into a framework for articulating the ethics of communication. Complex systems interacted only when systems were addressed in a language that their structure understood. Control became communication and communication was the emergent property of open, coupled systems interacting in recursive loops. Conversation was the mode of both self-regulation and entropic resistance, and without it life in the contingent universe would not persevere.

Coupling Complexity and the Case of Ocean Plastics

By the 1960s cybernetic thinking was disappearing from computer science.[40] High formalism and an emphasis on symbolic processing, precisely the kinds of cognitivist, determinist modes of knowledge that cybernetics and ecology pushed against, came to dominate the field. Today, a resurgence in cybernetic thinking is underway. Detectable in areas like robotics, biological computing, and machine learning, scientists and engineers are increasingly incorporating feedback loops between complex systems as a means of achieving systems performance in the world. Thus, revisiting ecological strands of cybernetics enlarges our imagination by showing how performative approaches change the stakes of critique and action, particularly in the face of twenty-first-century environmental crises. We ground this contribution by closing with an example of environmental change in order to illustrate what an eco-cybernetic approach to the problem might ~~look~~ perform like.

Every ocean in the world contains plastics. Ocean plastics are tiny (usually less than 5 millimeters), unevenly distributed in the water column, and spread across the oceans like a liquid smog.[41] In 2011, plastics outweighed plankton, and thus food stock, in the middle of the Pacific Ocean by six to one.[42] On beaches in Kauai, Hawaii, plastic washes ashore at a rate of 484 pieces a day.[43] Two-thirds of marine animals ingest ocean plastics; the chemicals associated with plastics become concentrated in their bodies and then through food chains.[44] In other words, we *know* that ocean plastics constitute an ecological crisis because they are changing how ecological systems work.

Yet ocean plastics are not legible as a single global phenomenon. There is no island of plastic to point to, let alone measure. Attempts to quantify plastic and paint a big picture have faced acute challenges. For example, a 2010 scientific report synthesizing all plastic collection studies conducted in the North Atlantic from 1986 to 2008 and based on more than sixty-four thousand pieces of plastic provided one of the most complete representations to date.[45] Yet the model did not indicate a stable progression of pollution over time. Rather, ocean plastics seemed to disappear and reappear. Thus, "despite a strong increase in discarded plastic, no trend was observed in plastic marine debris in the twenty-two-year data set . . . it is impossible to estimate the size of this sink."[46] Another recent report found that although we *know* (based on models and empirical data) that there are millions of tons of plastics, "[w]e can't account for 99% of the plastic that we have in the ocean."[47] Scientists hypothesize that fish are ingesting the plastic and that these plastics are then sinking, but "the pathway and ultimate fate of the missing plastic are as yet unknown."[48] For many scientists, policy makers and industry representatives, these results trigger calls for more data, more studies, and better models. In short, they call for bigger, better, clearer representations. But what call to action might result from an eco-cybernetic imagination?

DISCUSSANT: Do you really have anything in the nature of a meaningful measure of the environment in question?
BEER: I do not suppose I have to, and I do not care. You see, the thing is, if we preoccupy ourselves with transfer functions, and being able to define things, and the getting of meaningful measures, we are constructing a descriptive science—not a control science. [. . .] It seems to me that our conditions for the real control system are very, very simple. You take a line from a control box into the world, where there must be some effect, any effect; and you take a line out of the world and into the

control box, where there must be some effect. And this system will run to stability . . . we do not need to know what those transfer functions are.[49]

Following Beer, we could assume that the amount, location, and movement patterns of plastics in the ocean are unknown and perhaps even unknowable. They can remain black boxed. We can instead look at the inputs to and outputs from the ocean. We know from industry reporting that plastics production is increasing exponentially.[50] We also know that most ocean plastics come from land.[51] Thus, we can identify an input.

The ocean has long been regarded as the "ultimate sink," an "ecosystem service" that absorbs the detritus of production, allowing more to occur.[52] It is an infrastructure that allows for disposability, a way of making waste go away. This property is feedback. It fundamentally connects land-based economies to oceans such that together they constitute the oceanic environment. Many scientists and activists are calling for a *de*coupling of land and ocean. They argue that rather than build a scientific infrastructure for adequately representing the state of ocean plastics pollution, we should act to reduce the flow of plastics.[53] Interrupt the loop rather than open the black box. Many ocean plastic researchers are involved in banning microbeads in personal care products, plastic bag bans, and other legislation to reduce disposable plastics.[54]

Another example of eco-cybernetic thinking in marine plastics is the 2015 publication, "Plastic Waste Inputs from Land into the Ocean." In it, a group of American scientists estimates the amount of plastic entering oceans by "linking worldwide data on solid waste, population density, and economic status."[55] They argue that the "small size [of ocean plastics] renders this debris untraceable to its source and extremely difficult to remove from open ocean environments, suggesting that the most effective mitigation strategies must reduce inputs."[56] They stacked estimates of "mismanaged" waste that could easily escape into oceans from 192 coastal countries. Of course their numbers are estimates and approximations, because like ocean plastics, land-based waste and populations in developing countries are notoriously difficult to measure. Yet they generated a precise, rather than accurate, model that showed the countries with the highest impacts on ocean plastics. The authors conclude that, "industrialized countries can take immediate action by reducing waste and curbing the growth of single-use plastics."[57] This low-resolution model of inputs and outputs shows that plastics move in an ecosystemic fashion. The entire system cannot be defined in detail, nor should it be, as such an approach will miss inputs and possible outputs. The argument for action is to decouple industrial and oceanic systems that are reaching towards a plastic equilibrium.

Tapping back into a cybernetic imagination of ecological action can shift our strategies. Currently, most scholars and activists focus on problems of resolution, or how to understand and represent the mechanisms of ecological crisis. Thus, environmental and scientific disputes often revolve around questions of uncertainty:

How can we get a better picture, a clearer view, or bigger data? Exactly how much plastic is in the ocean? How can we become more certain that what we are seeing closely approximates what is actually happening in the world? Reframing the world as a contingent place filled with complex, interacting systems makes these concerns rather moot. We won't know and so uncertainty is no longer at stake. A cybernetic imagination of ecology might provide resources for worrying less about fidelity and resolution—the hallmarks of certainty—and moving towards comfort with not knowing and yet acting at the level of systems anyhow. We could turn to questions of what should and should not be coupled and whether there might be other kinds of regulating systems better equipped than either humans or computers to manage exceedingly complex problems.

Notes

1 Paul N. Edwards, *The Closed World: Computers and the Politics of Discourse in Cold War America* (Cambridge, MA: MIT Press, 1996), 1.
2 This model predates computers. See, for example, Thomas Richards, *The Imperial Archive: Knowledge and the Fantasy of Empire* (London:Verso, 1993).
3 Paul Edwards, *A Vast Machine: Computer Models, Climate Data, and the Politics of Global Warming* (Cambridge, MA: MIT Press, 2010).
4 We note that humanities approaches to representation, on the other hand, are characterized by thoroughgoing critique of knowledge transparency and profound recognition of textual polysemy, the play of meaning, and the role of subjectivity in textual formation. In the field of film studies, much of this debate unfolded in the 1970s in the pages of the British journal *Screen*.
5 Kurt Jax, "Holocoen and Ecosystem: On the Origin and Historical Consequences of Two Concepts," *Journal of the History of Biology*, 31, no. 1 (March 1, 1998): 113–142.
6 A.G. Tansley, "The Use and Abuse of Vegetational Concepts and Terms," *Ecology*, 16, no. 3 (July 1935): 284–307.
7 Ibid., 297, 299.
8 Frank Benjamin Golley, *A History of the Ecosystem Concept in Ecology: More Than the Sum of the Parts* (New Haven, CT:Yale University Press, 1993), 72. Note that most of these discussions were happening in the English-speaking worlds of the UK and the US at the time, while other parts of Europe were occupied with slightly different issues.
9 H.A. Gleason, "The Individualistic Concept of the Plant Association," *Bulletin of the Torrey Botanical Club*, 53, no. 1 (January 1926): 7–26.
10 Tansley, "The Use and Abuse of Vegetational Concepts and Terms," 300.
11 Ibid., 299.
12 See for example, Stephen Alfred Forbes, "The Lake as a Microcosm," *NHS Bulletin*, 15, no. 9 (November 1925), https://www.ideals.illinois.edu/handle/2142/45976.
13 Raymond L. Lindeman, "The Developmental History of Cedar Creek Bog, Minnesota," *American Midland Naturalist*, 25, no. 1 (January 1941): 101.
14 Lindeman, "The Trophic-Dynamic Aspect of Ecology," *Ecology*, 23, no. 4 (October 1, 1942): 399–417.
15 Ibid.
16 Bernard Patten and Eugene P. Odum, "The Cybernetic Nature of Ecosystems," *American Naturalist*, 118, no. 6 (December 1981): 890.
17 Lorraine Daston, "The Language of Strange Facts in Early Modern Science," in *Inscribing Science: Scientific Texts and the Materiality of Communication*, edited by Timothy Lenoir (Stanford, CA: Stanford University Press, 1998), 35.
18 Ibid., 35–36.

19 Norbert Wiener, *Cybernetics or Control and Communication in the Animal and the Machine* (New York, NY: John Wiley & Sons, 1948).
20 Arturo Rosenblueth, Norbert Wiener, and Julian Bigelow, "Behavior, Purpose, & Teleology," *Philosophy of Science*, 10 (1943): 18–24.
21 "Black box" is a term from Maxwell on thermodynamics that was annexed to cybernetics by Walter Ashby in his *Introduction to Cybernetics* to allow practitioners to construct systems descriptions without specifying their inner mechanisms.
22 Wiener was introduced to the term "homeostasis" by Walter Cannon, whose book *Wisdom of the Body* described how organisms self-regulate despite environmental fluctuations. In the late 1930s Wiener participated in a series of conversations organized by Cannon and Rosenblueth on the question of scientific method.
23 See Leon Brillouin, "Life, Thermodynamics, and Cybernetics," *American Scientist*, 37 (1949): 554–568.
24 Patten and Odum, "The Cybernetic Nature of Ecosystems," 894.
25 Bernard Patten, "An Introduction to the Cybernetics of the Ecosystem: The Trophic-Dynamic Aspect," *Ecology*, 40, no. 2 (1959): 221–231.
26 Norbert Wiener, *The Human Use of Human Beings: Cybernetics and Society* (New York, NY: Doubleday, 1950), 95.
27 Club of Rome, *The Limits to Growth: A Report for the Club of Rome's Project on the Predicament of Mankind*, edited by Donella H. Meadows (New York, NY: Universe Books, 1972).
28 Bateson characterized these systems as cybernetic because they had subsystems that were potentially regenerative, governing loops, and tendencies towards homeostasis. See Gregory Bateson, *Steps to an Ecology of Mind: Collected Essays in Anthropology, Psychiatry, Evolution, and Epistemology* (San Francisco, CA: Chandler, 1972), 446–453.
29 Mary Catherine Bateson, *Our Own Metaphor: A Personal Account of a Conference on the Effects of Conscious Purpose on Human Adaptation* (New York, NY: Alfred A. Knopf, 1972).
30 Ibid., 15.
31 Stafford Beer, "Holism and the Frou-Frou Slander: Opening Presidential Address at the Seventh Triennial International Congress of Cybernetics and Systems," in *How Many Grapes Went into the Wine: Stafford Beer on the Art and Science of Holistic Management*, edited by Roger Harnden and Allenna Leonard (West Sussex: Wiley, 1994), 17–18.
32 Beer was able to implement a large-scale cybernetic system (CYBERSYN) for controlling a nation's economy during Salvador Allende's presidency in Chile. On this brief but important experiment in socialist cybernetics, see Eden Medina, *Cybernetic Revolutionaries: Technology and Politics in Allende's Chile* (Cambridge, MA: MIT Press, 2011).
33 Beer, "Towards the Cybernetic Factory," in *How Many Grapes Went into the Wine*, 215.
34 Ibid.
35 Ibid., 215–216.
36 Andrew Pickering, *The Cybernetic Brain: Sketches of Another Future* (Chicago, IL: University of Chicago, 2011), 234.
37 Ibid.
38 Ibid., 235–236.
39 Martin Heidegger, "The Question Concerning Technology," in *Technology and Values: Essential Readings*, edited by Craig Hanks (Oxford: Wiley-Blackwell, 2010), 99–113.
40 The work credited with killing more cybernetic strains in computer science is Marvin Minsky and Seymour A. Papert, *Perceptrons: An Introduction to Computational Geometry* (Cambridge, MA: MIT Press, 1969). On the sociology of artificial intelligence see Mikel Olazaran, "A Sociological Study of the Official History of the Perceptron Controversy," *Social Studies of Science*, 26, no. 3 (1996): 611–659.
41 Richard Thompson et al., "Lost at Sea: Where Is All the Plastic?," *Science*, 304, no. 5672 (May 7, 2004): 838, doi:10.1126/science.1094559 (accessed April 2015). See also David

Barnes, Francois Galgani, Richard C. Thompson, and Morton Barlaz, "Accumulation and Fragmentation of Plastic Debris in Global Environments," *Philosophical Transactions of the Royal Society B: Biological Sciences*, 364, no. 1526 (July 27, 2009): 1985–1998, doi:10.1098/rstb.2008.0205 (accessed April 2015).

42 C.J. Moore, S.L. Moore, M.K. Leecaster, and S.B. Weisberg, "A Comparison of Plastic and Plankton in the North Pacific Central Gyre," *Marine Pollution Bulletin*, 42, no. 12 (December 2001): 1297–1300.

43 David Cooper and Patricia L. Corcoran, "Effects of Mechanical and Chemical Processes on the Degradation of Plastic Beach Debris on the Island of Kauai, Hawaii," *Marine Pollution Bulletin*, 60, no. 5 (May 2010): 650–654, doi:10.1016/j.marpolbul.2009.12.026 (accessed April 2015).

44 Murray R. Gregory, "Environmental Implications of Plastic Debris in Marine Settings—Entanglement, Ingestion, Smothering, Hangers-On, Hitch-Hiking and Alien Invasions," *Philosophical Transactions of the Royal Society B: Biological Sciences*, 364, no. 1526 (July 27, 2009): 2013–2025. See also Matthew Cole, Pennie Lindeque, Claudia Halsband, and Tamara S. Galloway, "Microplastics as Contaminants in the Marine Environment: A Review," *Marine Pollution Bulletin*, 62, no. 12 (December 2011): 2588–2597, and Yukie Mato, Tomohiko Isobe, Hideshige Takada, Haruyuki Kanehiro, Chiyoko Ohtake, and Tsuguchika Kaminuma, "Plastic Resin Pellets as a Transport Medium for Toxic Chemicals in the Marine Environment," *Environmental Science & Technology*, 35, no. 2 (2001): 318–324.

45 K.L. Law, S. Moret-Ferguson et al., "Plastic Accumulation in the North Atlantic Subtropical Gyre," *Science*, 329 (2010): 1185–1190.

46 Ibid., 1187.

47 Angus Chen, "Ninety-Nine Percent of the Ocean's Plastic Is Missing," *Consortium for Ocean Leadership*, July 2, 2014, http://oceanleadership.org/ninety-nine-percent-oceans-plastic-missing/ (accessed April 2015). See also Andrés Cózar, Fidel Echevarría, J. Ignacio González-Gordillo, Xabier Irigoien, Bárbara Úbeda, Santiago Hernández-León, Álvaro T. Palma, et al., "Plastic Debris in the Open Ocean," *Proceedings of the National Academy of Sciences* (June 30, 2014): 1–6, and Marcus Eriksen, Laurent C.M. Lebreton, Henry S. Carson, Martin Thiel, Charles J. Moore, Jose C. Borerro, Francois Galgani et al., "Plastic Pollution in the World's Oceans: More than 5 Trillion Plastic Pieces Weighing over 250,000 Tons Afloat at Sea," *PLoS ONE*, 9, no. 12 (December 10, 2014): e111913.

48 Cózar et al., "Plastic Debris in the Open Ocean," 5.

49 Harnden and Leonard, *How Many Grapes Went into the Wine*, 215–216.

50 PlasticsEurope, *Plastics—The Facts 2013*. PlasticsEurope, 2013: 10, http://www.plasticseurope.org/Document/plastics-the-facts-2013.aspx?FolID=2 (accessed April 2015).

51 S. Rech, V. Macaya-Caquilpán, J.F. Pantoja, M.M. Rivadeneira, D. Jofre Madariaga, and M. Thiel, "Rivers as a Source of Marine Litter—A Study from the SE Pacific," *Marine Pollution Bulletin*, 82, no. 1–2 (May 15, 2014): 66–75, doi:10.1016/j.marpolbul.2014.03.019 (accessed April 2015). See also C.J. Moore, G.L. Lattin, and A.F. Zellers, "Quantity and Type of Plastic Debris Flowing from Two Urban Rivers to Coastal Waters and Beaches of Southern California," *Journal of Integrated Coastal Zone Management*, 11, no. 1 (2011): 65–73.

52 Joel Tarr, "The Search for the Ultimate Sink: Urban Air, Land, and Water Pollution in Historical Perspective," *Records of the Columbia Historical Society, Washington, D.C.*, 51 (1984): 1–29, and Morgan Robertson, "Measurement and Alienation: Making a World of Ecosystem Services," *Transactions of the Institute of British Geographers*, 37, no. 3 (2012): 386–401.

53 Also see Max Liboiron, *Redefining Pollution: Plastics in the Wild* (New York, NY: New York University, 2013), http://gradworks.umi.com/35/53/3553962.html (accessed April 2015).

54 For example, Chelsea Rochman, Mark Anthony Browne, Benjamin S. Halpern, Brian T. Hentschel, Eunha Hoh, Hrissi K. Karapanagioti, Lorena M. Rios-Mendoza et al., "Policy: Classify Plastic Waste as Hazardous," *Nature*, 494, no. 7436 (2013): 169–171, and Carmen González, N. Machain, and C. Campagna, "Legal and Institutional Tools to Mitigate Plastic Pollution Affecting Marine Species: Argentina as a Case Study," *Marine Pollution Bulletin*, 92, no. 1–2 (2015): 125–133.

55 Jenna Jambeck, Roland Geyer, Chris Wilcox, Theodore R. Siegler, Miriam Perryman, Anthony Andrady, Ramani Narayan et al., "Plastic Waste Inputs from Land into the Ocean," *Science*, 347, no. 6223 (2015): 768.

56 Ibid., 768.

57 Ibid., 770.

15

THE INVISIBLE AXIS

From Polar Media to Planetary Networks

Peter Krapp

The polar regions of our planet were long considered inhospitable and inaccessible, but in the media age, they have become highly visible nonetheless. Representations of the Arctic and Antarctic provide us with historical, political, and ecological frames of reference for broader epistemological insights, from heroic seafaring to climate science, from *Nanook* (dir. Flaherty, 1922) or *Scott of the Antarctic* (dir. Frend, 1948) to works by DJ Spooky and Werner Herzog, and from travel documentaries in the era of adventurous exploration to installation art using live feed in the age of GPS and webcams.[1] While thinking in terms of globalization places the accent on how economic change brings about a homogenization that in turn impacts the cultural sphere, the concept of "the planetary" emphasizes constellations and positions that problematize knowledge of our whole world, with its rising sea levels, species extinctions, and climate change. This chapter traces the multiple cultural implications of two pivotal infrastructural shifts: a redeployment of military capacity towards exploration at the turn of the nineteenth to the twentieth century and, later, a shift from big science collaborations to using the power of the media to help foment a planetary perspective or world picture. Moreover, the sheer inaccessibility of the North and South Poles makes them a crucible for the persistent questions of access and data visualization that characterize the information age. Planetary discourse is thus neither simply the consequence nor a facile critique of the history of colonizing the planet; as its etymological roots suggest, the planetary is transient, capable of unmooring fixed positions but by the same token *expanding* the frame of reference.

It might seem that before we can discuss or even conceive of climate change, we need to be able to envision Earth as one habitat; that planetary consciousness only fully congeals with the availability of an image of our whole world from a distance. Taken on December 7, 1972, by the crew of Apollo 17 at a distance of about 28,000 miles, the so-called "Blue Marble" is one of the most trafficked images in human history, from postcards to political posters and from postage stamps to book covers, not to mention its cinematic appearances. Remarkably, it prominently features Antarctica. But even before that historic photograph, as this chapter will discuss, the invisible axis of the North and South Poles galvanized media attention; over a century ago, Shackleton financed his expeditions by selling image rights.

In providing an invisible axis as a condition of possibility for our world picture, polar media challenge the scale and dimensionality of how we define and comprehend the planet, encompassing both spatial orders of magnitude (from the nanometric to the astronomic) and temporal scales (from the milliseconds of neurological apperception to the millennia of geologic and ecological structure). And yet, our ever more high-tech ways of envisioning and representing the planet are, at the same time, underpinned by colonial histories in ways that directly impinge on the fate of the world in the Anthropocene.[2] For example, the persistent attraction of the idea of a tunnel or axis connecting the north and south polar regions is a key trope in how we visualize and explore, represent and navigate the Earth: whether in terms of magnetism, without which Columbus or Magellan would not have achieved what either did; or in terms of radio waves; or indeed in the dense and global trafficking of images that seek to represent the extremes of the planet to those who are trying to comprehend their place on it.

What is at stake in media studies of the planetary—beyond the reductive media topoi of McLuhan's global village, Innis's trade routes, Chardin's noosphere, Castells's information society, or de Kerckhove's telepresence, and beyond the diagnosis of an expanded technical reach and shrinking in scale of our lived experience—is not simply globalization as enabled by a political and technical infrastructure of digital information and thus of capital flows among computers, satellites, and mobile phones. While we might observe that the history of exploration and colonization of the farthest reaches of the planet has led directly to the conditions of climate change we confront today, thinking about planetary infrastructure entails more than merely envisioning a globe with cultural and cosmic layers. Rather, a full-fledged planetary consciousness requires us to recognize the struggle between colonializing space on, inside, above, and around our planet through media-technical world images, and the pivotal insight that this very same high-tech human involvement on the planet makes an irrevocable difference in the ecosystem.[3] To that extent, engaging with polar media is perhaps a little like falling down a rabbit hole that tunnels through the planet, traveling inexorably yet unpredictably along that invisible axis that promises to connect the far north and the most inaccessible south. Imagine Alice staring into the abyss of the rabbit hole, wondering not so much about orange marmalade, but about the time it takes to traverse the whole planet and pop out the other end.[4] Camille Flammarion, as late as 1909, argued in his notorious essay "A Hole through the Earth" that "the inhabitants of the earth are still very far from a full knowledge of the planet on which they dwell." Pointing out that such a hole had been stipulated by Plutarch, Maupertuis, Voltaire, Dante, and others, Flammarion sought to debunk the notion that the core of the planet is "liquid and incandescent," and proposed drilling a tunnel right through the planet that would accelerate transportation and communication and also provide a source of geothermal energy:[5]

> I have had the idea in mind for some time past of sinking a shaft into the earth for the express purpose of scientific exploration, descending as far below the surface as the utmost resources of modern science would permit.[6]

It is that colonial and entrepreneurial prospecting that this chapter seeks to describe before turning to a critical, reflexive notion of polar media and planetary networks.

From Hollow Earth Visions to Planetary Prospecting

One trajectory of the historical logic of polar media along this invisible but powerful axis that we ought to be able to excavate is the shift from the nineteenth century's pronounced emphasis on race to a growing concern with environmental factors—particularly with weather and the global meteorological consequences of melting polar ice caps in the course of the twentieth century and into the twenty-first. The racial tensions that used to be found in nearly every canonical text about the extreme North and far South now often yield to discussions foregrounding climate consciousness and global politics. In centuries past, racist conspiracy theories congealed around inchoate ideas of purity where the pole— elsewhere imagined as vortex, void, or whirlpool—was conceived as an entry point into a different realm: the geographical prize as a spiritual prize. Antarctica figured as a southern counterpart to the *Ultima Thule* of occult racist fantasies about the north. In 1818, John Cleves Symmes posited a tunnel from pole to pole. His search for *Symzonia*, the land of perfect whiteness and "abode of a race perfect in its kind" was held out in *Harper's* magazine to be a great motivation for Antarctic exploration. "I declare the earth is hollow and habitable within," Symmes asserted, arguing of the entire globe "that it is open at the poles."[7] In 1873, the *Atlantic Monthly* could still argue the feasibility of this theory; only the Amundsen and Scott expeditions in 1911–12 laid it to rest. Edgar Allan Poe was greatly fascinated with Symmes's hollow Earth idea, and arguably in *The Narrative of Arthur Gordon Pym of Nantucket* (1839) projected onto that global scale the tensions between North and South in the United States. Furthermore, Poe ends Pym's travel narrative with a fictive editor's framing note that associates the whiteout of the South Pole with the black cavern of the inverted Eden he calls Tsalal, along with speculations on black-on-white scribbles that supposedly denote shade and darkness in esoteric languages. Beyond the model of *Gulliver's Travels* or similar fare, darkness here is interpolated with and dissolved in snow and ice, reminiscent of the allegorical complications of the whiteness of Melville's whale, or indeed of the albatross in nineteenth-century literature.[8] At this intersection of the trajectories of race and ecology we also find *Frankenstein*—where the hubris of creating life meets its double in Walton's desire to "satiate my curiosity with the sight of a part of the world never before visited."[9] As Shelley writes,

> What may not be expected in a country of eternal light? I may there discover the wondrous power which attracts the needle and may regulate a thousand celestial observations that require only this voyage to render their seeming eccentricities consistent forever.[10]

Yet viewers of James Whale's classic horror films based on Shelley's *Frankenstein* realize that this landscape is, by the same token, one of eternal darkness, where proximity to one source of magnetism occludes the opposite, balancing pole.

If an emphasis on climate issues in planetary consciousness comes more to the fore as time goes on, this concentration must take into account the history of heroic exploration when the redeployment of military resources—from earlier seafaring explorations (particularly for the English Navy after the Napoleonic Wars) to the service of colonial and scientific expeditions—led to international collaborations on an unprecedented scale. The Austro-Hungarian expedition of 1872–74 led by Carl Weyprecht made some discoveries but lost the ship *Tegethoff*, although everyone on board was saved. Weyprecht saw that individual expeditions could only contribute in small ways and, at scientific conventions and conferences in 1875, he presented a more ambitious plan, proposing that polar research is of great importance for knowledge of nature and that expeditions have value only if they prepare the way for scientific exploration. He held that detailed topographic mapping is secondary, that the geographical pole has no more value than other high latitudes, and that—beyond isolated data series—the intensity of the phenomena to be studied should be the deciding factor on favorable places for observation.[11] Weyprecht promoted the idea of establishing stations for meteorological and magnetic measurements throughout a year. Around the same time, Georg von Neumann, director of the German Maritime Observatory, proposed similar ideas but focused on the Southern Hemisphere. After some discussion among meteorologists, an International Polar Commission was created in October 1879 at a meeting in Hamburg, in anticipation of the next major period of sunspot activity (1881–82). However, after only four countries signed up by August 1880, a meeting in Berne postponed investigation of solar-terrestrial relations. At a third meeting in St. Petersburg in August 1881, a research program including atmospheric electricity, meteorology, geomagnetism, and ocean currents was scheduled for the period from August 1882 through August 1883, with eleven nations supporting fourteen stations—twelve around the Arctic, one in South Georgia, and one at Cape Horn. Weyprecht died in 1881, but had inspired over seven hundred men to venture out on the largest coordinated series of scientific expeditions of the nineteenth century. Unfortunately the International Polar Commission dissolved, and each participating nation published its own results; the potential benefit of coordination envisioned by Weyprecht was lost. Nonetheless it is worth mentioning that meteorological research received major impulses from Swedish, Scottish, and German expeditions to Antarctica early in the twentieth century, when balloons recorded different aerial temperatures at different altitudes and allowed the testing of thermodynamic theories.

During World War I, media technologies—including but not limited to telegraph, radio, telephone lines, and electric power lines—experienced unexplained interruptions and contaminations, and the rise of the airplane and of new instruments for ocean navigation renewed interest in magnetic research around the poles. After World War I, new meteorological measurements with radiosondes to investigate upper air motion reinvigorated these discussions, and interest in trans-Arctic air routes (as well as the discovery of the "jet stream") created a demand for more detailed weather information. The organization of a second international polar year was suggested for 1932–33, and an international meteorological committee started working towards that goal in 1927.[12] A scientific conference in Copenhagen in 1929 pointed to expected practical applications in terrestrial magnetism, marine and aerial navigation, wireless telegraphy, and weather forecasts, resulting in the establishment of a Commission for the International Polar Year (IPY) and inviting the General Assembly of the International Union of Geodesy and Geophysics, who provided funding for instruments and publications.[13] Despite the political upheavals of the early 1930s worldwide, the Second IPY was scheduled for a period of minimal solar activity between August 1932 and August 1933. At a convention in Locarno, Switzerland, in October 1931, it was decided to establish twenty-seven stations supported by forty-four nations, but none of them in the Antarctic due to fiscal restraints. The main US scientific contributions were led by Admiral Byrd.

The aviator and naval officer Richard E. Byrd had long been a polar explorer. His controversial claim to have reached the North Pole from Spitsbergen (in the Svalbard archipelago) by airplane in 1926 may not be verifiable, since his typewritten report of June 22, 1926 to the National Geographic Society is not corroborated by sextant data from his journal entry of May 9, 1926 (which was erased yet still legible). But on November 28, 1929, he flew from the Ross Ice Shelf over the South Pole in a Ford Trimotor. Byrd shrewdly harnessed media popularity for his polar adventures. An eighty-two-minute (mostly silent) documentary film about this quest, *With Byrd at the South Pole* (1930), won best cinematography at the third Academy Awards—the first-ever Oscar for a documentary. Based on the acclaim that mission received, Byrd went on to lead four later Antarctic missions. As his autobiography documents, during his second sojourn he barely survived an extended polar winter, from March 22 to October 14, 1934, at an advance weather station where he lived alone for nineteen weeks.[14] His third expedition during 1939–40 had the full backing of the US government, and he was recalled into active military duty in 1940. Right after World War II, Byrd was chosen to lead what may still be the largest Antarctic expedition to date, Operation Highjump.

The US Navy assembled over forty-seven hundred men, a flagship and an aircraft carrier, thirteen support ships, six helicopters, six amphibious planes, two seaplane tenders, and fifteen other aircraft. This contingent arrived in Antarctica at the end of 1946 to carry out aerial surveys, train personnel and test equipment in low temperature conditions, and develop techniques for establishing, maintaining, and utilizing bases on ice.

After Operation Highjump, Byrd famously warned the media that the world was shrinking rapidly, that an attack on the United States over either of the poles by fast airplanes was possible, and that isolation or distance are no longer guarantees of safety.[15] Conspiracy theories have made much of the fact that early in 1939, a German expedition visited Dronning Maud Land (an Antarctic region claimed by Norway) to establish a base, and that between 1943 and 1945 Britain ran a Special Air Service Regiment (SAS) operation, code named Tabarin, against Germans on Antarctica. Given that much of Operation Highjump was classified, it is hardly surprising that science fiction and thriller writers and filmmakers should spin yarns about secret Antarctic Nazi bases, UFOs from the pole, and the like.[16] One extant conspiracy theory maintains that "Germany lost the European war in 1945 in order to win the South Polar one in 1947."[17] Some even believe that purportedly observed UFOs are not actually alien vessels but Nazi flying saucers. Indeed some advanced German aircraft designs did reach prototype stage in 1944 and 1945. However, it does not follow, as conspiracy theorists continue to allege, that leading Nazis were "able to escape the ruins of the Third Reich and continue their nefarious plans for world domination in the icy fastnesses of the Arctic and Antarctic."[18] By the same token, there can be no doubt that the power of flight changed not only the conduct of warfare but also the very accessibility of the polar regions, as well as their visibility and traversibility.

It is true that a German expedition under Alfred Ritschler in 1938 and 1939 viewed 230,000 square miles of Antarctica from the air and photomapped 135,000. But it is not verifiable that Operation Highjump or, for that matter, Operation Argus (three 1958 nuclear tests on Antarctica, yet nowhere near where the Germans had landed two decades earlier) were directed against any military targets on Antarctica. Nonetheless, the Byrd expedition revived hollow Earth fantasies in novels and movies.[19] To this day, cinema and literature enjoy propagating the fantasy that German U-boats and flying saucers were stationed in Antarctica. As historians point out, "activities that were classified have subsequently been declassified and it is no longer difficult to separate fact from fancy, despite the fact that many find it attractive not to do so."[20] Aside from using Antarctica (as well as Greenland) to stage training exercises for a possible conflict with the Soviet Union in the high Arctic, Operation Highjump was mainly about cartography, geology, meteorology, and the hydrological and electromagnetic conditions in the area. Nonetheless the topos remains popular, particularly in conjunction with Americans fighting Nazis. For instance, in Michael Chabon's 2000 novel *The Amazing Adventures of Kavalier & Clay*, Joe Kavalier joins the US Navy in December 1941, is stationed as a radioman at Kelvinator Station in Antarctica, and monitors the Germans in Queen Maud Land.[21] An Antarctic-German World War II connection is also perpetuated in numerous works of science fiction.[22] One of the most curious examples may be the 1934 German science fiction novel *A Star Fell from the Sky*, in which a German expedition to Antarctica deploys advanced technologies including jet propulsion and automatic steering to discover a meteorite full of gold, thus again giving rise to the suggestion that a secret Nazi base near the South Pole could be feasible.[23] Another remarkable novel from the year 1939 sends its protagonist five millennia back in time, where he meets "Antarkans"—it is unusual for pre–World War II stories to mention atomic bombs, as this one does well before the first such bomb was built. Presumably the US government ignored this science fiction so that the Manhattan Project could continue in secrecy.[24]

After World War II spurred the development of computational weather models and simulations that aided the nondestructive development and testing of explosives and rockets, discussions soon emerged about prospective internationally coordinated investigations to take place during a period of maximum solar activity from 1957–58. Ionospheric probes and rocket launches were proposed at a meeting of the Joint Commission of the Ionosphere of the International Council of Scientific Unions at Brussels in September 1950, and the World Meteorological Organization was invited to collaborate. The third polar year became the International Geophysical Year, facilitating investigations into solar radiation, aurora and cosmic radiation, geomagnetism, glaciology, gravity, ionospheric physics, meteorology, and seismology. During this year, measurements were taken not only in the Arctic and Antarctic, but also on the equator and in North America, Europe, and Oceania, with sixty-seven nations participating and about four thousand stations, many of which were newly set up for the occasion.

In Antarctica, fifty-five stations were set up by twelve nations, and the Commonwealth Trans-Antarctic Expedition (led by Vivian Fuchs and Edmund Hillary) crossed the continent by tractor via the South Pole, traversing 2,158 miles from Shackleton station in ninety-nine days and arriving at Scott station on March 2, 1958. This was the first expedition to reach the South Pole in forty-six years, preceded only by Amundsen and Scott's respective achievements in 1911 and 1912; it remained the only crossing of the Antarctic continent until Ranulph Fiennes matched it in 1981. Right after the end of the third International Polar Year, the Soviets also drove a tractor exploration from Komsomolskaya base 1,142 miles to the South Pole, arriving on December 26, 1959. Admiral Byrd once again commanded a US Navy operation in Antarctica during the International Geophysical Year, called Operation Deep Freeze I, establishing permanent bases at McMurdo Sound, the Bay of Whales, and the South Pole. Operation Deep Freeze I also established five new stations fronting the Indian, Pacific, and Atlantic oceans, having explored over a million square miles of unknown territory in Wilkes Land—the vast area of Antarctica where the US Exploring Expedition under Lt. Charles Wilkes had reported land in longitude 158° degree east on January 16, 1840, and then skirted the coast for 1,500 miles, thus establishing the existence of the Antarctic continent.

In this zeal for exploration and mapping, it is easy to lose sight of the fact that the geographically meaningful axis for South and North Polar adventures is also imaginary, ambiguous, and shifting. For one thing, the magnetic poles (and the magnetic north-south axis) are distinct from the more familiar geographic poles, inclining at roughly 12° from the geographic axis and therefore lying almost eight hundred miles from the geographic poles. Moreover, the South Magnetic Pole, where the magnetic field lines point vertically, continues to shift several miles northwest each year; while Shackleton's expedition in 1909 located it in the highlands of Victoria Land, in 1962 it was near the Adélie Coast, and by 2010 it was actually off the coast of Antarctica.[25] The fact that the South Geomagnetic Pole is very close to Vostok, the Russian Antarctic base—where a record low temperature of −127°F (−89°C) was recorded on August 24, 1960—has given rise to speculation (and some scientific experiments) regarding what might be buried there under 12,140 feet of ice.

While science fiction and fantasy literature as well as movies indulge in tales of bacterial or alien life under the ice, one of the most globally influential effects of Antarctic exploration in the Cold War period, amid all the highly telegenic media campaigns around heroic military and scientific exploits, turned out to be the decision to set aside territorial claims in Antarctica. Of course, it proved consequential for media history when the Soviet *Sputnik* satellite was launched on October 4, 1957, right when the Scientific Committee on Antarctic Research conference on rockets and satellites took place in Washington DC, triggering a Cold War arms race. Nevertheless, the third IPY also resulted in an international peace treaty, albeit one restricted to Antarctica. On December 1, 1959, the twelve countries whose scientists participated in the polar year signed the Antarctic Treaty, which entered into force in 1961, and has since been acceded to by many other nations; currently fifty nations have signed on. Providing that Antarctica shall be used for peaceful purposes only, and prohibiting territorial claims to all the land and ice south of latitude 60° south, the Treaty upholds freedom of scientific exploration and the exchange of observations and results.

A fourth, expanded International Polar Year (March 2007 through March 2009) was again an interdisciplinary initiative, devoted mainly to the effects of climate change. It also marked the 125th anniversary of the first IPY, the 75th anniversary of the second IPY, and the 50th anniversary of the third IPY, and involved over sixty thousand scientists from sixty-six nations. This history of the IPY goes hand in hand with milestones in polar and scientific media, from the heroic age of exploration to the launch of the global space age with *Sputnik* and *Vanguard*.

Tracing the history of the IPY also raises questions of scale in the context of communications media and collaborative initiatives. As Ray and Charles Eames demonstrated in their classic short *Powers of Ten* (1977), the translatability of abstract (but calculable and measurable) into concrete sensory data allows us to comprehend and incorporate conceptions of scale that are otherwise beneath or above the thresholds of our natural perception. This sort of scaling up or down is how microscopic, but also satellite and space imagery, can become appropriated and understood. And it is how we apprehend representations of the polar regions that would be otherwise invisible, inaccessible, and unimaginable.

Planetary Networks

The development of the "cybercartographic atlas" of Antarctica—a web-based reference for the entire Antarctic region—is an interesting example for the challenges and attractions that are part and parcel of a project to visualize and catalogue an entire continent for the productive use by people who, for the most part, have not been there and may not expect to set foot there.[26] The project encompasses not only online interactive visualization of maps built from distributed sources and updated with real-time data, but also collaborative maps used via telepresence.[27] Instead of topographic surveys and census information—both of which are hard to acquire for a vast ice desert of over five million square miles—the atlas amasses data from distributed networks of GPS and other sensor inputs.[28] In the context of this history and the laborious constitution of the cybercartographic atlas, we might perceive Google Earth as a step towards commodification and trivialization of the hard-won insights that, in the heroic and military ages of exploration, were paid with many deaths, and that were no easier to come by in the periods of polar exploration enhanced by science and media. The comparison sends us back to fundamental questions about the role of maps, not merely in exploring or navigating, nor even in establishing and disputing territorial claims, but as media that query the representation of the unknown defining our comprehension of the history and culture of Antarctica. Instead of seeing the continent merely as a giant science lab, thinking through the polar imaginary may help us understand broader ecosystemic interrelationships, including the hydrosphere of global ocean currents and the atmosphere, as we come to terms with climate change.

Antarctica is the planet's highest continent, averaging more than a mile above sea level, with the Vinson Massif in the Ellsworth Mountains its highest peak at 16,860 feet. The continent stores seventy percent of the planet's fresh water—an estimated seven million cubic miles of ice, most of the world's supply, up to 14,000 feet thick on the Hollick-Kenyon plateau, pressing the continent down (as it does in Greenland). As *National Geographic* magazine calculated in 1963, if all the ice were to melt, it would raise sea levels globally by two hundred fifty feet.[29] Consulting the *RealClimate* website, browsing data compiled by the World Meteorological Organization, or consulting NASA reports, it becomes plain that the ice changes around Earth's frozen caps, and that sea levels are rising.[30] The United Nations Educational, Scientific and Cultural Organization (UNESCO) urges us to consider species extinction, referring to its universal declaration on cultural diversity.[31] This concern for life on the planet should not merely extend to husky heroics, melodrama displaced upon the sacrifice or consumption of man's best friend, or the chronicling of the humanoid waddling of penguins. The past hundred years saw such a radical transformation in life on the planet's crust, with extreme climates, population growth, deforestation, ocean acidification, and species extinction, that another geologic era has been established: what ecologist Eugene Stoermer and atmospheric chemist Paul Crutzen in 2000 called the Anthropocene.[32]

In the face of such massive geological features and dramatic changes, we may ask: at what scale is Earth understood, at what scale are Antarctica and the high Arctic comprehensible, and how do we visualize and narrate their appeal? Umberto Eco discussed the theoretical possibility of a 1:1 map—it would need to be coextensive, yet not transparently covering the territory; it would need to faithfully reproduce all features, functioning as a semiotic tool:

> The map cannot be a transparent sheet in any way fixed over the territory on which the reliefs of the territory itself are projected point by point; for in that case any extrapolation carried out on the map would be carried out at the same time on the territory beneath it, and the map would lose its function as a maximum existential graph.[33]

In this extended meditation upon Borges's celebrated short story, "On Exactitude in Science," Eco concludes that an opaque map covering the territory would necessarily change the ecological equilibrium of the territory itself, thus depicting the territory differently from its actual state. Suspending the map does not solve the issue, since it would have an atmospheric impact. Yet a transparent, permeable, extended, and adjustable map would eliminate the semiotic function Eco requires: "It would be functional as sign only in the presence of its own referent: residing on the map, the subjects could not tend the territory, which would deteriorate, making the map unfaithful."[34]

One might argue that this "deterioration" of the territory is indeed already in progress in the way our mediascape in the Northern Hemisphere has been representing Antarctica. Not even the cybercartographic project can circumvent the paradox of Eco's essay: a cybercartographic data feed that fully shows what Antarctica is would need to include the sources of such data as they are acquired and compiled, since the remote sensing, weather stations, vehicles, and settlements that produce the data points are all part of Antarctica.[35] This would invite a bad infinity of representing the territory and its maps. Following Eco, a 1:1 map inevitably reproduces the territory unfaithfully; or, in the very realization of the map, the territory becomes irreproducible. At its point of maximum representation in interactive multimedia, cinema, literature, photography, data visualizations, and so forth, Antarctica itself becomes imperceptible. This is the problem facing

Buckminster Fuller's fluid geography kit; his "dymaxion projection" world map so prominently featured Antarctica, but none of his clients knew what to do with that "empty" continent.

This conundrum of irreproducibility may also become a challenge or inspiration for works of art and data-rich media representations. The Pavilion of the Bahamas at the 2013 Venice Biennial, for instance, took up the ineffability of the poles as a theme. Artist Tavares Strachan, who in 2006 had FedEx ship a 4.5-ton block of ice from a subarctic river in Alaska to a primary school in the Bahamas, where solar energy was used to keep it frozen, has also developed what he calls the "Bahamas Aerospace and Sea Exploration Center" in Nassau. For "Polar Eclipse," his 2013 installation in Venice, he refers to Matthew Alexander Henson's 1912 autobiography, which recounts the role Henson (1866–1955) and four Inuit guides played in Robert Peary's "discovery" of the North Pole on April 6, 1909. The African American Henson explored the Arctic with Peary for twenty-two years, but Peary credited neither him nor the mission's Inuit members with significant contributions. However, Inuit oral tradition accords the guides a crucial role in the historical exploits of Peary, and posterity also honored Henson's contributions. Henson died in 1955 at the age of eighty-eight and was buried at Woodlawn Cemetery (a National Historic Landmark in The Bronx, New York), but in 1988, his remains were exhumed and reburied at Arlington National Cemetery, near the grave of Admiral Peary. The Matthew Henson Earth Conservation Center in Washington, DC, is named for him. Tavares Strachan's fourteen-channel video installation "Standing Alone" (2013) reenacts Peary and Henson's 1909 expedition while simultaneously exploring issues of identity, belonging, and globalization:

> Instead of aggrandizing himself as a victorious explorer able to reenact Matthew Henson's presumed 1909 "discovery" of the North Pole, which in reality is only a conceptual point, Strachan highlights the absurdity and even the impossibility of doing so. While his flag in "Standing Alone" might appear to be demarcating the North Pole, it is only connected to the constantly moving ice shelf covering portions of the Arctic Ocean, and thus is as unanchored as the artist standing beside it.[36]

Documenting the difficulty of putting any marker on the "conceptual point" that is the North Pole (the ice shelf covering it sometimes moves miles within minutes), Strachan also invited forty Bahamian children to Venice where they performed a traditional Inuit song, which Strachan claims is "untranslatable." The Biennial exhibit also featured a block of North Pole ice as well as a "cloned" block of ice (fabricated to contain a similar composition of chloride, sodium, sulfate, magnesium, calcium, potassium, bromide, zinc, etc.), displayed alongside and inviting reflection about their relation to each other as well as the viewer's relation to either of them.

DJ Spooky's *Terra Nova: The Antarctic Suite*, a multimedia performance work commissioned by German National Radio, transforms his encounter with a rapidly changing continent in December 2007 and January 2008 into seventy minutes of music compiled from field recordings. Grounded in "music composed from the different geographies that make up the land mass" and his DJ craft, the project seeks to reflects an environment that is only minimally illustrated by the accompanying visuals from the Getty collection of Antarctic images.[37] It differs in its reflexivity from what might be the world's most widely seen PowerPoint presentation, namely Al Gore's *An Inconvenient Truth* (2006). As Kathryn Yusoff states, an "increasing vigilance and reflexivity to representational and metaphoric practices in the constructions of environments" is necessary to cope with the technological commodification of ecosystem services.[38]

And of course, the ultimate irony of our remote-sensing data center world is that the huge server farms for our planetary cloud-computing providers are among the biggest culprits of heat dissipation and energy consumption (see Brennan in this volume). Or as Thomas Pynchon's latest protagonist, Gabriel Ice, has it in *Bleeding Edge* (2013):

> "More and more servers together in the same place putting out levels of heat that quickly become problematic unless you spend the budget on A/C. Thing to do," Ice proclaims, "is to go north, set up server farms where heat dissipation won't be so much of a problem, take your power from renewables like hydro or sunlight, use surplus heat to help sustain whatever communities grow up around the data centers. Dome communities across the Arctic tundra. My geek brothers! the tropics may be OK for cheap labor and sex tours, but the future is out there on the permafrost, a new geopolitical imperative—gain control of the supply of cold as a natural resource of incomputable worth, with global warming, even more crucial."

Notably, Pynchon's narrative voice contextualizes right away:

> There is something creepily familiar about this go-north argument. By a corollary of Godwin's law valid only on the Upper West Side, Stalin's name, like Hitler's, is 100% certain to enter a discussion of any length, and Maxine now recalls Ernie telling her about the genocidal Georgian and his plans back in the 1930s for colonizing the Arctic with domed cities and armies of young technicians, otherwise known, Ernie was always careful to point out, as forced labor, bringing out the multimedia emphasis his 78 rpm album of *The Attractive Schoolgirl of Zazhopinsk*, an obscure opera from the purge era, strangled Russian bass-tenor duets invoking steppes of ice, thermodynamic night.[39]

As always, fiction never lags far behind the facts, and indeed often forms the leading edge. In April 2007, Internet pioneer Vint Cerf (in the 1970s co-designer of the TCP/IP protocol, and more recently Google's "chief Internet evangelist") appeared at the University of California, Irvine (where I work), and explicitly called for data centers in the polar regions. Speaking to audiences at Calit2, a state-funded Institute for Science and Innovation focused on Telecommunications and Information Technology, Cerf pointed out that in terms of infrastructure, the North and South Poles are well situated to connect via satellite relays or ocean cable to areas of the globe that have not been fully connected to the Internet. And indeed, earlier in 2013 it was revealed that there may well be a chain of data centers and listening posts all along the coast of Antarctica, as marked on a map featured on one of the PowerPoint slides that National Security Agency (NSA) contractor Edward Snowden took and allowed the news media to publish.[40]

Technologies of measuring and representing, exploring and navigating, remote sensing and telecommunication have produced a planetary imaginary in the high-tech crucible of polar media. The very remoteness and darkness of the Arctic and Antarctic is pictured, broadcast, webcast . . . and former Soviet icebreakers that threatened to become unsupportable on dwindling budgets are turned over to private enterprise as lucrative tourist vessels venturing into extreme regions. Indeed, media mogul Howard Hughes's obsession with the Cold War thriller *Ice Station Zebra* (1968) still looms large: the US National Academy of Sciences maintained as recently as 2008 that "polar regions play key roles in understanding impacts of ever-changing space weather on technologies for modern communication and power distribution."[41] While the most remote polar regions become convenient nodes for planetary networks, the abstraction of the planet as infrastructure shifts the register of observation and representation. The cover of the 1969 *Whole Earth Catalog* showed Earth as seen from space, as does the cover of the famous 1972 report on *The Limits to Growth* by the pioneering think tank, the Club of Rome, as well as Al Gore's famous slides from 2006. As astronaut Eugene Cernan, the last human to walk on the moon, mused:

> When you get 250,000 miles from earth and look back, the earth is really beautiful. You can see the roundness, you can see from the snowcap of the North Pole to the ice caps of the South Pole. You can see across continents. You look for the strings that are holding this earth up and you look for the fulcrum and it doesn't exist.[42]

When the American minimalist composer Terry Riley was commissioned by the Kronos Quartet (and curator of the NASA Art Program Bertram Ulrich) to use audio tapes of cosmic phenomena recorded by the University of Iowa Physics Department, he also made use of a very similar quote from Neil Armstrong, the first astronaut to walk on the moon: "You see from pole to pole and across oceans and continents. You can watch it turn, and there's no strings holding it up. And it's moving in a blackness that is almost beyond conception."[43] In the original photo, the south polar ice cap appears on top and the north at the bottom because of the orientation in which the astronauts were traveling. Stewart Brand, publisher of *The Whole Earth Catalog* between 1968 and 1972 and intermittently thereafter, had been lobbying NASA since 1966 to release a photograph of the entire Earth. But despite the fact that the 1972 photo features Antarctica so prominently, the desire for an emblematic image of planetary consciousness has not contributed as much as Brand and others had hoped to a changed attitude towards modifying the very industrial infrastructure that provides us with such images. The 1995 film *Apollo 13* (dir. Ron Howard) shows that same Apollo 17 photograph of Earth countless times, but every time Tom Hanks look out the window, he catches the same view of Antarctica; in Hollywood's safe vision, the planet does not even rotate around its invisible axis.

Notes

1 Two recent books are great compendia of polar media. See Andrea Polli and Jane D. Marsching, *Far Field: Digital Culture, Climate Change, and the Poles* (New York, NY: Intellect Press, 2011), and E. Glasberg, *Antarctica as Cultural Critique: The Gendered Politics of Scientific Exploration and Climate Change* (London: Palgrave Macmillan, 2012).
2 Paul Crutzen and Eugene Stoermer, "The Anthropocene," *International Geosphere-Biosphere Programme Newsletter*, 41 (May 2000): 17–18, http://www.igbp.net/publications/globalchangemagazine/globalchangemagazine/globalchangenewslettersno4159.5.5831d9ad13275d51c098000309.html.
3 Mary Louise Pratt, *Imperial Eyes: Travel Writing and Transculturation* (London: Routledge, 1992).
4 Lewis Carroll, *Alice's Adventures in Wonderland* (London: Macmillan, 1865).
5 Indeed, scientists in 2014 should not have been so surprised when satellite imagery and ice-penetrating radars pulled by snowmobiles aided them in the discovery of a hidden valley buried under West Antarctic ice that appears to be bigger than the Grand Canyon—two miles deep, two hundred miles long, and fifteen miles wide. See William Pentland, "Massive Hole Discovered under Antarctica, Bigger Than the Grand Canyon," *Forbes*, January 15, 2015, http://www.forbes.com/sites/williampentland/2014/01/15/massive-hole-discovered-under-antarctica-bigger-than-the-grand-canyon/.
6 Camille Flammarion, "A Hole through the Earth," *Strand Magazine*, 38 (1909): 349–355.
7 John Cleves Symmes, *Symzonia, A Voyage of Discovery* (London: J. Seymour, 1818). Compare J. McBride, *The Symmes Theory of Concentric Spheres* (Cincinnati, OH: Morgan Lodge and Fisher, 1828), and R. Rucker, *The Hollow Earth: The Narrative of Mason Algiers Reynolds of Virginia* (New York, NY: William Morrow, 1990).
8 Here one thinks first of Samuel Taylor Coleridge's poem *The Rime of the Ancient Mariner* (1798), but the albatross figures similarly in Mary Shelley's *Frankenstein* (1818) and Charles Baudelaire's *Les Fleurs du mal* (1857).
9 Mary Shelley, *Frankenstein, or the Modern Prometheus* (New York, NY: Signet, 1963), 194. Coleridge and Poe personify the pole as an animal lover, inhuman but nonetheless (selectively) protecting life.
10 Mary W. Shelley, *Frankenstein or, The Modern Prometheus* (Boston and Cambridge: Sever, Francis & Co., 1869), 15.
11 F.W.G. Baker, "The First International Polar Year, 1882–83," *Polar Record*, 21, no. 132 (1992): 275–285. Compare Carl Weyprecht, "Vortrag," in *Tageblatt der 48. Versammlung Deutscher Naturforscher und Ärzte zu Graz vom 18–24. September 1875* [*Publications of the 48th Assembly of German Natural Scientists and Medical Researchers, September, 18–24, 1875*]. Graz, 39–42.

12 V. Laursen, "The Second International Polar Year (1932/33)," *WMO Bulletin*, 31, no. 3 (1982): 214–222.
13 Henrik G. Cannegieter, "The History of the International Meteorological Organization 1872–1951," *Annalen der Meteorologie*, 1 (1963): 1–280.
14 Richard E. Byrd, *Alone* (New York, NY: G.P. Putnam, 1938). Byrd lived alone between March 28 and August 11.
15 David A. Kearns, "Operation Highjump: Task Force 68," in *Where Hell Freezes Over: A Story of Amazing Bravery and Survival* (New York, NY: Thomas Dunne, 2005). Compare with Halford Mackinder, "The Geographic Pivot of History," in *The Geopolitics Reader*, edited by Gearóid Ó. Tuathail, Simon Dalby, and Paul Routledge (London: Routledge, 1998), 27–31.
16 "Government Document Archive," The Black Vault, http://www.theblackvault.com/m/articles/view/Operation-Highjump.
17 "The Secret Expedition to the Nazi South Polar Base," German UFO Chatter, http://www.germanufochatter.com/Nazi-South-Polar-Base/index.html.
18 Alan Baker, *Invisible Eagle* (New York, NY: Virgin, 2000).
19 See, for instance, Raymond W. Bernard, *The Hollow Earth: The Greatest Geographical Discovery in History Made by Admiral Richard E. Byrd in the Mysterious Land beyond the Poles—The True Origin of the Flying Saucers* (New York, NY: Bell, 1964).
20 Colin Summerhayes and Peter Beeching, "Hitler's Antarctic Base: The Myth and the Reality," *Polar Record*, 43, no. 224 (2007): 1–21.
21 Michael Chabon, *The Amazing Adventures of Kavalier & Clay* (New York, NY: Random House, 2000).
22 See Clive Cussler, *Valhalla Rising* (New York, NY: Putnam, 2001), and several other titles by the same author. Other recent popular polar fiction deploying Nazi myths includes William Dietrich, *Ice Reich* (New York, NY: Warner, 1998), Mick Farren, *Underland* (New York, NY: Doherty, 2002), and Richard P. Henrick, *Ice Wolf* (New York, NY: Harper, 1994).
23 Hans Dominik, *A Star Fell from the Sky* (Leipzig: von Hase & Koehler, 1934). Coincidentally, Dominik was a boarding school student of science fiction pioneer Kurd Lasswitz; see Laurence Rickels, *Psy Fi* (Minneapolis: University of Minnesota Press, 2002), 179f., on Dominik's science fiction from the 1930s.
24 E. and O. Binder, *The Lords of Creation* (New York, NY: Munsey, 1939).
25 The magnetic North Pole is several hundreds of miles south of the geographic North Pole, which itself changes slightly due to a wobble in Earth's axis. The same factors affect the relation between the magnetic South Pole and the geographic one.
26 Peter Pulsifer et al., "The Development of the Cybercartographic Atlas of Antarctica," *Cybercartography: Theory and Practice* (Amsterdam: Elsevier, 2005), 461–490.
27 See Glenn Brauen, "Designing Interactive Sound Maps Using Scalable Vector Graphics," *Cartographica* (special issue on cybercartography), 41, no. 1 (March 2006): 59–71; and Peter L. Pulsifer, Sébastien Caquard, and D.R. Fraser Taylor (eds.), *Toward A New Generation of Community Atlases—The Cybercartographic Atlas of Antarctica* (New York, NY: Springer, 2007).
28 An important extension of my argument here would need to pursue GIS-based knowledge politics, for instance along the lines of Sarah Elwood and Agnieszka Leszczynski, "New Spatial Media, New Knowledge Politics," *Transactions of the Institute of British Geographers* (Royal Geographical Society), 38 (2012): 544–559.
29 Atlas Plate 65, *National Geographic* magazine (February 1963).
30 The Center for Astrophysical Research in Antarctica offers a virtual tour of the South Pole at http://astro.uchicago.edu/cara/vtour/pole.
31 See the Arctic Climate Impact Assessment (ACIA), November 2004, http://www.acia.uaf.edu, as well as "Realclimate: Climate Science from Climate Scientists," http://realclimate.org; "NASA Eyes Ice Changes Around Earth's Frozen Caps," NASA,

http://www.nasa.gov/vision/earth/lookingatearth/icecover.html; "UNESCO Universal Declaration on Cultural Diversity," (UNESCO, 2002), http://unesdoc.unesco.org/images/0012/001271/127160m.pdf; "Impacts of the Kyoto Protocol on U.S. Energy Markets & Economic Activity," Energy Information Administration (EIA), http://www.eia.doe.gov/oiaf/kyoto/kyotorpt.html; Fred Pearce, "Post-Kyoto Climate Negotiations Look Troubled," New Scientist, http://www.newscientist.com/article.ns?id=dn6816.

32 See note 2.
33 Umberto Eco, "On the Impossibility of Drawing a Map of the Empire on a Scale of 1 to 1," *How to Travel with a Salmon and Other Essays* (New York, NY: Harcourt & Brace, 1982), 95–106.
34 Ibid. See Jorge Luis Borges, "On Exactitude in Science," *A Universal History of Infamy* (London: Penguin, 1975), first published as "Del rigor en la ciencia," *Los Anales de Buenos Aires*, 1, no. 3 (March 1946): 53, which itself borrows its motif from Lewis Carroll's 1893 book *Sylvie and Bruno Concluded* (London: Macmillan), 169.
35 The cybercartographic project has received more media attention after being mentioned by former US Vice President Al Gore in a speech on "The Digital Earth: Understanding Our Planet in the 21st Century," January 31, 1998, at the California Science Center in Los Angeles; see http://www.digitalearth.gov/VP199B0131.html.
36 Strachan exhibition blurb, Venice Biennal 2013.
37 See Paul D. Miller, *The Book of Ice* (New York, NY: Subliminal Kid, 2011).
38 Kathryn Yusoff, "Excess, Catastrophe, and Climate Change," *Environment and Planning D: Society and Space*, 9 (2009): 1010–1029.
39 Thomas Pynchon, *Bleeding Edge* (New York, NY: Penguin, 2013), 310–311.
40 On Snowden and the materials he released to the public, see Luke Harding, *The Snowden Files: The Inside Story of the World's Most Wanted Man* (New York, NY: Vintage, 2014).
41 ICSU, "A Framework for the International Polar Year 2007–2008," http://www.ipy.wnoz.us.edu.pl/strony/IPY_framework.pdf.
42 Eugene Cernan, "Apollo 17 Astronauts Weigh Space Challenge," *Lowell Sun* (March 29, 1973), 26.
43 Terry Riley (2002). *Sun Rings*, http://kronosquartet.org/projects/detail/sun_rings.

NOTES ON CONTRIBUTORS

Shane Brennan is a Ph.D. Candidate in the Department of Media, Culture, and Communication at New York University. He is currently working on a dissertation project about the visual and cultural politics of solar energy in three US environments.

Alenda Y. Chang is Assistant Professor at the University of California, Santa Barbara, having received her Ph.D. from the Department of Rhetoric at the University of California, Berkeley. With a multidisciplinary background in biology, literature, and film, she combines eco-critical theory with the analysis of contemporary media. Current projects include *Playing Nature*, an ecological approach to computer and video games, as well as new and ongoing research in documentary, sound and media studies, and environmental literature. She also maintains the *Growing Games* blog, a resource for game studies, environmental humanities, and ecomedia scholars, and has recently published work in *Ant Spider Bee*, *The Information Society*, *qui parle*, and *Interdisciplinary Studies in Literature and Environment.*

Sean Cubitt is Professor of Film and Television and co-Head of the Department of Media and Communications at Goldsmiths, University of London. His publications include *Timeshift: On Video Culture*, *Videography: Video Media as Art and Culture*, *Digital Aesthetics*, *Simulation and Social Theory*, *The Cinema Effect*, *EcoMedia*, and his most recent book, *The Practice of Light* (MIT 2014). He has recently co-edited *Rewind: British Video Art of the 1970s and 80s*, *Relive: Media Art History*, *Ecocinema: Theory and Practice*, *Ecomedia: Key Issues* (Earthscan/Routledge 2015), and the open-access anthology *Digital Light* (fibreculture 2015). He is the series editor for Leonardo Books at MIT Press. His current research is on media technologies, environmental impacts of digital media, and on media arts and their history.

Jennifer Gabrys is Reader in the Department of Sociology at Goldsmiths, University of London, and Principal Investigator on the ERC-funded project, "Citizen Sense." Her publications include *Digital Rubbish: A Natural History of Electronics* (University of Michigan Press 2011) and *Program Earth: Environmental Sensing Technology and the Making of a Computational Planet* (University of Minnesota Press, 2016). Her work can be found at http://citizensense.net and http://jennifergabrys.net.

Bishnupriya Ghosh is Professor of English at the University of California, Santa Barbara, where she teaches postcolonial theory, literature, and global media studies. Author of the recent *Global Icons: Apertures to the Popular* (2011), she is working on two monographs: *The Unhomely Sense: Spectral Cinemas of Globalization*, a book on contemporary European and South Asian global cinemas, and *The Virus: Touch: Living with Epidemics*, a comparative study of HIV/AIDS epidemic media from the United States, South Africa, and South Asia.

Minori Ishida is Associate Professor of contemporary visual culture at Niigata University. Her research investigates interdisciplinary approaches to visual culture across a range of manga and cinema, as a milieu of communication between producer and consumer. Her publications include *Secret Education: Prehistory of Yaoi and Boys Love (Hisoyakana kyōiku: Yaoi bōizu rabu zenshi)* (2008) and "You of a Profile: Keiji SADA (Yokogao no Kimi: SADA Keiji)" (2010).

Peter Krapp is Professor and Chair of the Department of Film and Media Studies at the University of California, Irvine, where he also holds courtesy appointments in the Departments of English and of Informatics, and contributes to an interdisciplinary doctoral program in Visual Studies. His research interests include media history and cultural memory, secret communications, and the theory and history of games and simulations. He is the author of *Noise Channels: Glitch and Error in Digital Culture* (University of Minnesota Press 2011) and *Déjà Vu: Aberrations of Cultural Memory* (University of Minnesota Press 2004), and a co-editor of *Medium Cool*, a special issue of the *South Atlantic Quarterly* (Duke University Press 2002), and the forthcoming international *Handbook Language Culture Communication* in the long-standing series "Handbooks of Linguistics and Communication Science" (Berlin: De Gruyter, 2016).

Max Liboiron is an Assistant Professor of culture and technology at Memorial University of Newfoundland. Her research focuses on how harmful, invisible, emerging phenomena such as disasters, toxicants, and marine plastics become manifest in science and activism, and how these methods of representation relate to action. Liboiron's major public projects include managing the *Discard Studies* blog, a public forum for a variety of audiences interested in waste and pollution, and directing the Civic Laboratory for Environmental Action Research (CLEAR), a citizen science laboratory for environmental monitoring. Prior to her position at Memorial, Liboiron was a postdoctoral fellow at both Northeastern

University's Social Science Environmental Health Research Institute (SSEHRI) and at Intel's Science and Technology Center for Social Computing. She holds a Ph.D. in Media, Culture, and Communication from New York University.

Colin Milburn is Gary Snyder Chair in Science and the Humanities and Professor of English, Science and Technology Studies, and Cinema and Digital Media at the University of California, Davis. His research focuses on the intersections of science, literature, and media technologies. He is the author of *Nanovision: Engineering the Future* (Duke University Press 2008) and *Mondo Nano: Fun and Games in the World of Digital Matter* (Duke University Press 2015). At UC Davis, he directs the ModLab, an experimental research laboratory for digital humanities.

Rahul Mukherjee is the Dick Wolf Assistant Professor of Television and New Media Studies in the Cinema Studies program at University of Pennsylvania. He is currently working on two collaborative projects: one concerned with the circulation of locally produced music videos through microSD cards in India, and the other exploring Information and Communication Technologies usage in Zambia. Mukherjee's book project combines science studies, media theory, and cultural studies to examine environmental effects of media infrastructures.

Jussi Parikka is Professor of Technological Culture and Aesthetics at the Winchester School of Art, University of Southampton, and Docent at University of Turku, Finland. He has written on media theory and media archaeology, the biopolitics of network culture, and the connections between media arts and cultural theory. His books include *Digital Contagions* (Peter Lang 2007), *The Spam Book* (Hampton Press 2009, with Tony Sampson), *Insect Media* (University of Minnesota Press 2010), *Media Archaeology* (University of California Press 2011, with Erkki Huhtamo), and *What Is Media Archaeology?* (Polity 2012). Parikka's most recent book, *A Geology of Media* (University of Minnesota Press) came out in 2015.

Erica Robles-Anderson is an Assistant Professor in the Department of Media, Culture, and Communication at New York University. Robles-Anderson focuses on the role media technologies play in the production of space. In particular, she concentrates on configurations that enable a sense of public, collective, or shared experience, especially through the structuring of visibility and gaze. Trained as both an experimental psychologist and a cultural historian, she has employed a range of methodologies to explore definitions and forms of media in the built environment. She is currently writing a book about the twentieth-century transformation of Protestant worship space into a highly mediated, spectacular megachurch. Prior to her position at NYU she was a Research Fellow in New Media and Architecture in joint affiliation with the Department of Culture and Media and the Humanities and Technology Laboratory (HUMlab) at the University of Umeå in Sweden. Robles-Anderson holds a Ph.D. in Communication from Stanford University.

Amy Rust is Assistant Professor in the Department of Humanities & Cultural Studies at the University of South Florida, where her research investigates meetings of technology, violence, and media ecology. In addition to a recently completed manuscript, *"Passionate Detachment": Technologies of Vision and Violence in American Cinema, 1967–1974*, Rust has published work in *Cinema Journal*, *Quarterly Review of Film & Video*, and *Films for the Feminist Classroom*.

John Shiga is Assistant Professor in the School of Professional Communication at Ryerson University. He completed his Ph.D. in the School of Journalism and Communication at Carleton University and was a Postdoctoral Fellow in the Department of Art History and Communication Studies at McGill University. He has published on digital audio, intellectual property, media theory, and interspecies communication. His current project is a cultural history of sonar, which explores the role of underwater listening technologies in the transformation of scientific, military, and popular understandings of the ocean.

Nicole Starosielski is Assistant Professor in the Department of Media, Culture, and Communication at New York University. Her research focuses on the global distribution of digital media and the relationships between media, society, and aquatic ecologies. She is author of *The Undersea Network* (Duke University Press 2015), an exploration of the histories, environments, and cultures of transoceanic cable systems, and co-editor of *Signal Traffic: Critical Studies of Media Infrastructure* (University of Illinois Press 2015).

Hunter Vaughn is Associate Professor of Cinema Studies at Oakland University. His research interests include film-philosophy, media and the environment, and identity studies. His first book, *Where Film Meets Philosophy: Godard, Resnais, and Experiments in Cinematic Thinking* (Columbia University Press 2013), reconciles the phenomenology of Merleau-Ponty with Deleuze's image philosophy to construct a foundational film-philosophy according to a theory of subject-object relations. His current research focuses on an ecomaterialist alternate narrative of Hollywood cinema.

Janet Walker is Professor and Chair of Film and Media Studies at the University of California, Santa Barbara, where she is also affiliated with the Environmental Media Initiative of the Carsey-Wolf Center. A specialist in documentary film, trauma and memory, and media and environment, her books include *Trauma Cinema: Documenting Incest and the Holocaust* (University of California Press 2005) and *Documentary Testimonies: Global Archives of Suffering* (Routledge 2010, co-edited with Bhaskar Sarkar). Walker co-chairs the Media and the Environment scholarly interest group of the Society for Cinema and Media Studies. She is currently writing about site-specific documentary film and other geolocative technologies.

INDEX